Proterozoic Geology of the Southern Rocky Mountains

Edited by

Jeffrey A. Grambling
Department of Geology
University of New Mexico
Albuquerque, New Mexico 87131

Barbara J. Tewksbury
Department of Geology
Hamilton College
Clinton, New York 13323

SPECIAL PAPER
235
1989

Published by The Geological Society of America, Inc.
3300 Penrose Place, P.O. Box 9140, Boulder, Colorado 80301

GSA Books Science Editor Campbell Craddock

Printed in U.S.A.

Library of Congress Cataloging-in-Publication Data

Proterozoic geology of the southern Rocky Mountains / edited by
Jeffrey A. Grambling, Barbara J. Tewksbury.
p. cm.—(Special papers ; 235)
Papers presented at a 2-day symposium, held in Durango, Colorado, May 1984, in conjunction with the Rocky Mountain Section Meeting of the Geological Society of America.
Bibliography: p.
Includes index.
ISBN 0-8137-2235-7
1. Geology, Stratigraphic—Pre-Cambrian—Congresses. 2. Geology--Rocky Mountains Region—Congresses. I. Grambling, Jeffrey A. II. Tewksbury, Barbara J. III. Geological Society of America. Rocky Mountain Section. IV. Series: Special papers (Geological Society of America) ; 235.
QE653.P797 1989
551.7'15'0978—dc19 89-1092
CIP

Cover Photo: Proterozoic rocks of northern New Mexico are separated into two tectonometamorphic plates by an extensive, subhorizontal, south-verging shear zone. Rocks overlying the shear zone preserve peak metamorphic conditions near the aluminum silicate triple point; rocks beneath the shear zone attained peak metamorphism in the lower granulite facies. The upper tectonic plate was metamorphosed as the lower plate thinned and the shear zone evolved.

The cover shows oriented photomicrographs from three different structural positions superimposed on a schematic structural cross-section. (a) Kyanite-andalusite-sillimanite schist, with biotite, muscovite, hematite and quartz, characteristic of specimens collected well above the shear zone. Minerals record minimal deformation; andalusite (gray) and sillimanite (red needles) are intergrown epitaxially, sharing a common c-axis. Kyanite is orange. (b) Kyanite-sillimanite-muscovite quartzite collected within the ductile shear zone that separates the two plates. Kyanite is orange; thin needles include both kyanite and sillimanite; muscovite is blue to green. Aluminum silicates are elongate in the extension lineation; asymmetry of the kyanite-muscovite "tectonic fish" records the top-to-the-south sense of shear. (c) Foliated and lineated sillimanite-potassium feldspar gneiss from granulite-grade rocks below the shear zone. Sillimanite is blue, potassium feldspar shows cross-hatched twinning, and quartz is colorless.

North is to the right, and the long edge of the field of view is 2.1 mm across in each photo. For additional discussion, see J. A. Grambling, M. L. Williams, R. F. Smith, and C. K. Mawer, this volume. Photomicrographs by C. K. Mawer and J. A. Grambling.

10 9 8 7 6 5 4 3 2

Contents

Preface

A two-day symposium on "Proterozoic Geology of the Southern Rocky Mountains" was held in Durango, Colorado, May 1984, in conjunction with the Rocky Mountain Section Meeting of the Geological Society of America. The goal of the symposium was to foster communication among workers actively studying the Early and Middle Proterozoic rocks in that region. This book is an outgrowth of the symposium.

This volume, containing eleven chapters, addresses the ca. 1,500-km-wide Proterozoic orogenic belt that lies south of the Archean craton of western North America. The orogenic belt is exposed in Wyoming, Colorado, Utah, New Mexico, Arizona, and possibly Nevada; it extends farther east in the subsurface. It is composed of meta-igneous and metasedimentary rocks, the former having yielded U-Pb zircon ages of between 1800 Ma and 1600 Ma. The lower Proterozoic rocks have been intruded by younger plutons, most of which have given U-Pb zircon ages of between 1500 Ma and 1300 Ma, although a few are several hundred million years younger. This volume provides an overview of the lithology, geochemistry, geochronology, deformation, and petrology of the orogenic belt.

The eleven chapters treat different geographic areas and are presented in order from northeast to southwest. The first paper (by R. S. Houston, E. M. Duebendorfer, K. E. Karlstrom, and W. R. Premo) examines the transition from Archean to Proterozoic rocks in southern Wyoming. The next three studies summarize U-Pb zircon geochronology. Premo and W. R. Van Schmus report on the geochronology of southeastern Wyoming and adjacent Colorado; M. E. Bickford and R. D. Shuster address the Gunnison and Salida areas of Colorado; and Bickford, R. L. Cullers, Shuster, Premo, and Van Schmus treat rocks from the Wet Mountains and southern Front Range of Colorado. The fifth study, by B. J. Tewksbury, summarizes the Proterozoic geology of the Needle Mountains, southwestern Colorado, and is followed by J. D. Collier's discussion of middle Proterozoic plutonic rocks in the same area. Four chapters consider the Proterozoic rocks of New Mexico. The first, by J. A. Grambling, M. L. Williams, R. F. Smith, and C. K. Mawer, hypothesizes that a vast area with kyanite-andalusite-sillimanite mineral assemblages formed as a result of synmetamorphic crustal extension. R. A. Wobus introduces the geology of Cimarron Canyon northeast of Taos, New Mexico. J. M. Robertson and K. C. Condie examine the major and trace element geochemistry of the Pecos greenstone belt, a volcanic-plutonic terrane in the southern Sangre de Cristo Mountains. D. G. Brookins and A. Majumdar consider the Rb-Sr systematics and geochemistry of the 1450 Ma Sandia granite. The volume closes with a chapter by K. E. Karlstrom which documents the presence and significance of Early Proterozoic structures in a part of Arizona.

Three general problems caught our interest as we edited the chapters in this volume: the geographic distribution of U-Pb zircon dates, the nature of lithologic contacts within the tectonic belt, and the significance of the middle Proterozoic granitic rocks that have intruded the lower Proterozoic units across most of the belt.

Several workers have attempted to synthesize the Proterozoic tectonic history of the belt using the assumption that rock ages become younger toward the south. U-Pb zircon dates from plutonic and volcanic rocks, as presented in this volume, challenge that assumption: rocks from all parts of the tectonic belt seem to show a similar range of U-Pb zircon dates. Premo and Van Schmus report dates of 1780 Ma to 1670 Ma in southern Wyoming and northern Colorado, and cite one date of 1625 Ma. Bickford and Shuster report dates from west-central Colorado that range from 1780 Ma to 1670 Ma. U-Pb zircon dates from east-central Colorado, described by Bickford and others, fall between 1705 Ma and 1665 Ma, with two samples possibly as young as 1620 Ma. Meta-igneous rocks in northern and central New Mexico have yielded U-Pb zircon dates from 1765 Ma to 1650 Ma, based upon references cited in Grambling and others. The observation that all regions seem to show similar, overlapping ranges of U-Pb zircon dates must be incorporated into future models of the Proterozoic tectonic evolution of the mountain belt.

Lithologic contacts in Proterozoic rocks of the region traditionally have been interpreted as either depositional or intrusive. The chapters by Tewksbury, Grambling and others, and Karlstrom challenge that view. These studies show that at least some lithologic contacts represent layer-parallel faults and shear zones. Although the amounts of offset across such structures are not well constrained, Grambling and others have identified one contact that separates rocks which differ by 6 kbar in peak metamorphic pressures; rocks on opposite sides of that contact show substantially different P-T paths. The tectonic nature of these contacts opens the possibility that structures in other parts of the tectonic belt are more complex than has been recognized.

Although middle Proterozoic plutonic rocks in the southern Rocky Mountains generally have been considered "anorogenic" or post-tectonic (cf., Collier, this volume), some plutons of this age in both Colorado and New Mexico, as described in this volume, are moderately to intensely foliated (see chapters by Bickford and others, Brookins and Majumdar, and Grambling and others). The foliations are solid-state deformational fabrics, and in places the foliated middle Proterozoic plutons are separated by shear zones from supracrustal rocks lacking any contact metamorphic overprint. One is left with the suspicion that a major component of ductile deformation occurred more recently than 1500 Ma in areas as widely separated as central Colorado and central New Mexico.

The volume illustrates previously unknown complexities of the Proterozoic tectonic belt. It also emphasizes that, although the belt is spectacularly exposed across an area several times the size of New England (U.S.A.), large areas have not been examined except at reconnaissance scale. If this volume serves no other purpose, we hope that it stimulates interest in the Proterozoic geology of the region.

We thank the following reviewers for providing prompt and useful technical reviews: John Aleinikoff, Barbara Barreiro, S. Chaudhuri, Jack Ellingson, Dave Frishman, Mike Garcia, Judith Hannah, Tom Krogh, Paul Lindberg, Mike McCallum, Bob Moench, Sharon Mosher, Mike O'Neill, William Olszewski, Carol Simpson, George Snyder, Frank Spear, Jon Spencer, Howard Stensrud, and Bob Wintsch.

Jeffrey A. Grambling
Barbara J. Tewksbury
June, 1988

Dedication to Ogden L. Tweto

This volume is dedicated to the memory of Ogden L. Tweto—expert on the geology and mineral deposits of Colorado and the Southern Rockies, and inspirational leader to many younger colleagues and friends. For more than 45 years he was identified with the U.S. Geological Survey, and for much of this time he was deeply involved in the geology of Colorado.

Tweto's long involvement with the geology of the Southern Rockies culminated with the preparation of the Geologic Map of Colorado published in 1979. This map stands as a monument to his remarkable contributions to our present understanding of the complex geology of this important area.

As an outgrowth of the Colorado map, Tweto embarked on studies of the larger features of the geology of the Southern Rocky Mountains; the results include a geologic map and accompanying report on the Proterozoic basement, a basement configuration map of Colorado, a tectonic map and report on the Rio Grande rift system in Colorado, and a regional synthesis of the tectonism of west-central Colorado. These syntheses chronicle Tweto's remarkable ability to assemble and synthesize regional geologic data into coherent and unifying concepts.

Tweto was a modest man of great integrity, and he had a truly uncommon amount of common sense. His wisdom was always available to those who sought his counsel, and is reflected in the merits of the collection of papers in this volume.

P. K. Sims

Geological Society of America
Special Paper 235
1989

A review of the geology and structure of the Cheyenne belt and Proterozoic rocks of southern Wyoming

R. S. Houston
Department of Geology and Geophysics, University of Wyoming, Laramie, Wyoming 82071
E. M. Duebendorfer
Department of Geoscience, University of Nevada, Las Vegas, Nevada 89154
K. E. Karlstrom
Department of Geology, Northern Arizona University, Flagstaff, Arizona 86011
W. R. Premo
U.S. Geological Survey, MS 963, Box 25046, Denver Federal Center, Denver, Colorado 80225

ABSTRACT

The Colorado Proterozoic province is separated from Archean rocks of the Wyoming province by a major structural boundary, the Cheyenne belt. Proterozoic rocks south of the Cheyenne belt are exposed in the Sierra Madre, Medicine Bow Mountains, and Laramie Range of southern Wyoming. They consist of metavolcanic units, metagraywacke, pelitic schist and gneiss, amphibolite, and felsic to mafic intrusive rocks that locally resemble rocks of central Colorado. North of the Cheyenne belt, Archean granite and gneiss of the Wyoming craton are overlain by a Late Archean and Early Proterozoic supracrustal sequence that contains quartzite, metadolomite, phyllite, and subordinate metavolcanic rocks. The eugeoclinal character of the metamorphic rocks south of the Cheyenne belt contrasts sharply with the dominantly siliciclastic supracrustal rocks north of the Cheyenne belt. Although specific sequences south of the belt have not yet been correlated between the Sierra Madre, Medicine Bow Mountains, and Laramie Range, similarities in age, lithology, and major element chemistry suggest that they are part of a single geologic terrane.

Macroscopic structure and microscopic kinematic indicators within the Cheyenne belt suggest that accretion of the Proterozoic rocks of northern Colorado to the Archean Wyoming craton was accomplished primarily by large-scale thrusting. Following accretion of individual thrust blocks, the boundary zone was steepened by folding and reactivated locally during a period of strike-slip movement. Presence of similar lithologies and shear zones south of the Cheyenne belt suggests that the southern margin of the Wyoming craton may have been a long-lived zone of crustal accretion.

INTRODUCTION

A major structural boundary between the Archean Wyoming craton to the north and the Proterozoic rocks of northern Colorado to the south was first recognized in the Medicine Bow Mountains in southeastern Wyoming by Houston and McCallum (1961). Originally named the Mullen Creek–Nash Fork shear zone for exposures in the Medicine Bow Mountains, the boundary has since been recognized in the Sierra Madre to the west and may be present in the Richeau Hills to the east, across a total distance of over 200 km (Fig. 1). In the Laramie Range, the Middle Proterozoic Laramie Anorthosite and the Sherman Granite occupy the projected position of the boundary (Fig. 1), but its location is reasonably well constrained by Nd isotopic data (Geist and others, 1987). In recognition of its regional extent, structural complexity, diverse character, and tectonic importance, the

Houston, R. S., Duebendorfer, E. M., Karlstrom, K. E., and Premo, W. R., 1989, A review of the geology and structure of the Cheyenne belt and Proterozoic rocks of southern Wyoming, *in* Grambling, J. A., and Tewksbury, B. J., Proterozoic geology of the southern Rocky Mountains: Boulder, Colorado, Geological Society of America Special Paper 235.

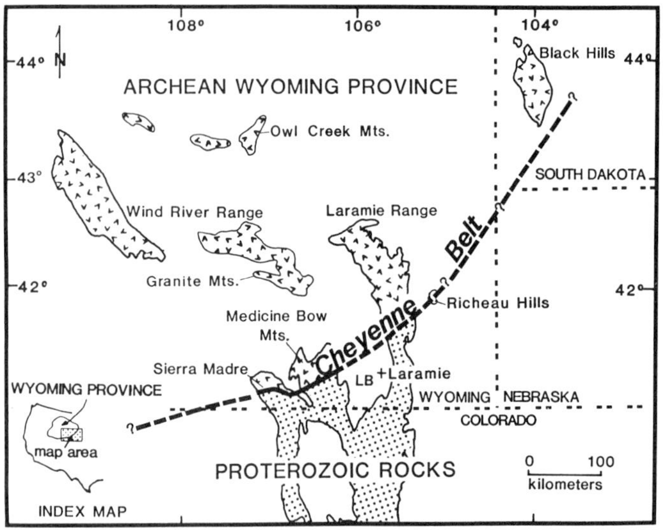

Figure 1. Index map showing location of Cheyenne belt in southern Wyoming LB = Laramie Basin (modified after Duebendorfer and Houston, 1987a).

boundary was renamed the Cheyenne belt by Houston and others (1979).

The intent of this review paper is to describe briefly the general rock types and structures within and south of the Cheyenne belt in the Sierra Madre and Medicine Bow Mountains of southern Wyoming and to establish the northern boundary of the Colorado Proterozoic province (Bickford and others, 1986), where Tweto did much of his work. The reader is referred to more detailed papers by Hills and Houston (1979), Karlstrom and Houston (1984), Johnson and others (1984), and Duebendorfer and Houston (1986, 1987a) for evidence used to define the margin and formulate a model for its development.

Recent geochronologic work continues to indicate that Archean rocks are absent south of the Cheyenne belt. Results from U-Pb zircon geochronology (Premo and Van Schmus, 1982; Premo, 1984a, b) are consistent with those of Hills and Houston (1979) and Hedge (in Karlstrom and others, 1981) north of the Cheyenne belt, but substantially refine earlier work of Hills and others (1968) and Divis (1976) south of the Cheyenne belt. The oldest reliable data obtained south of the Cheyenne belt is 1,791 ± 15 Ma on a relatively undeformed metavolcanic rock of the Green Mountain Formation of the south-central Sierra Madre (Premo and Van Schmus, this volume). Rocks older than 1,800 Ma have not been recognized south of the Cheyenne belt, and Sm-Nd studies support the earlier conclusion of DePaolo (1981) that crustal residence times for rocks of the Colorado Proterozoic province do not exceed 1,800 Ma (Nelson and DePaolo, 1985). Dates on Proterozoic rocks south of the Cheyenne belt are summarized in Table 1.

COCORP reflection profiles across the Laramie Range and Laramie Basin show reflectors interpreted by Allmendinger and others (1982) as shear zones striking northeast and dipping 55° southeast. They suggested that these shear zones may mark the boundary between Archean and Proterozoic rocks. Bouguer gravity profiles across the Cheyenne belt show a southward gradient with gravity values to the north as much as 100 milligals higher than values to the south. These data are consistent with a thicker and/or more felsic crust south of the Cheyenne belt, but Johnson and others (1984) favor the former interpretation as it is more consistent with results of refraction studies (Prodehl and Pakiser, 1980).

PRECAMBRIAN ROCKS NORTH OF THE CHEYENNE BELT

Late Archean (2,900 to 2,500 Ma) rocks north of the Cheyenne belt include a greenstone belt in the Laramie Mountains (Graff and others, 1982) and metasedimentary rocks that overlie crystalline basement in the Sierra Madre, Medicine Bow Mountains, and Hartville Uplift. The metasedimentary successions contain mature quartzite, metagraywacke, slate, phyllite, and iron formation and are transitional in lithology between typical Archean greenstone-belt assemblages and Early Proterozoic, quartz-rich miogeoclinal sequences (Karlstrom and others, 1981).

Early Proterozoic metasedimentary successions are exposed in the Sierra Madre and the Medicine Bow Mountains (Fig. 2). The Snowy Pass Supergroup of the Medicine Bow Mountains is a quartzite-dominated miogeoclinal succession approximately 10,000 m thick (Houston and others, 1968; Karlstrom and others, 1981). These sedimentary rocks record transgressive depositional conditions from dominantly fluvial, through shallow marine (deltaic), to deeper water marine deposition. The sequence is cut by numerous tholeiitic gabbro sills and dikes. A small, K-feldspar-poor, plagioclase-rich intrusion, which is spatially associated with and interpreted as a differentiate of the mafic dikes that cut part of the succession (Karlstrom and others, 1981), yielded a Rb-Sr whole-rock date of 2,000-2,150 Ma (Hedge, personal communication, 1982). The Snowy Pass Supergroup may record an Early Proterozoic rifting event reflected by the tholeiitic intrusive activity (Karlstrom, 1981; Karlstrom and others, 1983; Karlstrom and Houston, 1984).

PROTEROZOIC METAMORPHIC ROCKS SOUTH OF THE CHEYENNE BELT

Proterozoic metamorphic rocks south of the Cheyenne belt include metavolcanic rocks, metagraywacke, metapelite, amphibolite, hornblende gneiss, calc-schist, and marble. Where protoliths can be determined, rocks south of the Cheyenne belt are eugeoclinal in contrast to the dominantly miogeoclinal rocks to the north. The character of the basement upon which rocks south of the Cheyenne belt were deposited is not known. Due to metamorphism and deformation, stratigraphic relationships of these rocks have not been determined. Several ages or rock successions may be present.

TABLE 1. SUMMARY OF DATES ON PROTEROZOIC ROCKS FROM SOUTHERN WYOMING*

Locality and rock type or unit	Method	Age (Ma)	Interpretation	Reference†
Laramie Mountains, Wyoming				
Syenite in Laramie Anorthosite Syenite Complex (four localities)	U-Pb-Pb zir	1435 ± 15	P	1
Sherman Granite	Pb-Pb zir	1420 ± 20	P	1
Sherman Granite	Rb-Sr rk	1430 ± 60	P	1
Medicine Bow Mountains, Wyoming				
Keystone Quartz Diorite	U-Pb zir	1781 ± 7	P	2
Horse Creek Granite	U-Pb zir	1777 ± 7	P	2
Rambler Granite	Rb-Sr rk	1730 ± 15	P	3
Horse Creek Granite	Rb-Sr rk	1725 ± 120	P	3
Pegmatite M.77	Rb-Sr mu	1620 ± 40	P or T	4
Pegmatite M.35	Rb-Sr ml	1565 ± 40	P or T	4
Pegmatite M.75	Rb-Sr ml	1510 ± 40	P or T	4
Sherman Granite	Rb-Sr rk	1385 ± 30	P	4
Sierra Madre, Wyoming				
Green Mountain Formation (metavolcanic)	U-Pb zir	1791 ± 15	P	2
Encampment River Granodiorite	U-Pb zir	1779 ± 5	P	2
Sierra Madre Granite	U-Pb zir	1749 ± 8	P	2
Sierra Madre Granite	U-Pb zir	1763 ± 6	P	2
Big Creek Gneiss	U-Pb zir	1618 ± 22	T	2
White quartz monzonite	U-Pb zir	1627 ± 4	P or T	2
Sierra Madre Granite	Rb-Sr rk	1645 ± 50	T	5
Sierra Madre Granite	Rb-Sr rk	1635 ± 20	T	3
Elkhorn Mountain Gabbro	U-Pb zir	1774 ± 2	P	6
Elkhorn Mountain Gabbro	U-Pb zir	1769 ± 6	P	6
Elkhorn Mountain Gabbro	U-Pb zir	1768 ± 8	P	6

*P = age of primary igneous event; T = age of thermal event or thermally disturbed age; rk = whole-rock isochron; ml = mineral isochron; zir = zircon; mu = muscovite.
†References: 1 = Subbarayudu (1975); 2 = Premo and Van Schmus (this volume); 3 = Hills and Houston (unpublished data); 4 = Hills and others (1968); 5 = Divis (1976); 6 = Pallister and Aleinikoff (1987).

Medicine Bow Mountains

The dominant rock in the south-central Medicine Bow Mountains is dark gray, quartz-biotite-andesine gneiss (Fig. 2). The gneiss interfingers southward with a heterogeneous succession of mafic, pelitic, and calcareous gneisses. The latter succession crops out in a 5- to 8-km-wide band that extends entirely across the southern Medicine Bow Mountains (Houston and others, 1968). Based on lithology, it is probably volcanic and sedimentary in origin; however, no primary textures or structures are preserved. This sequence locally displays remarkably regular layering defined by alternating orthoamphibolite (tholeiitic basalt) and quartzofeldspathic gneiss (granite), which may represent original contacts transposed during deformation associated with development of the Cheyenne belt (Houston and others, 1968). Amphibolite ranges in thickness from one to several meters and locally crosscuts the gneiss, indicating that the amphibolite represents deformed and recrystallized dikes or sills of undetermined age.

The most northerly block within the Cheyenne belt proper is a sequence of amphibolite-grade gneiss, which in the eastern Medicine Bow Mountains, includes laminated quartzofeldspathic gneiss, migmatitic hornblende augen gneiss, and amphibolite (metagabbro; Fig. 3). Gneisses are granodioritic or tonalitic and the augen gneiss falls within the calc-alkaline field on several major-element plots (Duebendorfer, 1986). The laminated quartzofeldspathic gneiss may have a mixed igneous and sedimentary protolith. Concordant bodies of amphibolite are interpreted as sills or dikes emplaced prior to major deformation and metamorphism associated with development of the Cheyenne belt. In the western Medicine Bow Mountains, a bimodal succession of alternating tholeiitic orthoamphibolite and feldspathic orthogneiss (alkali granite) of undetermined age is present.

South of the northern gneissic block (Fig. 3) is a migmatitic gneiss, sillimanite-garnet schist, and hornblende gneiss terrane, which probably represents an original sequence of graywacke, impure siltstone, shale, and volcanogenic sedimentary rocks of intermediate to mafic composition. Mineral assemblages are diagnostic of the sillimanite–K-feldspar zone of the amphibolite facies (Duebendorfer, 1986). Associated granitic sills, dikes, and irregular bodies are corundum-normative and highly peraluminous, indicating that the igneous component of the migmatites may owe its origin to partial melting of the pelitic metasedimentary rocks. These are the highest grade rocks recognized south of the Cheyenne belt in Wyoming.

Sierra Madre

Metavolcanic rocks south of the Cheyenne belt are better preserved in the Sierra Madre (Fig. 2). The least deformed metavolcanic rocks, exposed in the Fletcher Park and Huston Park

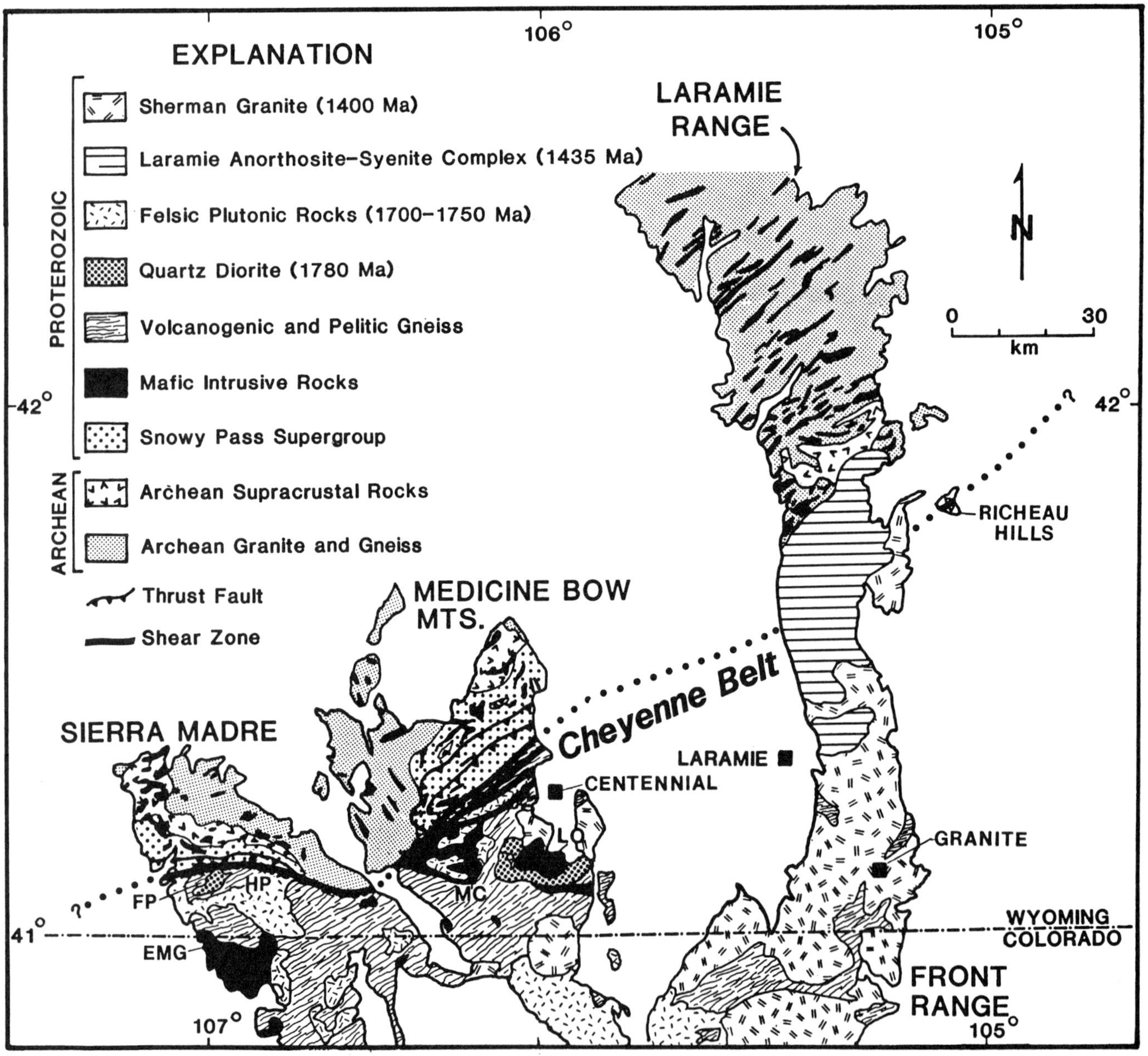

Figure 2. Generalized geologic map of Precambrian rocks, Sierra Madre, Medicine Bow Mountains, and Laramie Range, Wyoming. LO = Lake Owen Mafic Complex, MC = Mullen Creek Mafic Complex, EMG = Elkhorn Mountain Gabbro, FP = Fletcher Park area, HP = Huston Park area (modified after Duebendorfer and Houston, 1987a).

areas (FP and HP of Fig. 2), exhibit relict flow structure, pillows, eutaxitic structure, and recognizable lapilli and are interpreted as metamorphosed flows, tuffs, breccias, and agglomerates (Schmidt, 1983; Condie and Shadel, 1984). Mafic and felsic volcanic rocks occur in about equal proportions, similar to typical bimodal basalt-rhyolite assemblages (Divis, 1976). However, the rocks display calc-alkaline differentiation trends (Schmidt, 1983). Local preservation of primary features such as pillows establishes a marine origin for at least part of the assemblage, but felsic agglomerates may have been deposited subaerially. Although stratigraphy is difficult to document, two mafic-to-felsic cycles have been recognized in the Fletcher Park area; others may be present but unrecognized due to structural complexities (Houston and others, 1984).

The volcanic sequences of Fletcher Park and Huston Park are separated from supracrustal rocks of the remainder of the Sierra Madre by Proterozoic granitic intrusions; however, metamorphic rocks of the southwest Sierra Madre near the Colorado border (Fig. 2) are probably fine-grained correlatives of the Fletcher Park rocks (Swift, 1982). In this area, deformation and amphibolite-grade metamorphism have destroyed original textures and structures, and it is difficult to demonstrate conclusively a volcanic origin for these rocks.

Volcanogenic rocks of the south-central and southeastern Sierra Madre resemble those of the southern Medicine Bow Mountains. A large circular mass of amphibolite, approximately 10 km in diameter, in the south-central Sierra Madre locally displays textures such as recognizable lapilli suggestive of a volcanic origin but is more highly recrystallized and more mafic than metavolcanic rocks of the Fletcher Park area. North and east of the amphibolite, metamorphic rocks consist of biotite gneiss, calc-schist, marble, amphibolite, hornblende gneiss, sillimanite gneiss, and amphibole-garnet gneiss. These rocks probably represent a mixed volcano-sedimentary succession similar to that of the southern Medicine Bow Mountains, with which they are nearly contiguous.

Premo and Van Schmus (this volume) report an age of 1,791 ± 15 Ma for zircon from a felsic metavolcanic rock from

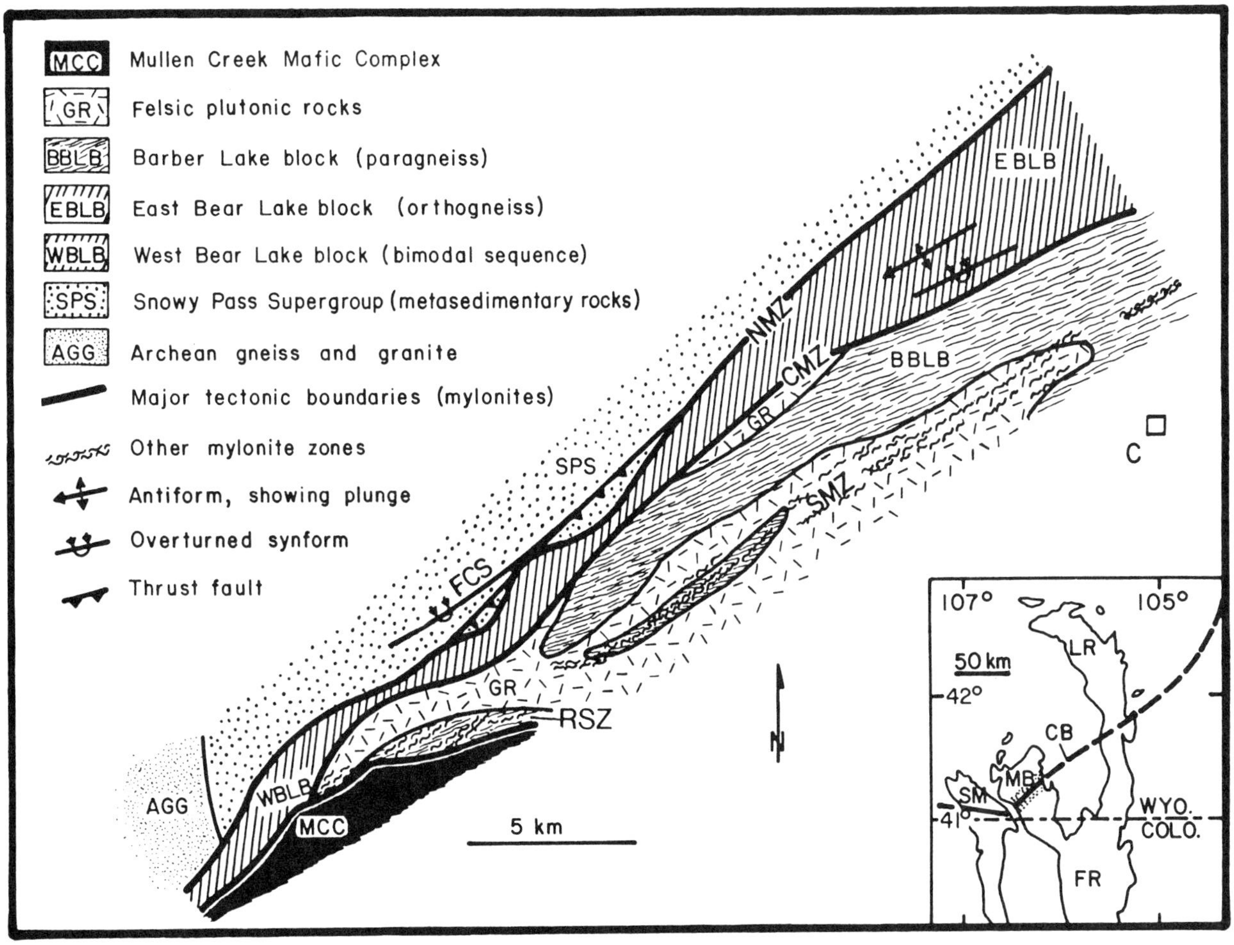

Figure 3. Highly generalized geologic map of the Cheyenne belt in the Medicine Bow Mountains showing tectonic blocks and major structures. NMZ = north mylonite zone, CMZ = central mylonite zone, SMZ = south mylonite zone, FCS = French Creek syncline, C = town of Centennial. Inset: Location of Cheyenne belt in southeastern Wyoming. Stippled pattern = area of geologic map, SM = Sierra Madre, MB = Medicine Bow Mountains, LR = Laramie Range, FR = Front Range (modified after Duebendorfer and Houston, 1987a).

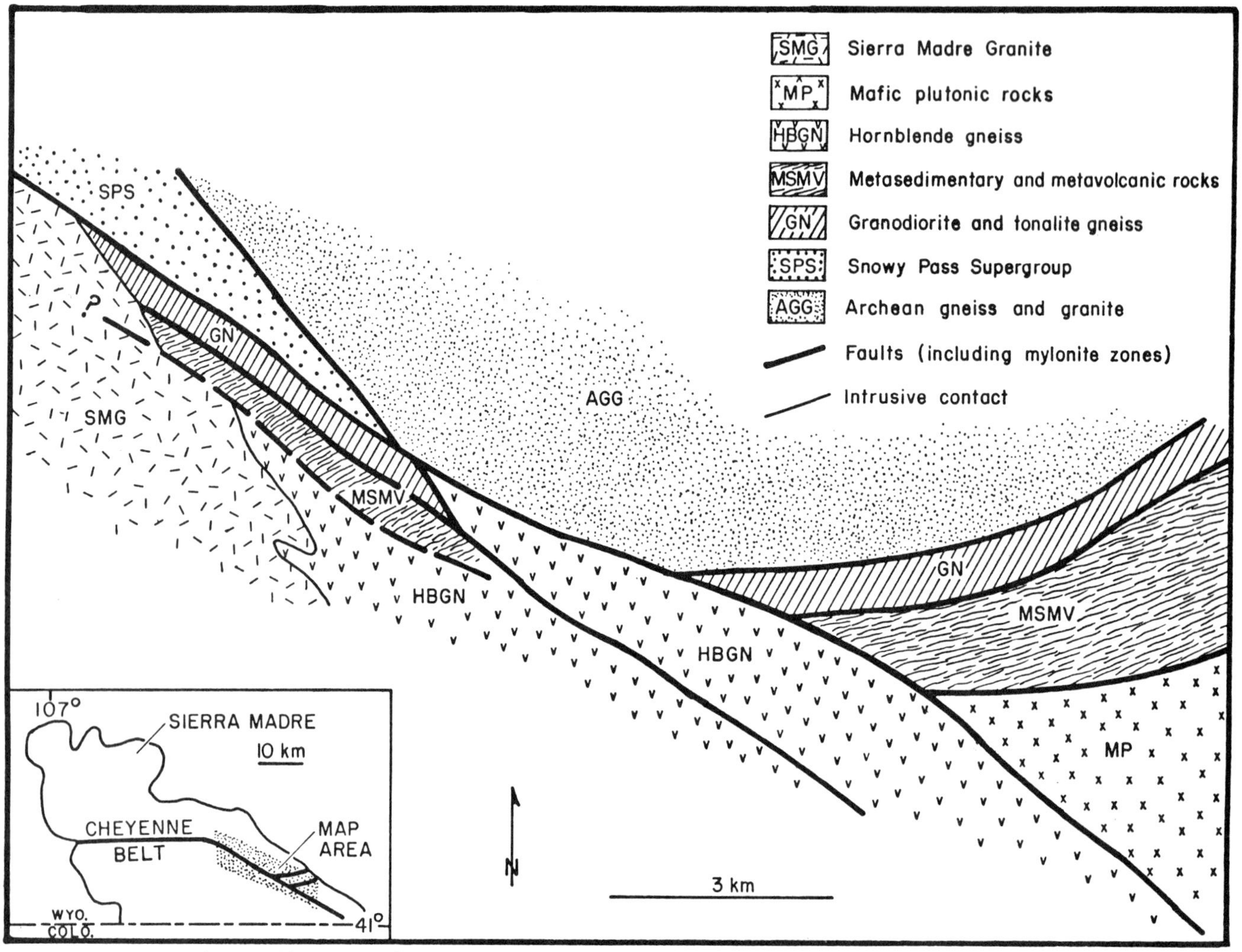

Figure 4. Highly generalized geologic map of the Cheyenne belt in the eastern Sierra Madre showing tectonic blocks and major structures.

the metavolcanic sequence. Zircons from the Encampment River Granodiorite, which may be comagmatic with the metavolcanic rocks, yielded an age of 1,779 ± 5 Ma. Zircons from the Big Creek Gneiss yield an age of 1,618 ± 22 Ma (Premo and Van Schmus, this volume), which may represent the age of a high-grade thermal event.

Four distinct, but narrow, lithotectonic blocks are present within the Cheyenne belt in the eastern Sierra Madre (Fig. 4). From north to south, these include Archean granite gneiss and associated supracrustal rocks; compositionally banded granodiorite and tonalite gneiss of undetermined age; a block containing sillimanite-garnet schist and hornblende gneiss; and a hornblende gneiss–mafic plutonic terrane. These rocks are lithologically similar to rocks exposed within the Cheyenne belt in the Medicine Bow Mountains, are bounded by major mylonite zones, and appear in the same north-south progression as do the blocks in the Medicine Bow Mountains. These observations support the inference that the Cheyenne belt, as a lithologic and structural feature, is correlative between the two ranges.

Laramie Mountains

Metamorphic rocks are exposed locally in the Laramie Mountains south of the projected position of the Cheyenne belt. North and southwest of Granite, Wyoming (Fig. 2), metavolcanic rocks very similar to those of the central and southwest Sierra Madre and Centennial Ridge (west of Centennial, Wyoming) in the Medicine Bow Mountains are exposed in a north-northeast–striking strip. These metavolcanic rocks have well-preserved relict textures and structures, and range from basalt to rhyodacite, but contain a higher proportion of felsic rocks than the volcanic sequences of the Sierra Madre or Medicine Bow Mountains. These rocks extend south to within 3 to 5 km of the Colorado border where they grade into quartz-biotite-plagioclase gneiss with some interlayered hornblende gneiss, biotite-microcline-quartz gneiss, and amphibolite (Darton and others, 1910). This latter sequence appears largely metasedimentary, but probably includes a metaigneous component represented by amphibolite and biotite-microcline-quartz gneiss.

Patches of metamorphic rocks, largely pelitic schists, are present in the south-central Laramie Mountains (Fig. 2), and metasedimentary rocks also crop out in the Richeau Hills (Figs. 1 and 2; Mueller, 1982). The metasedimentary rocks of the Richeau Hills are intruded by granites of possible Archean age (G. L. Snyder, personal communication, 1985).

None of the metamorphic rocks of the southern Laramie Mountains has been dated, but based on lithologic similarities the metavolcanic rocks probably correlate with those of the Sierra Madre and Medicine Bow Mountains, and thus may be Early Proterozoic. The age of the metasedimentary rocks is undetermined.

PROTEROZOIC PLUTONIC ROCKS SOUTH OF THE CHEYENNE BELT

Mafic plutonic rocks

Three large mafic bodies crop out south of the Cheyenne belt in southern Wyoming and northern Colorado (Fig. 2). In the Medicine Bow Mountains, these include the Lake Owen Mafic Complex and the Mullen Creek Mafic Complex, two large (60 to 125 km^2) layered gabbroic complexes. A date of 1,777 ± 4 Ma by U-Pb determination on zircons has been obtained from the Horse Creek Granite, which cuts the Mullen Creek Mafic Complex (Premo and Van Schmus, this volume).

The Lake Owen Mafic Complex of the eastern Medicine Bow Mountains has been tilted to expose as much as 4.5 km of interlayered norite, gabbro, troctolite, and anorthosite. It is divided into three major cycles, which are distinguished on the basis of differences in lithology, igneous structures, and major-element compositional trends (Houston and Orback, 1976). The Mullen Creek Mafic Complex of the western Medicine Bow Mountains is compositionally similar to the Lake Owen Mafic Complex, but is highly disrupted and invaded by granite sills, dikes, and irregular pods. Since the granite is largely confined to the Mullen Creek Complex and is older than most granite south of the Cheyenne belt, it may be comagmatic with the rocks of the complex.

A body of non-layered gabbro, the Elkhorn Mountain Gabbro (Snyder, 1980), crops out in a 125-km^2 area in the Sierra Madre and adjacent parts of the Park Range in northern Colorado. Snyder reported a U-Pb zircon age of 1,781 Ma (Hedge, in Snyder, 1980) for this body; Pallister and Aleinikoff (1987) report overlapping U-Pb zircon ages of 1,774 ± 2 Ma, 1,769 ± 6 Ma, and 1,768 ± 8 Ma for intrusive rocks that cut, and inclusions within, the Elkhorn Mountain Gabbro. The Elkhorn Mountain Gabbro, which includes hornblende-clinopyroxene-biotite gabbro, hornblende-biotite gabbro, hornblende gabbro, and porphyritic feldspar amphibolite (Snyder, 1980), is compositionally more homogeneous than either the Lake Owen or Mullen Creek Mafic Complexes. Medium grain size and diabasic texture suggest emplacement at relatively shallow tectonic levels. According to Snyder (personal communication, 1984), the average composition of eight samples from the Elkhorn Mountain Gabbro is that of high-Al basalt.

Smaller bodies of mafic intrusive rock ranging in width from a few to nearly 1,000 m are common south of the Cheyenne belt in all three ranges. Metagabbro dominates, but olivine metagabbro, metadiorite, and metapyroxenite are also present. Most mafic bodies have been largely or completely converted to amphibolite.

The tectonic significance of the mafic complexes remains obscure. Their chemical and mineralogical characteristics, and occurrence near the Cheyenne belt, suggest that they may be tectonically emplaced remnants of the lower parts of magmatic arcs (Hills and Houston, 1979; Karlstrom and Houston, 1984; Pallister and Aleinikoff, 1987).

Intermediate plutonic rocks

The Encampment River Granodiorite in the Sierra Madre and the Keystone Quartz Diorite in the Medicine Bow Mountains are distinctive plutonic rocks with similar mineralogic and chemical characteristics. The Encampment River Granodiorite intrudes metavolcanic rocks of the central Sierra Madre and contains abundant metavolcanic inclusions. The Keystone Quartz Diorite contains abundant highly altered inclusions that are probably mafic volcanic rocks (Houston and others, 1968). The Encampment River Granodiorite has been dated by U-Pb zircon at 1,779 ± 5 Ma (Premo and Van Schmus, this volume), and the Keystone Quartz Diorite has yielded an age of 1,781 ± 7 Ma (Premo and Van Schmus, this volume). Smaller intermediate plutonic bodies containing abundant metavolcanic xenoliths are present in the Laramie Mountains south of Granite, Wyoming. Complex cross-cutting relationships between the metavolcanic, mafic, and intermediate plutonic rocks suggest that they may represent a broadly comagmatic suite. This is supported by the geochronological and geological data, which suggest that the metavolcanic rocks dated at 1,791 ± 15 Ma, mafic complexes dated at about 1781 Ma, and intermediate plutonic rocks dated at 1,781 ± 7 Ma are genetically related. If so, the intermediate and mafic intrusive rocks may represent plutons associated with the lower parts of island arcs (represented by the metavolcanic rocks) sutured to the southern margin of the Wyoming craton along the Cheyenne belt (Hills and Houston, 1979; Pallister and Aleinikoff, 1987). Overall geologic relationships suggest a deeper level of exposure in the southern Medicine Bow Mountains and the southeastern Sierra Madre than elsewhere. This might explain the higher proportion of mafic rocks and the greater intensity of deformation and metamorphism in these areas.

Felsic plutonic rocks

Granite and minor quartz monzonite cut metavolcanic and metasedimentary rocks, mafic plutonic rocks, and rocks of the granodiorite-tonalite suite. The felsic intrusive rocks make up about 50 percent of the outcrop area of the Sierra Madre, where they are referred to as the Sierra Madre Granite (Divis, 1976). Felsic plutonic rocks have also been recognized in the Medicine

Bow Mountains. These include two-mica, peraluminous granite, and leucogranite of Mullen Creek, which extends northeast-southwest in a discontinuous band across the Medicine Bow Mountains (Fig. 3; Duebendorfer, 1986). A small body, the Rambler Granite, is present in the south-central Medicine Bow Mountains.

These intrusive bodies generally have sharp contacts with the country rock. Most of the felsic intrusive rocks are faintly foliated to massive and contain xenoliths that show an early deformational fabric (discussed below). This relationship suggests that they largely postdate that deformational event (discussed below); however, they locally exhibit strong foliation parallel to that of the country rock. Snyder and Hedge (1978) suggest that the Buffalo Pass Pluton in northern Colorado is syntectonic in the sense that it may have been emplaced during the late stages of the principal deformation episode. The granites of Mullen Creek exhibit two lineations, suggesting that they are late synkinematic with respect to the early deformational event but pre-date later strike-slip deformation (see below).

Felsic plutonic rocks in the Sierra Madre (Sierra Madre Granite) and the Medicine Bow Mountains have been dated by U-Pb zircon methods and yield ages ranging from 1,763 ± 6 Ma to 1,749 ± 8 Ma (Premo and Van Schmus, this volume). Thus, on both geologic and geochronologic grounds, there is a well-deformed episode of emplacement of late synkinematic (with respect to the early inferred thrusting event) felsic intrusions in southern Wyoming and northern Colorado at approximately 1,750 Ma.

Post-kinematic plutonic rocks

A later magmatic event, apparently confined to the Sierra Madre, is represented by a white, coarse-grained quartz monzonite and associated pegmatite. A sample of this quartz monzonite, which intrudes both the Sierra Madre Granite and the supracrustal rocks, yielded a U-Pb zircon date of 1,627 ± 4 Ma (Premo and Van Schmus, this volume). Both the zircon morphology and the peraluminous character (Divis, 1976) suggest an origin by partial melting of supracrustal rocks and perhaps pre-existing igneous rocks.

The Laramie Anorthosite is a major massif-type anorthosite exposed in the central Laramie Range (Fig. 2). It occurs in two masses, a large, north-striking, antiformal body exposed over a 760-km^2 area and a smaller (60 km^2), east-northeast–trending body. Two positive magnetic anomalies in the Laramie Basin west of the exposed anorthosite suggest that it extends beneath Phanerozoic cover 30 km to the southwest. The anorthositic core of the larger body is rimmed by noritic anorthosite (Hagner, 1968) and bordered on the northwest by a large, crescent-shaped mass of syenite (Newhouse and Hagner, 1957). Gravity (Hodge and others, 1973) and seismic data (Allmendinger and others, 1982) indicate that the anorthosite is a 4- to 5-km-thick slab, possibly floored by granite. Studies of the crystallization conditions of ferrosyenite associated with the anorthosite (Bochensky and Frost, 1982) suggest an emplacement depth of 7 to 9 km. Syenite that is probably comagmatic with the Laramie Anorthosite has been dated as 1,435 ± 15 Ma (U-Pb zircon, Subbarayudu, 1975).

The Sherman Granite and equivalent bodies crop out in the Laramie Range, the eastern Medicine Bow Mountains, and the Park and Front Ranges of northern Colorado (Fig. 2). The Sherman Granite is distinguished from older granites by its coarse grain size and weak, though usually discernible, primary flow foliation. Facies of the Sherman Granite include a medium-grained quartz monzonite, a porphyritic granite with large microcline phenocrysts, and a pink coarse-grained granite with crystals that average 5 to 6 mm in diameter. The coarse-grained granite constitutes 80 to 90 percent of the exposures. Contacts between the Sherman Granite and country rocks are typically sharp and irregular. Inclusions are common but not abundant; most are elongate blocks of altered amphibolite with nearly vertical long axes. The Sherman Granite in the Laramie Range has been dated at 1,430 ± 60 Ma (Rb-Sr whole rock, Subbarayudu, 1975); granite in the Medicine Bow Mountains has been dated at 1,385 ± 30 Ma (Rb-Sr whole rock, Hills and others, 1968).

STRUCTURE OF THE CHEYENNE BELT

In the Medicine Bow Mountains, where it is best exposed, the Cheyenne belt is a 1- to 7-km-wide zone of penetrative ductile deformation that includes parts of three, possibly four, tectonic blocks (Fig. 3; McCallum, 1964; Duebendorfer and Houston, 1986, 1987a). These blocks are separated by 50- to 200-m-wide, northeast-striking, subvertical to steeply southeast-dipping mylonite zones. The blocks exhibit successively higher grades of metamorphism to the south, from the chlorite zone of the greenschist facies in the north to the sillimanite–K-feldspar zone of the amphibolite facies in the south. Increases in metamorphic grade occur abruptly across the mylonite zones, suggesting south-side-up juxtaposition of blocks that originated at different crustal levels. This interpretation is strengthened further by the presence of an inverted metamorphic gradient in the supracrustal rocks directly north of the Cheyenne belt in the eastern Medicine Bow Mountains (Duebendorfer, 1988).

Within the Sierra Madre, the Cheyenne belt can be divided into an eastern, northwest- to northeast-trending segment (Fig. 4) and a western, east-west-trending segment. The eastern segment, like its counterpart in the Medicine Bow Mountains, consists of three mylonite zones separating four distinct tectonic blocks, which may be broadly correlative with rocks in the Medicine Bow Mountains. The zone of deformation, which includes the mylonite zones as well as penetratively deformed and undeformed areas within individual tectonic blocks, ranges from 0.5 to 2.0 km in width. The western segment consists largely of a single shear zone 20 to 100 m wide. With the exception of localized pods of ductilely deformed gneiss, rocks of this segment are cataclastic. In addition, supracrustal rocks north of the shear zone are weakly deformed, with relict bedding clearly evident. This contrasts with the strongly transposed character of correlative rocks north of the shear zone in the Medicine Bow Mountains. These observations collectively suggest that much of the ductile deformation zone within the eastern Sierra Madre was removed by a later brittle event and that the present east-west trend may not reflect the initial orientation of the Cheyenne belt in the western Sierra Madre.

The blocks and mylonite zones in both ranges exhibit evidence for a complex polyphase deformation history involving development of penetrative fabrics, mylonite, and cataclasite. Two main events involving ductile deformation have been documented (Duebendorfer and Houston, 1986, 1987a, b), an early phase of north-northwest–directed thrusting and a later phase of distributive, dominantly dextral strike-slip.

Early thrusting (D_1)

Structures associated with D_1 include a steeply dipping mylonitic fabric; a strong, subvertical, penetrative mineral elongation lineation; and variably plunging, southwest- to northeast-trending folds (Duebendorfer and Houston, 1986, 1987a, b). Directly north of the Cheyenne belt in the Medicine Bow Mountains, rocks of the Snowy Pass Supergroup are folded into a macroscopic, overturned, steeply northeast-plunging syncline, the south limb of which is cut by a southeast-dipping thrust fault that deletes at least 3 km of section (Fig. 3). North of the northernmost mylonite zone of the Cheyenne belt in both ranges, mesoscopic structures—including steeply plunging isoclinal folds, mullion structures in quartzites, and a subvertical mineral elongation lineation—are strongly developed in both Archean crystalline and associated supracrustal rocks.

Rocks within the Cheyenne belt south of the northern mylonite zone in the western Medicine Bow Mountains and eastern Sierra Madre also exhibit a penetrative, subvertical mineral elongation lineation defined by minerals such as hornblende, which are characteristic of the amphibolite facies. In the eastern Medicine Bow Mountains, subhorizontal compositional banding, which contains mesoscopic rootless isoclines, is deformed into macroscopic and mesoscopic, open to tight, upright to northwest-verging folds that plunge gently southwest. Near the north and central mylonite zones of the Cheyenne belt in the eastern Medicine Bow Mountains (Fig. 3), this subhorizontal fabric is transposed into parallelism with the subvertical foliation of the bounding mylonite zones. Transposition is accompanied by increased fold amplitude, marked hinge-line arcuation, and steepened fold hinges. This change in fold geometry is coincident with, and restricted to, the vicinity of the major mylonite zones.

The mylonite zones, which bound tectonic blocks in both ranges, are dominated by a strong planar fabric that is parallel to the foliation in the immediately adjacent rocks. Where not overprinted by later structures, a steeply plunging mineral lineation is evident. Folds are rare, occurring only locally as rootless hinges within the mylonitic foliation. Cataclasis, generally restricted to narrow (<10 m) fault zones, locally obscures the mylonitic foliation.

The parallelism of planar and linear fabric elements across the entire Cheyenne belt suggests that these structures developed in response to the same deformational event. The argument is more compelling when one considers that, with increasing distance north of the Cheyenne belt, regional foliation lies at high angles to the subvertical fabric of the Cheyenne belt. This implies that pre-existing fabrics have been rotated due to high strains associated with mylonite development. This is further substantiated by steepening of fold axes and associated hinge-line arcuation near the mylonite zones. The following observations of macroscopic structure support a hypothesis of early thrusting as a mechanism for juxtaposition of contrasting terranes across the Cheyenne belt: (1) The steeply plunging mineral elongation lineation associated with the Cheyenne belt indicates predominantly dip-slip movement. (2) Northwest-verging folds in the eastern part of the Cheyenne belt in the Medicine Bow Mountains are suggestive of northwest-directed thrusting. (3) Juxtaposition along mylonite zones of blocks representing successively deeper crustal levels (i.e., increase in metamorphic grade to the south) is consistent with northwest-directed thrusting. (4) Systematic study of microscopic sense-of-shear indicators in rocks adjacent to and within the Cheyenne belt in both ranges indicates a south-side-up shear sense (Duebendorfer and Houston, 1987a, b). (5) Quartz fabric analysis on mylonitic quartzites from both the Medicine Bow Mountains and the Sierra Madre indicates north-directed tectonic transport (Duebendorfer, 1986; Duebenforder and Houston, 1987b). Equilibrium mineral assemblages associated with development of principal fabric elements indicate deformation under low- to middle-amphibolite-facies conditions. This estimate is consistent with garnet-biotite geothermometry on synkinematic minerals, which indicates that deformation occurred at temperatures of at least 475°C (Duebendorfer, 1986; Duebendorfer and Houston, 1987a).

At present, planar fabrics within the Cheyenne belt are subvertical to steeply southeast dipping. We favor low-angle thrusting followed by rotation of the mylonite zones to the present steep orientation rather than high-angle reverse faulting to explain the observed juxtaposition for the following reasons: (1) Large displacements along the Cheyenne belt are implied by the juxtaposition of markedly dissimilar rocks formed at different crustal levels. (2) Bedding plane faults exhibiting lateral ramp geometries cut the Snowy Pass Supergroup and are interpreted as thrust faults (Karlstrom and Houston, 1984). (3) It is difficult to envision a mechanism by which high-angle reverse faulting could produce penetrative strains over a zone as wide as the Cheyenne belt. (4) Juxtaposition of contrasting age provinces, one floored by Archean basement and the other by Proterozoic basement, cannot be explained by vertical movements alone. (5) Based on geochronological evidence, Karlstrom (1981) and Karlstrom and Houston (1984) postulated several hundred kilometers of displacement along the Cheyenne belt. This magnitude of displacement is clearly inconsistent with high-angle reverse faulting. Steepening of originally low-angle fabrics during continued postcollisional convergence has been proposed for the Alps (Roeder,

1973, 1977), the northern Sierra Nevada (Moores and Day, 1984; Day and others, 1985), and other orogenic belts (Burke and others, 1975), and may be typical of convergent zones.

On the basis of U-Pb zircon geochronology (Premo and Van Schmus, this volume) on the lake synkinematic Sierra Madre Granite, we infer that the main phase of thrusting occurred prior to 1,750 Ma.

Late strike-slip (D_2)

Within the Medicine Bow Mountains, the early, penetrative, subvertical mineral lineation is overprinted by a nonpenetrative, localized, subhorizontal slickenside lineation defined by synkinematic chlorite and epidote, suggesting deformation at conditions characteristic of the greenschist facies. These are associated with kink folds, which have a consistent z-shaped down-plunge profile and locally break into minor ductile faults that exhibit consistent dextral offset. Retrograde assemblages (chlorite and epidote) associated with these ductile shears indicate development under lower greenschist-facies conditions. The steeply plunging nature and consistent dextral asymmetry of the folds suggest that they developed in a zone of dextral strike-slip translation. Microscopic kinematic indicators also indicate a dextral shear sense (Duebendorfer and Houston, 1987b). Strike-slip faulting in the Medicine Bow segment of the Cheyenne belt appears to have been localized along the northern and southern mylonite zones.

Nonpenetrative subhorizontal lineations, similar to those of the Medicine Bow segment, are present, but less well developed in the eastern segment of the Cheyenne belt in the Sierra Madre. Dextral kink folds have not been recognized within the Sierra Madre; however, macroscopic and microscopic kinematic indicators in pre-D_2 granitoids suggest dextral shear (Duebendorfer and Houston, 1987b).

The age of the strike-slip event is bracketed only between the inferred date of thrusting (1,750 Ma) and the age of the undeformed Sherman Granite (1,385 to 1,420 Ma).

SUMMARY

Rocks south of the Cheyenne belt in Wyoming comprise a heterogeneous assemblage of metasedimentary and metavolcanic units intruded by several generations of felsic to mafic plutonic rocks. The eugeoclinal character of the southern metamorphic assemblage contrasts sharply with the miogeoclinal nature of the supracrustal rocks north of the Cheyenne belt.

Although specific sequences south of the Cheyenne belt in the Sierra Madre, Medicine Bow Mountains, and the Laramie Range have not yet been definitively correlated, similarity in lithologic character, age, and major element geochemistry (where available) suggest that these ranges constitute part of a single geologic province.

Geochronological evidence from plutonic rocks, combined with geologic observations, suggest the following history for rocks south of the Cheyenne belt in southern Wyoming. (1) An early, volcano-plutonic event involving development of early (1,790 Ma) arc volcanics and associated calc-alkaline granodiorites and quartz diorites on non-Archean, probably oceanic, crust. The roots of these arc volcanic rocks may be represented by the large mafic complexes south of the Cheyenne belt. (2) Deformation of the above sequences, probably related to thrusting (D_1) along the Cheyenne belt between 1,780 and 1,750 Ma. (3) Late-synkinematic granitic plutonism (1,750 Ma) that may represent melts derived from crustal thickening during thrusting. These rocks record the latest deformation associated with thrusting. (4) Dextral strike-slip movement along the Cheyenne belt. (5) Post-kinematic intrusion of quartz monzonite and pegmatite. The temporal relationship between (4) and (5) is undetermined.

Macroscopic structure and microscopic kinematic indicators within the Cheyenne belt suggest that juxtaposition of this southern province against the Archean Wyoming province was accomplished primarily by large-scale thrusting. Following accretion of individual blocks, the boundary zone was steepened and reactivated locally during a period of distributive strike-slip movement. It is not clear whether the two deformations represent separate, distinct tectonic events or different stages of a single protracted event perhaps related to oblique collisions, as suggested by Duebendorfer and Houston (1987a). Presence of similar lithologies and shear zones south of the Cheyenne belt in southern Wyoming and northern Colorado suggests that the southern margin of the Wyoming craton may have been a long-lived zone of crustal accretion.

ACKNOWLEDGMENTS

Field research was supported by National Science Foundation Grants #EAR-8400227, #EAR-8615736, and #EAR-8796347 to R. S. Houston, S. B. Smithson, and E. M. Duebendorfer. Geochronological research was supported by #EAR 79-19544, #EAR 81-18234, and #EAR 83-03110 to W. R. Van Schmus. We thank J. A. Grambling, M. E. McCallum, and an anonymous reviewer for their helpful comments.

REFERENCES CITED

Allmendinger, R. W., Brewer, J. A., Brown, L. D., Kaufman, S., Oliver, J. E., and Houston, R. S., 1982, COCORP profiling across the Rocky Mountain Front in southern Wyoming; Part 2, Precambrian basement structure and its influence on Laramide deformation: Geological Society of America Bulletin, v. 93, p. 1253–1263.

Bickford, M. E., Van Schmus, W. R., and Zietz, I., 1986, Proterozoic history of the midcontinent region of North America: Geology, v. 14, p. 492–496.

Bochensky, P., and Frost, B. R., 1982, Contact metamorphism produced by the Laramie anorthosite complex, Morton Pass, Wyoming: Geological Society of America Abstracts with Programs, v. 14, p. 303.

Burke, K., Dewey, J. F., and Kidd, W.S.F., 1975, Dominance of horizontal movements, arc, and microcontinental collisions during the later permobile regime, *in* Windley, B. F., ed., The early history of the Earth: London, Wiley Interscience, p. 113–129.

Condie, K. C., and Shadel, C. A., 1984, An early Proterozoic volcanic arc succession in southeastern Wyoming: Canadian Journal of Earth Science, v. 21, p. 415–427.

Darton, N. H., Blackwelder, E., and Siebenthal, C. E., 1910, Geologic atlas of the United States, Laramie–Sherman folio, Wyoming: U.S. Geological Survey, 17 p.

Day, H. W., Moores, E. M., and Tuminas, A. C., 1985, Structure and tectonics of the northern Sierra Nevada: Geological Society of America Bulletin, v. 96, p. 436–450.

DePaolo, D. J., 1981, Neodymium isotopes in the Colorado Front Range and crust; Mantle evolution in the Proterozoic: Nature, v. 291, p. 193–196.

Divis, A. F., 1976, Geology and geochemistry of the Sierra Madre Range, Wyoming: Colorado School of Mines Quarterly, v. 71, 127 p.

Duebendorfer, E. M., 1986, Structure, metamorphism, and kinematic history of the Cheyenne belt, Medicine Bow Mountains, southeastern Wyoming [Ph.D. thesis]: Laramie, University of Wyoming, 323 p.

—— , 1988, Evidence for an inverted metamorphic gradient associated with a Precambrian suture, southern Wyoming: Journal of Metamorphic Geology, v. 6, p. 41–63.

Duebendorfer, E. M., and Houston, R. S., 1986, Kinematic history of the Cheyenne belt, Medicine Bow Mountains, southeastern Wyoming: Geology, v. 14, p. 171–174.

—— , 1987a, Proterozoic accretionary tectonics at the southern margin of the Archean Wyoming craton: Geological Society of America Bulletin, v. 98, p. 554–568.

—— , 1987b, Microstructure and quartz fabric analysis, Cheyenne belt, Wyoming: Geological Society of America Abstracts with Programs, v. 19, p. 648.

Geist, D. J., Frost, C. D., and Frost, B. R., 1987, Delineation of a hidden Archean–Proterozoic suture; Nd isotopes in the Sherman Granite: Geological Society of America Abstracts with Programs, v. 19, p. 674.

Graff, P. J., Sears, J. W., Holden, G. S., and Hausel, W. D., 1982, Geology of the Elmers Rock greenstone belt, Laramie Range, Wyoming: Geological Survey of Wyoming Report of Investigations 14, p. 23.

Hagner, A. F., 1968, The titaniferous magnetite deposit at Iron Mountain, Wyoming, *in* Ore deposits of the United States, 1933–1967 Graton–Sales Volume: American Institute of Mining and Metallurgical Engineering, v. 1, p. 665–680.

Hills, F. A., and Houston, R. S., 1979, Early Proterozoic tectonics of the central Rocky Mountains, North America: Contributions to Geology, v. 17, p. 89–109.

Hills, F. A., Gast, P. W., Houston, R. S., and Swainbank, I., 1968, Precambrian geochronology of the Medicine Bow Mountains of southeastern Wyoming: Geological Society of America Bulletin, v. 79, p. 1757–1984.

Hodge, D. S., Owen, L. B., and Smithson, S. B., 1973, Gravity interpretation of the Laramie anorthosite complex, Wyoming: Geological Society of America Bulletin, v. 84, p. 1451–1464.

Houston, R. S., and McCallum, M. E., 1961, Mullen Creek–Nash Fork shear zone, Medicine Bow Mountains, southeastern Wyoming [abs.]: Geological Society of America Special Paper 68, p. 91.

Houston, R. S., and Orback, C. J., 1976, Geologic map of the Lake Owen Quadrangle, Albany County, Wyoming: U.S. Geological Survey Map GQ–1034, scale 1:24,000.

Houston, R. S., and others, 1968, A regional study of rocks of Precambrian age in that part of the Medicine Bow Mountains lying in southeastern Wyoming, with a chapter on the relationship between Precambrian and Laramide structure: Wyoming Geological Survey Memoir 1, 167 p.

Houston, R. S., Karlstrom, K. E., Hills, F. A., and Smithson, S. B., 1979, The Cheyenne belt; A major Precambrian crustal boundary in the western United States: Geological Society of America Abstracts with Programs, v. 11, p. 446.

Houston, R. S., Schmidt, T. G., and Lane, M. E., 1984, Mineral resource potential and geologic map of the Huston Park roadless area, Carbon County, Wyoming: U.S. Geological Survey Map MF–1637, scale 1:24,000.

Johnson, R. A., Karlstrom, K. E., Smithson, S. B., and Houston, R. S., 1984, Gravity profiles across the Cheyenne belt, a Precambrian crustal suture in southeastern Wyoming: Journal of Geodynamics, v. 1, p. 445–472.

Karlstrom, K. E., 1981, Late Archean and Early Proterozoic geologic history of metasedimentary rocks of the Medicine Bow Mountains, Wyoming [Ph.D. thesis]: Laramie, University of Wyoming, 162 p.

Karlstrom, K. E., and Houston, R. S., 1984, The Cheyenne belt; Analysis of a Proterozoic suture in southern Wyoming: Precambrian Research, v. 25, p. 415–446.

Karlstrom, K. E., Houston, R. S., Flurkey, A. J., Collidge, C. M., Kratochvil, A. L., and Sever, D. K., 1981, A summary of the geology and uranium potential of Precambrian conglomerates in southeastern Wyoming: Bendix Field Engineering Corporation Report DJBX–139–81, 541 p.

Karlstrom, K. E., Flurkey, A. J., and Houston, R. S., 1983, Stratigraphy and depositional setting of Proterozoic metasedimentary rocks in southeastern Wyoming; Record of an Early Proterozoic Atlantic-type cratonic margin: Geological Society of America Bulletin, v. 94, p. 1257–1294.

McCallum, M. E., 1964, Petrology and structure of the Precambrian and post-Mississippian rocks of the east-central portions of the Medicine Bow Mountains, Albany and Carbon Counties, Wyoming [Ph.D. thesis]: Laramie, University of Wyoming, 164 p.

Moores, E. M., and Day, H. W., 1984, Overthrust model for the Sierra Nevada: Geology, v. 12, p. 416–419.

Mueller, R. E., 1982, The Cheyenne belt, southeastern Wyoming; Part 1, Descriptive geology and petrography [M.S. thesis]: Laramie, University of Wyoming, 98 p.

Nelson, B. K., and DePaolo, D. J., 1985, Rapid production of continental crust 1.7 to 1.9 b.y. ago; Nd isotopic evidence from the basement of the North American mid-continent: Geological Society of America Bulletin, v. 96, p. 746–754.

Newhouse, W. H., and Hagner, A. F., 1957, Geologic map of anorthosite areas, southern part of the Laramie Range, Wyoming: U.S. Geological Survey Mineral Investigations Field Studies Map, MF–119, scale 1:63,360.

Pallister, J. S., and Aleinikoff, J. N., 1987, Gabbroic plutons south of the Cheyenne belt; Underpinnings of an Early Proterozoic continental margin arc: Geological Society of American Abstracts with Programs, v. 19, p. 325.

Premo, W. R., 1984a, U-Pb zircon geochronology of some Precambrian rocks in the Sierra Madre Range, Wyoming [M.S. thesis]: Lawrence, University of Kansas, 106 p.

—— , 1984b, U-Pb zircon geochronology of Early Proterozoic plutonism in N. Colorado and SE Wyoming: Geological Society of America Abstracts with Programs, v. 16, p. 251.

Premo, W. R., and Van Schmus, W. R., 1982, U-Pb zircon geochronology of the Sierra Madre Range, Wyoming: Geological Society of America Abstracts with Programs, v. 14, p. 346.

Prodehl, C., and Pakiser, L. C., 1980, Crustal structure of the southern Rocky Mountains from seismic measurements: Geological Society of America Bulletin, v. 91, p. 147–155.

Roeder, D. H., 1973, Subduction and orogeny: Journal of Geophysical Research, v. 78, p. 5005–5024.

—— , 1977, Continental convergence in the Alps: Tectonophysics, v. 40, p. 339–350.

Schmidt, T. G., 1983, Precambrian metavolcanic rocks and associated volcanogenic mineral deposits of the Fletcher Park and Green Mountain areas, Sierra Madre, Wyoming [M.S. thesis]: Laramie, University of Wyoming, 113 p.

Snyder, G. L., 1980, Geologic map of the northernmost Park Range and southernmost Sierra Madre, Jackson and Routt Counties, Colorado: U.S. Geological Survey Miscellaneous Investigations Series I–1113, scale 1:48,000.

Snyder, G. L., and Hedge, C. E., 1978, Intrusive rocks northeast of Steamboat Springs, Park Range, Colorado: U.S. Geological Survey Professional paper 1041, 42 p.

Subbarayudu, G. V., 1975, The Rb/Sr isotope composition and the origin of the Laramie anorthosite-syenite complex, Laramie Range, Wyoming [Ph.D. thesis]: Buffalo, State University of New York, 109 p.

Swift, P. N., 1982, Precambrian metavolcanic rocks and associated volcanogenic mineral deposits of the southwestern Sierra Madre, Wyoming [M.S. thesis]: Laramie, University of Wyoming, 61 p.

Manuscript Accepted by the Society September 27, 1988

Printed in U.S.A.

Geological Society of America
Special Paper 235
1989

Zircon geochronology of Precambrian rocks in southeastern Wyoming and northern Colorado

Wayne R. Premo* and W. R. Van Schmus
Department of Geology, University of Kansas, Lawrence, Kansas 66045

ABSTRACT

Archean gneisses and Early Proterozoic metasedimentary rocks of the Wyoming Province are separated from Proterozoic eugeoclinal metamorphic rocks by a major east-west–trending shear zone called the Cheyenne belt. U-Pb zircon ages of Archean tonalites north of the Cheyenne belt denote an intrusive event at 2,700 Ma. Detrital zircons from Proterozoic metasedimentary rocks north of the Cheyenne belt define an apparent age of 2,450 Ma for the source rock, similar to an age of 2,430 Ma obtained for a local granite. A metagabbro plug, which intruded the metasedimentary rocks about 2,100 Ma, constrains their deposition within this 350 m.y. period. Ages for key units just south of the Cheyenne belt in Wyoming delineate at least three magmatic events at 1,780; 1,750; and 1,625 Ma. Ages for large plutons in the northern Colorado area define pulses of granodioritic to granitic intrusions at approximately 1,720 and 1,670 Ma.

A U-Pb zircon age of 1,792 ± 15 Ma for a Proterozoic metavolcanic rock in the Sierra Madre is greater than ages reported for other Proterozoic metavolcanic rocks in the U.S. Rockies. However, ages for Proterozoic plutons in southeastern Wyoming are similar to other ages for plutonism and volcanism for rocks exposed in the central Colorado Rockies and are coeval with suturing of Proterozoic crust with the Archean Wyoming Province along the Cheyenne belt. Although at present the accretionary history for these Early Proterozoic rocks is not well understood, it is evident that there exists a progressive decrease in age for volcanism as well as plutonism from north to south.

INTRODUCTION

Archean and Early Proterozoic rocks of the Wyoming craton are separated from Proterozoic orogenic rocks by the Cheyenne belt (Houston and others, 1979), a discontinuous, east-west–trending, southeast-dipping (Allmendinger and others, 1982) Precambrian shear zone (Houston and others, 1968; Hills and Houston, 1979; Fig. 1). The zone is exposed in the Sierra Madre and the Medicine Bow Mountains of southeastern Wyoming, and marks the southern boundary of the Wyoming Province. Movement along the Cheyenne belt apparently occurred during at least two episodes, an early period of thrusting and a later period of dextral, strike-slip motion (Houston and others, 1979; Karlstrom and others, 1981, Karlstrom and Houston, 1984; Duebendorfer and Houston, 1986). Previous interpretations of the evolution of the rocks along the Cheyenne belt have been based on K-Ar and Rb-Sr ages (Hills and others, 1968; Divis, 1976; 1977). This study attempts to establish a more precise chronologic framework based on U-Pb zircon geochronology. Most ages are from rocks of the Sierra Madre, but additional results from other parts of southeastern Wyoming and northern Colorado are reported. Previous workers have interpreted these rocks as a well-preserved continental margin/volcanic arc collisional complex (Hills and Houston, 1979; Karlstrom and others, 1981; Condie, 1982). Understanding the timing of plutonism, deformation, and metamorphism are critical to interpretation of the collisional events.

*Present address: U.S. Geological Survey, MS 963, Box 25046, Denver Federal Center, Denver, Colorado 80225.

Premo, W. R., and Van Schmus, W. R., 1989, Zircon geochronology of Precambrian rocks in southeastern Wyoming and northern Colorado, *in* Grambling, J. A., and Tewksbury, B. J., Proterozoic geology of the southern Rocky Mountains: Boulder, Colorado, Geological Society of America Special Paper 235.

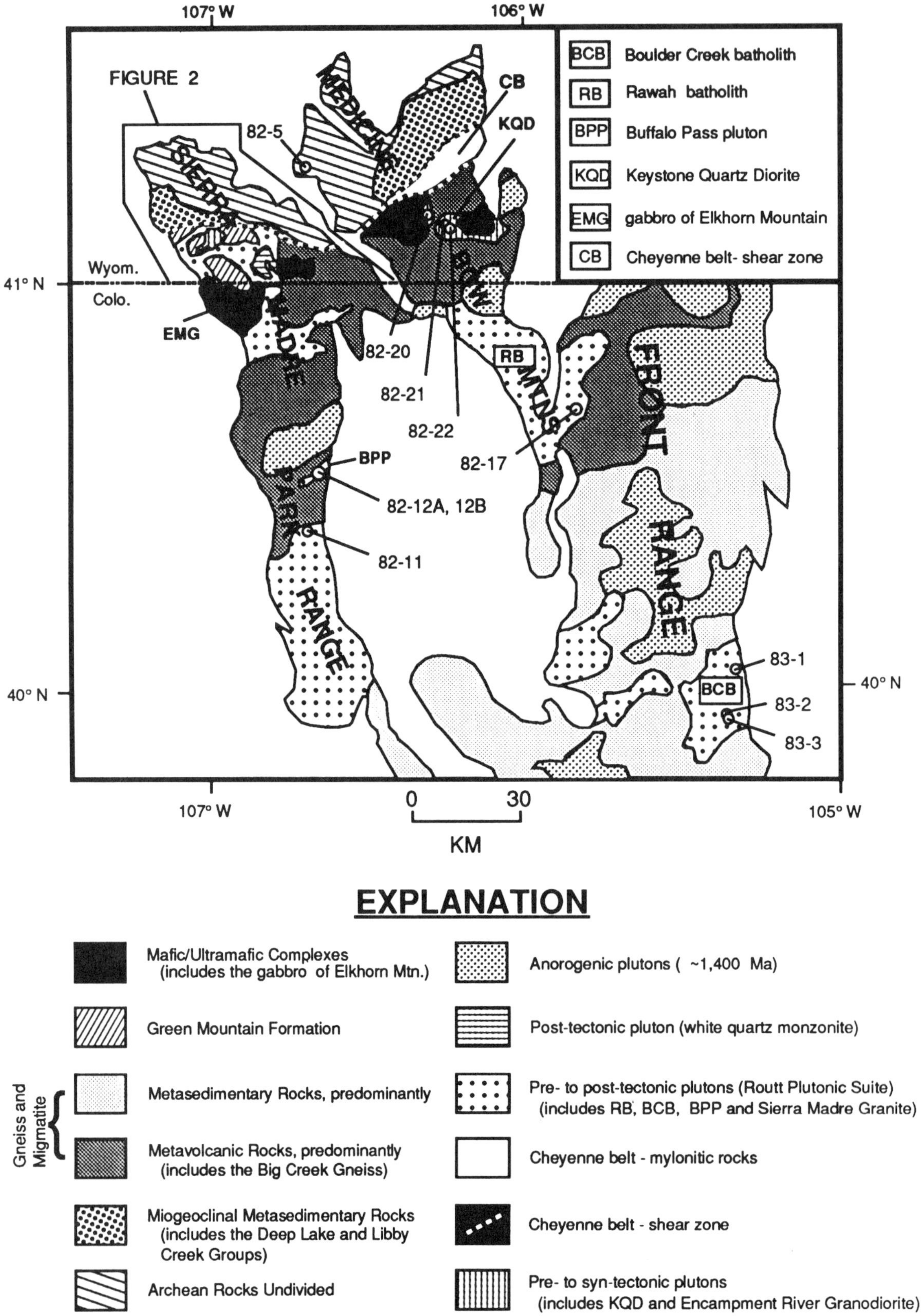

Figure 1. Generalized geologic map of Precambrian rocks in the southeastern Wyoming–northern Colorado region showing zircon sample locations (open circles) outside of the Sierra Madre. Geology from Tweto (1979) and Love and Christiansen (1985).

GEOLOGIC SETTING

Previous geological work in the area (Hedge and others, 1967; Peterman and others, 1968; Hills and others, 1968; Houston and others, 1968, 1977; McCallum and others, 1975; Divis, 1976; Snyder and Hedge, 1978; Graff, 1979) was recently summarized by Hills and Houston (1979). Karlstrom and others (1981, 1983) drew together detailed descriptions and interpretations of Archean and Early Proterozoic rocks north of the Cheyenne belt. It is from these studies that most of the summary below was derived. Proper names for units are from Karlstrom and others (1981) for rocks north of the belt and from Divis (1976) for rocks south of the belt.

Archean rocks north of the Cheyenne belt

The Archean rocks exposed along the southern edge of the Wyoming craton are subdivided into three major units: (1) a gneissic basement complex, (2) the Phantom Lake Metamorphic Suite, and (3) deformed plutons of tonalite and granite (Karlstrom and others, 1981, 1983; Fig. 2). The basement complex consists of heterogeneous gray, biotite-plagioclase gneiss; hornblende gneiss; and quartzofeldspathic gneiss with minor pods of quartzite, amphibolite, schist, and marble. Contacts are reported to be gradational, and structures are generally complex and poorly understood.

The Phantom Lake Metamorphic Suite unconformably overlies the basement complex and consists of quartzite, conglomerate, phyllite, metavolcanic rocks, and minor marble. Although these rocks are intensely deformed and metamorphosed to amphibolite facies, some sedimentary structures are preserved.

Granitic plutons of late Archean age intrude the basement complex and the Phantom Lake Metamorphic Suite, and are often complexly folded along with the enclosing rocks. Karlstrom and others (1981) recognized three types of pluton: (1) pink, coarse- to medium-grained, foliated, tonalitic orthogneiss; (2) gray, coarse-grained, foliated, tonalite-granodiorite (Spring Lake Granodiorite), which intrudes the Phantom Lake Metamorphic Suite in the central Sierra Madre; and (3) red, pink, or white, coarse- to medium-grained, potassic, granitic gneiss (Baggot Rocks Granite). Although previously reported radiometric ages for the Baggot Rocks Granite range from latest Archean to earliest Proterozoic (2,550 to 2,350 Ma; Hills and Houston, 1979), this unit is discussed herein with the Archean rocks.

Proterozoic rocks north of the Cheyenne belt

The Deep Lake and Libby Creek Groups are metasedimentary rocks of Early Proterozoic age that unconformably overlie Archean rocks north of the Cheyenne belt (Figs. 1 and 2). Their aggregate stratigraphic thickness may be as much as 10 km. In the Sierra Madre, the older Deep Lake Group is approximately 3 km thick and composed mainly of quartzites, phyllites, and conglomerates, interpreted as having been deposited in an intracratonic basin (Karlstrom and others, 1981, 1983). The Libby Creek Group consists of approximately 2 km of orthoquartzite, phyllite, metacarbonate, and metachert, interpreted as a passive continental margin sequence (Graff, 1979; Karlstrom and others, 1981, 1983).

Both the Deep Lake and Libby Creek Groups have been strongly deformed. Fold axes are subhorizontal and east-west trending, and bedding is steeply dipping to vertical. In the Sierra Madre, the rocks are imbricated by north-directed thrusts and cut by shear faults. Metamorphism never exceeded the staurolite zone of the amphibolite facies, and primary sedimentary structures are commonly preserved.

Abundant mafic sills, dikes, and stocks, some several hundred meters wide, intrude metasedimentary rocks north of the Cheyenne belt (Fig. 2). Most are tholeiitic and intrude only the Deep Lake Group and Phantom Lake Metamorphic Suite. Other small bodies of metagabbro, transitional between tholeiitic and calc-alkalic rocks, intrude rocks of the Libby Creek Group (Karlstrom and others, 1981).

Rocks south of the Cheyenne belt

Rocks south of the Cheyenne belt can be subdivided into three major suites: (1) gneiss and migmatite (including the Big Creek Gneiss of the Sierra Madre) that grade into lower amphibolite-facies metasedimentary rocks to the south-southeast in the Colorado Front Range; (2) interlayered, mafic and felsic metavolcanic and volcaniclastic rocks metamorphosed to greenschist or lower amphibolite facies (Green Mountain Formation) in the south-central Sierra Madre; and (3) metaplutonic rocks of various compositions (Figs. 1 and 2).

Gneisses and migmatites. A heterogeneous group of gneisses and migmatitic rocks make up much of the Proterozoic terrane south of the Cheyenne belt in southern Wyoming and northern Colorado (Fig. 1). These include hornblende gneiss, amphibolite, quartz-biotite-andesine gneiss, calc-schist, pelitic marble, orthogneiss, migmatite, and pegmatite (Hills and Houston, 1979; Karlstrom and Houston, 1984). Although primary textures and structures are largely obliterated, the dominant lithologies are suggestive of a mixed volcanic-sedimentary origin (Hills and Houston, 1979; Houston and others, this volume). In the part of the Sierra Madre where zircon samples were collected, these rocks have been called the Big Creek Gneiss (Fig. 2). The Big Creek Gneiss consists of high-grade, medium-grained, gray to blue-gray, interlayered mafic and felsic gneiss (Ridgley, 1971; Hills and Houston, 1979). The mineral assemblages in some samples (pyroxene, hornblende, orthoclase, andesine, and quartz) are consistent with upper amphibolite- to lower granulite-grade metamorphism. Two periods of deformation of the Big Creek Gneiss were recognized by Ridgley (1971); the first produced passive folding with axes trending approximately S70°W, parallel to the Cheyenne belt trend, and a second phase produced flexural folds.

Green Mountain Formation. The Green Mountain Formation, modified from Divis (1976), consists of metavolcanic

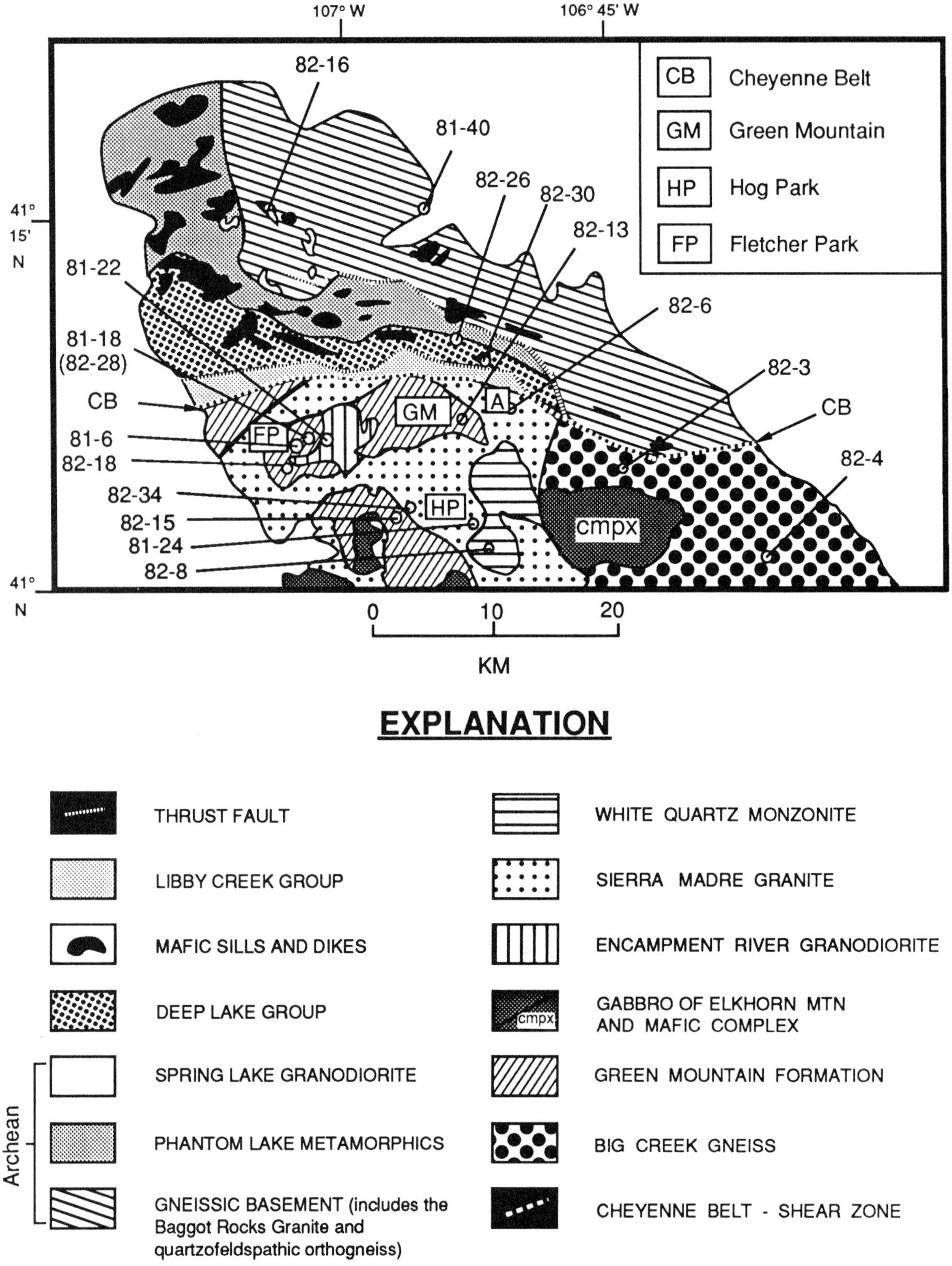

Figure 2. Generalized geologic map of Precambrian rocks in the Sierra Madre, southern Wyoming, showing zircon sample locations (open circles). The white square labeled "A" marks the approximate location of exposed Sierra Madre Granite crosscutting sheared rocks (CB) of the Libby Creek Group (see text for explanation). Also, note that deformed plutons of the Baggot Rocks Granite and the quartzofeldspathic orthogneiss are not distinguishable from the Archean gneissic basement because of their complex field relationship. The geology was compiled from many sources including Divis (1976), Houston and Ebbett (1977), Graff (1978), Swift (1982), and Karlstrom and others (1981).

and metasedimentary rocks exposed in roof pendants mainly in the Fletcher Park, Green Mountain, and Hog Park areas of the Sierra Madre (Fig. 2). The most abundant units in the Green Mountain Formation are interlayered mafic and felsic metavolcanic schist and gneiss, metagraywacke, breccia, volcanic conglomerate, and metatuff. Minor iron formation, exhalite, metapelite, and quartzite have also been reported (Divis, 1976; Swift, 1982; Condie and Shadel, 1984). A pervasive foliation in the Fletcher Park and Green Mountain pendants strikes approximately east-west and dips between 50° and 80° south, roughly parallel to the Cheyenne belt (Lackey, 1965). The Green Mountain Formation is sheared and truncated by the Cheyenne belt. Rocks in the three roof pendants differ in metamorphic grade and texture; in the Fletcher Park pendant they are of greenschist facies, whereas those in the Green Mountain and Hog Park pendants are higher grade, mostly sillimanite-muscovite zone of the amphibolite facies (Merry, 1963; Lackey, 1965; Swift, 1982).

Plutonic rocks. Plutons south of the Cheyenne belt are of four major types: (1) small, pre- to syn-tectonic, mafic to intermediate plutons exposed in the Sierra Madre and the Medicine Bow Mountains; (2) pre- to post-tectonic, massive, red granite plutons exposed mainly in the southwestern Sierra Madre; (3) pre- to post-tectonic batholiths of granitic composition (mostly quartz monzonite) in northern Colorado; and (4) a small, post-tectonic pluton of quartz monzonite that crops out in the south-central part of the Sierra Madre (Figs. 1 and 2). Younger plutons, approximately 1,450 to 1,400 Ma, are not included in this study.

Small, pre- to syn-tectonic plutons, just south of the Cheyenne belt, range in composition from hornblende-clinopyroxene-biotite gabbro to granodiorite. In the Sierra Madre, the Encampment River Granodiorite, a gray, medium- to coarse-grained, well-foliated pluton, clearly intrudes rocks of the Green Mountain Formation (Fig. 2). Metamorphic foliation in the Encampment River Granodiorite has attitudes similar to foliation in the Green Mountain Formation, suggesting that the two were subjected to the same deformational event (Lackey, 1965). The Keystone Quartz Diorite (KQD; Fig. 1), a similar medium- to coarse-grained, foliated pluton, is exposed in the east-central portion of the Medicine Bow Mountains. Houston and others (this volume) observed that these two rock units are similar in mineralogy and chemistry; both are closely associated with more mafic rock assemblages (Hills and Houston, 1979; Fig. 1). It has been suggested that the Green Mountain metavolcanic and volcaniclastic metasedimentary rocks and associated plutons of the southeastern Wyoming region are cogenetic and represent portions of a deformed oceanic island arc complex (Houston and others, 1979; Hills and Houston, 1979; Karlstrom and others, 1981; Swift, 1982; Condie, 1982; Condie and Shadel, 1984; Karlstrom and Houston, 1984).

Pre- to post-tectonic plutons of massive red granite are exposed mainly in the Sierra Madre but also in the Medicine Bow Mountains and parts of the Park Range in Colorado (Figs. 1 and 2). The Sierra Madre Granite forms a batholith exposed in the west-central Sierra Madre where it intrudes the Green Mountain Formation and the Encampment River Granodiorite (Lackey, 1965; Swift, 1982). The granite is red, coarse- to medium-grained, foliated to massive, and characterized by textural and compositional variations. The Sierra Madre Granite does not display a tectonic foliation, suggesting that it post-dates the major deformational event that affected the older units. The granite crosscuts the Cheyenne belt in the area of Copper Creek (marked A, Fig. 2), where it intruded sheared rocks of the Slaughterhouse Formation of the Libby Creek Group.

A small body of medium- to coarse-grained granitic gneiss, the Buffalo Pass pluton (BPP; Fig. 1) is exposed in the Park Range in northern Colorado (Snyder and Hedge, 1978). Similar small plutons of granite also crop out in the Medicine Bow Mountains (Houston and Ebbett, 1977; Hills and Houston, 1979).

In northern Colorado, batholiths of mainly quartz monzonite intrude the gneissic complex (Fig. 1). A large unnamed pluton is exposed in the southern Park Range. Although quartz monzonite is probably the dominant composition, these rocks range from quartz diorite through granite (Snyder, 1980a). The Rawah batholith (RB; Fig. 1) in the northwestern Front Range and southern Medicine Bow Mountains is composed mainly of granite and quartz monzonite together with more mafic phases (McCallum and others, 1975; McCallum and Hedge, 1976). In the east-central Front Range, the Boulder Creek batholith (BCB; Fig. 1) is composed of dominantly quartz monzonite and granodiorite (Gable, 1969, 1972), although border zones and isolated areas of quartz diorite and amphibolite have also been reported (Stern and others, 1971). Numerous smaller bodies of similar mineralogy, texture, and age are mapped throughout the northern Front Range (Hedge and others, 1967; Peterman and others, 1968; Barovich, 1986). These plutons contain abundant xenoliths, exhibit foliation, and are interpreted to have been emplaced during the second of two major phases of deformation and metamorphism that characterize the Proterozoic terrane (Braddock, 1966; Peterman and others, 1968).

A small, irregular pluton of white quartz monzonite is exposed around the Hog Park area (Fig. 2). The rock is medium-grained, massive to foliated, and displays primary igneous textures. It clearly intrudes the Sierra Madre Granite (Merry, 1963; Davis, 1976). The presence of muscovite and xenocrystic garnet is consistent with the peraluminous chemistry of this rock (Divis, 1976) and suggests an origin by partial melting of pelitic sedimentary rocks (Chappell and White, 1974).

ANALYTICAL PROCEDURES

Field reconnaissance and sampling depended on geological descriptions and mapping from many sources (Merry, 1963; Lackey, 1965; Ridgley, 1971; Stern and others, 1971; Divis, 1976; Tweto, 1976; Houston and Ebbett, 1977; Snyder and Hedge, 1978; Snyder, 1980a, 1980b, 1980c; Karlstrom and others, 1981). Most of the samples are from the Sierra Madre, because the rocks there are well preserved.

TABLE 1. SUMMARY OF CONCORDIA INTERCEPT AGES AND LOCATION OF SAMPLES

Sample Number	Unit and Location	N	P	Model	Age (± 2σ)	L.I.
I. Archean and earliest Proterozoic plutons						
81-40	Quartzofeldspathic orthogneiss 41°15.9'N., 106°55.4'W.	4	25	2	2,683 ± 6	110
82-05	Baggot Rocks Granite 41°16.4'N., 106°43.9'W.	4	34	1	2,429 ± 4	190
82-16	Spring Lake Granodiorite 41°16.1'N., 107°5.2'W.	6	<1	2	2,710 ± 10	248
II. Miogeoclinal suite (Deep Lake Group)						
82-26	Magnolia Formation (detrital) 41°10.6'N., 106°53.2'W.	4	30	1	2,451 ± 9	145
82-30	Metagabbro (post-Deep Lake Group) 41°9.6'N., 106°51.5'W.	4	75	1	2,092 ± 9	195
III. Sierra Madre orogenic suite						
81-06	Rhyolite porphyry, Fletcher Park 41°6.1'N., 107°3.2'W.	5	79	1	1,780 ± 6	103
81-18	Green Mountain Formation 41°6.1'N., 107°1.9'W.	1	--	-	1,742 (Pb-Pb)	
81-22	Encampment River Granodiorite 41°6.2'N., 107°1.2'W.	4	57	1	1,779 ± 5	100
81-24	Sierra Madre Granite, N. Hog Park 41°2.1'N., 106°51.7'W.	5	<1	2	1,744 ± 14	125
82-06	Sierra Madre Granite, Dead Horse Pk. 41°8.6'N., 106°52.6'W.	4	57	1	1,763 ± 6	62
82-13	Green Mountain Formation 41°7.3'N., 106°53.3'W.	6	<1	2	1,792 ± 15	211
82-15	Green Mountain Formation 41°2.2'N., 106°55.2'W.	1	--	-	1,653 (Pb-Pb)	
82-18	Green Mountain Formation 41°5.0'N., 1074.2'W.	1	--	-	1,777 (Pb-Pb)	
82-28	Green Mountain Formation 41°6.1'N., 107°2.1'W.	1	--	-	1,679 (Pb-Pb)	
82-34	Sierra Madre Granite, W. Hog Park 41°7.1'N., 106°54.0'W.	5	23	2	1,749 ± 8	47
IV. Sierra Madre metamorphic suite and late plutons*						
82-03	Big Creek Gneiss, N. Blackhall Mtn. 41°4.9'N., 106°43.5'W.	5	85	1	(<1,684 ± 5)	−123
82-04	Big Creek Gneiss, Pearl, Colorado 41°0.6'N., 106°34.3'W.	4	<1	2	(>1,618 ± 22)	−614
82-08	White quartz monzonite 41°1.5'N., 106°51.6'W.	5	64	1	1,627 ± 4	43
V. Orogenic rocks from elsewhere in SE Wyoming and N Colorado						
82-11	Quartz diorite, Rabbit Ears Pass Park Range, Colorado 40°20'N., 106°44'W.	6	<1	2	1,735 ± 8	12
82-12A	Granite, Buffalo Pass, Park Range 40°31'N., 106°44'W	4	10	2	1,746 ± 6	137
82-12B	Pegmatite, Buffalo Pass (same location as 12A)	5	<1	2	1,765 ± 8	241

TABLE 1. SUMMARY OF CONCORDIA INTERCEPT AGES AND LOCATION OF SAMPLES (continued)

Sample Number	Unit and Location	N	P	Model	Age (± 2σ)	L.I.
V. Orogenic rocks from elsewhere in SE Wyoming and N Colorado (continued)						
82-17	Quartz monzonite, Rawah Batholith, Northern Front Range, Colorado 40°41'N., 105°48'W.	4	17	2	1,720 ± 8	68
82-20	Granite, Horse Creek Medicine Bow Mts., Wyoming 41°10.5'N., 106°16.5'W.	5	95	1	1,777 ± 4	59
82-21	Keystone Quartz Diorite, Medicine Bow Mts., Wyoming 41°10'N., 106°15'W.	5	15	2	1,781 ± 7	143
82-22	Red granite, Keystone, Medicine Bow Mts., Wyoming 41°8'N., 106°15'W.	5	<1	2	1,744 ± 18	178
83-01	Boulder Creek Granodiorite Front Range, Colorado 40°2'N., 105°18.9'W.	4	64	1	1,714 ± 6	56
83-02	Boulder Creek quartz monzonite, Front Range, Colorado 39°56.6'N, 105°20.9'W.	6	1	2	1,672 ± 8	132
83-03	Boulder Creek quartz monzonite 39°56.4'N., 105°20.8'W.	4	89	1	1,670 ± 4	156

Notes: N = number of fractions; P = probability for fit (in percent); Model refers to models of Ludwig (1980) for models 1 and 2; Ages are upper intercepts, given in Ma, with uncertainties at 2-sigma. L.I. = lower intercept age given in Ma. See text for further details. *zircons contain inherited cores; see text for discussion.

Zircon yields ranged between 5 and 1,000 mg, and were split into fractions using a magnetic separator. Lesser- to non-magnetic fractions were collected and labeled according to their magnetic susceptibilities. Designations used in this chapter refer to the side angle at which the fractions were collected with the magnetic separator set at 1.5 amp. Fractions were hand-picked from the less magnetic fractions of some samples. Zircon fractions for analysis usually weighed between 2 and 6 mg. Subsequent chemical procedures used for extraction of U and Pb for isotope dilution analysis are slightly modified from Krogh (1973).

Isotopic analyses were obtained using a 22.5-cm sector, Nier-type, single-filament mass spectrometer with programmable magnetic field scanning and digital output. Individual isotope composition and isotope dilution aliquots were loaded using silica-gel and phosphoric acid. Analytical Pb blanks were between 1 and 2 ng. U blanks were less than 0.5 ng. All Pb isotope data were corrected using modern (204):(206):(207):(208) ratios of (1.0):(18.6):(15.8):(39.0) for blank common Pb. Assumed initial common Pb isotope compositions were derived from the Stacey and Kramers (1975) two-stage model for the approximate age of the sample. Uncertainties for Pb/U ratios range between ± 1.0 to 1.5 percent at the two sigma level. Radiogenic $^{207}Pb/^{206}Pb$ ratios are uncertain to ±0.1 to 0.2 percent at 2 sigma, resulting in a corresponding uncertainty of ±0.2 to 0.3 percent for individual Pb-Pb ages. Ages were calculated using decay constants from Steiger and Jager (1977). Concordia intercept ages were determined using the regression treatment of Ludwig (1980) with correlated errors; uncertainties are reported at the 2-sigma confidence level. The resultant concordia upper intercept ages are interpreted as representative of the time of crystallization for each igneous rock analyzed, except where explained otherwise (i.e., the Big Creek Gneiss, white quartz monzonite, and the Magnolia Formation). The U-Pb ages and their uncertainties are summarized in Table 1. U-Pb data are given in Table 2. Sample locations are listed in Table 1 and shown in Figs. 1 and 2.

RESULTS AND INTERPRETATION

Samples were selected in order to: (1) define the age of Archean plutons north of the Cheyenne belt, (2) define age limits for deposition of Early Proterozoic-age epicratonic strata north of the belt, (3) identify the oldest rocks south of the belt and the time span for arc volcanism, and (4) establish ages for plutonic rocks south of the belt in order to provide constraints on the timing of major tectonic events.

TABLE 2. ANALYTICAL DATA FOR ZIRCON FRACTIONS

Fraction[1]	Concentrations[2]		Observed[3]	Radiogenic Ratios[4]				Ages[5]		
	U (ppm)	Pb (ppm)	$^{206}Pb/^{204}Pb$	$^{207}Pb/^{206}Pb$	$^{208}Pb/^{206}Pb$	$^{206}Pb/^{238}U$	$^{207}Pb/^{235}U$	$^{206}Pb/^{238}U$	$^{207}Pb/^{235}U$	$^{207}Pb/^{206}Pb$
81-6										
HP	275	87	790	0.10867	0.0952	0.2941	4.406	1,662	1,714	1,777
NM(0)	305	93	1,825	0.10840	0.0948	0.2870	4.290	1,627	1,691	1,773
M(0)	337	101	1,660	0.10841	0.0956	0.2795	4.177	1,588	1,670	1,773
M(1)	378	111	1,410	0.10835	0.0953	0.2744	4.099	1,562	1,654	1,772
M(2)	412	118	960	0.10818	0.0985	0.2624	3.914	1,502	1,617	1,769
81-18										
Total	370	77	360	0.10661	0.1248	0.1747	2.568	1,038	1,292	1,742
81-22										
NM(0)	294	78	900	0.10795	0.0969	0.2454	3.652	1,415	1,561	1,765
M(0)	378	97	680	0.10751	0.1046	0.2288	3.391	1,328	1,502	1,758
M(2)	518	115	500	0.10680	0.1124	0.1916	2.821	1,130	1,361	1,746
M(3)	574	119	410	0.10618	0.1196	0.1724	2.523	1,025	1,279	1,735
81-24										
HP	300	84	430	0.10611	0.0560	0.2546	3.725	1,462	1,577	1,734
NM(1)	431	109	870	0.10520	0.0508	0.2400	3.482	1,387	1,523	1,718
M(1)	506	117	930	0.10484	0.0495	0.2191	3.168	1,277	1,449	1,712
M(3)	628	136	750	0.10476	0.0545	0.2009	2.901	1,180	1,382	1,710
M(4)	654	136	680	0.10438	0.0544	0.1921	2.764	1,133	1,346	1,704
81-40										
NM(0)	395	168	1,610	0.18147	0.1619	0.3592	8.987	1,978	2,337	2,666
M(0)	411	173	1,530	0.18113	0.1584	0.3555	8.878	1,961	2,326	2,663
M(2)	499	178	1,110	0.17971	0.1642	0.2968	7.353	1,675	2,155	2,650
M(4)	557	170	860	0.17847	0.1717	0.2498	6.148	1,438	1,997	2,639
82-3										
NM(0)	375	102	3,100	0.10353	0.0223	0.2782	3.972	1,582	1,628	1,688
M(0)	612	164	3,185	0.10370	0.0278	0.2711	3.876	1,546	1,609	1,691
M(1)	639	167	2,170	0.10387	0.0382	0.2611	3.739	1,495	1,580	1,695
M(3)	1,093	267	2,415	0.10415	0.0365	0.2435	3.496	1,405	1,526	1,699
M(3,4)AA	1,461	368	3,530	0.10381	0.0218	0.2558	3.662	1,469	1,563	1,693
82-4										
NM(-1)	515	138	2,170	0.10026	0.0271	0.2771	3.831	1,577	1,599	1,629
M(-1)	609	165	2,430	0.09985	0.0298	0.2738	3.769	1,560	1,586	1,621
M(0)	894	233	1,790	0.10167	0.0395	0.2620	3.673	1,500	1,566	1,655
M(1)	589	151	800	0.10136	0.0408	0.2528	3.533	1,453	1,535	1,649
82-5										
NM(-1)	262	128	2,220	0.15652	0.1934	0.4106	8.860	2,324	2,218	2,418
M(-1)	296	139	2,120	0.15651	0.1951	0.3953	8.530	2,147	2,289	2,418
M(0)	341	152	1,351	0.15572	0.2050	0.3664	7.867	2,013	2,216	2,410
M(1)	426	171	1,025	0.15440	0.2127	0.3248	6.915	1,813	2,101	2,395
82-6										
NM(1)	406	131	500	0.10760	0.0789	0.2830	4.198	1,606	1,674	1,759
M(1)	483	144	1,060	0.10756	0.0831	0.2771	4.109	1,577	1,656	1,758
M(2)	534	155	935	0.10758	0.0864	0.2656	3.940	1,519	1,622	1,759
M(3)	616	172	750	0.10732	0.0907	0.2516	3.723	1,447	1,576	1,755

TABLE 2. ANALYTICAL DATA FOR ZIRCON FRACTIONS (continued)

Fraction[1]	Concentrations[2]		Observed[3]	Radiogenic Ratios[4]				Ages[5]		
	U (ppm)	Pb (ppm)	$^{206}Pb/^{204}Pb$	$^{207}Pb/^{206}Pb$	$^{208}Pb/^{206}Pb$	$^{206}Pb/^{238}U$	$^{207}Pb/^{235}U$	$^{206}Pb/^{238}U$	$^{207}Pb/^{235}U$	$^{207}Pb/^{206}Pb$
82-8										
NM(0)	248	67	880	0.10000	0.0955	0.2468	3.402	1,422	1,505	1,624
M(0)	361	91	500	0.09972	0.0909	0.2211	3.040	1,288	1,417	1,619
M(1)	494	112	410	0.09946	0.0841	0.1951	2.675	1,149	1,322	1,614
M(2)	674	137	365	0.09937	0.0814	0.1717	2.352	1.021	1,228	1,612
M(3)	776	145	360	0.09905	0.0838	0.1579	2.157	945	1,167	1,607
82-11										
NM(0)	543	149	4,445	0.10616	0.0543	0.2737	4.006	1,560	1,636	1,734
M(0)	657	177	5,495	0.10628	0.0501	0.2690	3.942	1,536	1,622	1,737
M(1)	687	179	1,935	0.10579	0.0577	0.2526	3.685	1,452	1,568	1,728
M(2)	892	219	1,850	0.10631	0.0558	0.2393	3.508	1,383	1,529	1,737
M(3) c	1,081	248	1,480	0.10615	0.0597	0.2197	3.216	1,280	1,461	1,734
M(3) f	970	215	1,815	0.10597	0.0568	0.2151	3.142	1,256	1,443	1,731
82-12A										
NM(0)	367	100	4,310	0.10625	0.0568	0.2708	3.968	1,545	1,628	1,736
M(0)	464	119	3,485	0.10583	0.0512	0.2539	3.704	1,458	1,572	1,729
M(1)	617	141	2,720	0.10506	0.0446	0.2269	3.286	1,318	1,478	1,715
M(2)	851	158	1,560	0.10392	0.0422	0.1820	2.607	1,078	1,303	1,695
82-12B										
NM(0)	358	101	1,660	0.10657	0.0623	0.2741	4.028	1,562	1,640	1,741
M(0)	461	117	1,695	0.10590	0.0605	0.2461	3.594	1,419	1,548	1,730
M(1)	547	128	1,170	0.10499	0.0558	0.2234	3.235	1,300	1,466	1,714
M(2)	861	171	810	0.10234	0.0522	0.1875	2.645	1,108	1,313	1,667
M(4)	1,490	244	515	0.09911	0.0592	0.1477	2.018	888	1,122	1.607
82-13										
NM(1)a	304	111	215	0.10852	0.1095	0.2749	4.113	1,566	1,657	1,775
NM(1)b	300	109	216	0.10815	0.1084	0.2741	4.087	1,561	1,652	1,769
M(1)	348	132	165	0.10848	0.1215	0.2630	3.934	1,505	1,621	1,774
M(2)	398	156	126	0.10708	0.1338	0.2470	3.647	1,423	1,560	1,750
M(3)	478	194	102	0.10688	0.1505	0.2326	3.428	1,348	1,511	1,747
M(4)	495	206	89	0.10675	0.1639	0.2226	3.276	1,296	1,475	1,745
82-15										
Total	463	73	210	0.10159	0.0832	0.1486	2.082	893	1,143	1,653
82-16										
NM(-2)	83	47	1,060	0.18491	0.2210	0.4621	11.782	2,449	2,587	2,698
M(-2)	91	51	1,450	0.18482	0.2035	0.4581	11.674	2,431	2,579	2,697
M(-1)	94	50	4,810	0.18457	0.1972	0.4444	11.310	2,371	2,549	2,694
M(0)	102	53	1,280	0.18319	0.2064	0.4253	10.742	2,284	2,501	2,682
M(1)	116	57	1,815	0.18346	0.1953	0.4047	10.238	2,191	2,456	2,684
M(2)	103	53	1,010	0.18351	0.1976	0.4153	10.509	2,239	2,481	2,685
82-17										
NM(1)	835	211	1,900	0.10489	0.0676	0.2435	3.522	1,405	1,532	1,712
M(1)	918	220	1,600	0.10449	0.0749	0.2276	3.279	1,322	1,476	1,705
M(2)	1,001	213	1,670	0.10420	0.0755	0.2027	2.912	1,190	1,385	1,700
M(3)	1,097	208	1,795	0.10403	0.0735	0.1818	2.607	1,077	1,303	1,697

TABLE 2. ANALYTICAL DATA FOR ZIRCON FRACTIONS (continued)

Fraction[1]	Concentrations[2]		Observed[3]	Radiogenic Ratios[4]				Ages[5]		
	U (ppm)	Pb (ppm)	$^{206}Pb/^{204}Pb$	$^{207}Pb/^{206}Pb$	$^{208}Pb/^{206}Pb$	$^{206}Pb/^{238}U$	$^{207}Pb/^{235}U$	$^{206}Pb/^{238}U$	$^{207}Pb/^{235}U$	$^{207}Pb/^{206}Pb$
82-18										
Total	52	16	132	0.10864	0.1748	0.2697	4.039	1,539	1,642	1,777
82-20										
NM(0)	1,362	379	5,235	0.10835	0.0644	0.2735	4.086	1,558	1,651	1,772
M(0)	823	222	5,350	0.10836	0.0657	0.2645	3.951	1,513	1,624	1,772
M(1)	908	229	4,310	0.10813	0.0673	0.2465	3.674	1,420	1,566	1,768
M(2)	1,053	240	3,080	0.10787	0.0686	0.2207	3.283	1,286	1,477	1,764
M(3)	1,193	242	2,300	0.10751	0.0711	0.1953	2.895	1,150	1,381	1,758
82-21										
NM(-1)	306	87	2,690	0.10856	0.0651	0.2797	4.187	1,590	1,672	1,776
M(-1)	309	85	1,530	0.10876	0.0728	0.2624	3.935	1,502	1,621	1,779
M(0)	324	85	2,670	0.10860	0.0663	0.2545	3.812	1,462	1,595	1,776
M(1)	327	82	1,930	0.10837	0.0668	0.2421	3.617	1,398	1,553	1,772
M(2)	354	85	1,825	0.10835	0.0645	0.2307	3.446	1,338	1,515	1,772
82-22										
NM(0)	2,722	755	1,260	0.10587	0.1002	0.2568	3.748	1,473	1,582	1,730
M(0)	3,610	885	1,630	0.10409	0.1071	0.2275	3.265	1,322	1,473	1,698
M(1)	4,704	1,012	1,460	0.10346	0.1126	0.1981	2.826	1,165	1,362	1,687
M(2)	4,393	913	1,025	0.10318	0.1098	0.1888	2.686	1,115	1,325	1,682
M(3)	5,908	1,145	1,350	0.10282	0.1061	0.1790	2.537	1,061	1,283	1,676
82-26										
NM(0)	216	97	1,790	0.15809	0.2738	0.3609	7.867	1,987	2,216	2,435
M(0)	206	89	6,025	0.15753	0.2802	0.3510	7.624	1,939	2,188	2,429
M(1)	327	137	4,185	0.15740	0.2867	0.3355	7.281	1,865	2,147	2,428
M(2)	353	141	1,860	0.15686	0.2919	0.3129	6.767	1,755	2,081	2,422
82-28										
Total	1,382	225	575	0.10300	0.0644	0.1586	2.252	949	1,198	1,679
82-30										
NM(0)	356	146	1,330	0.12876	0.2119	0.3448	6.121	1,910	1,993	2,081
M(0)	314	124	2,400	0.12867	0.2113	0.3370	5.980	1,872	1,973	2,080
M(1)	413	159	2,230	0.12835	0.2136	0.3276	5.797	1,827	1,946	2,075
M(2)	408	154	1,860	0.12819	0.2184	0.3189	5.636	1,784	1,922	2,073
82-34										
HP	268	75	390	0.10677	0.0766	0.2484	3.656	1,430	1,562	1,745
NM(0)	271	70	640	0.10640	0.0742	0.2366	3.471	1,369	1,521	1,739
M(0)	369	92	525	0.10654	0.0798	0.2216	3.255	1,290	1,470	1,741
M(1)	399	99	435	0.10632	0.0817	0.2151	3.153	1,256	1,446	1,737
M(2)	464	107	485	0.10630	0.0815	0.2024	2.966	1,188	1,399	1,737
83-1										
NM(-2)	560	130	800	0.10420	0.1076	0.2078	2.986	1,217	1,404	1,700
M(-2)	551	120	630	0.10397	0.1096	0.1916	2.747	1,130	1,341	1,696
M(-1)	667	133	475	0.10356	0.1118	0.1702	2.430	1,013	1,252	1,689
M(0)	746	131	375	0.10315	0.1131	0.1456	2.071	876	1,139	1,682

TABLE 2. ANALYTICAL DATA FOR ZIRCON FRACTIONS (continued)

Fraction[1]	Concentrations[2]		Observed[3]	Radiogenic Ratios[4]				Ages[5]		
	U (ppm)	Pb (ppm)	$^{206}Pb/^{204}Pb$	$^{207}Pb/^{206}Pb$	$^{208}Pb/^{206}Pb$	$^{206}Pb/^{238}U$	$^{207}Pb/^{235}U$	$^{206}Pb/^{238}U$	$^{207}Pb/^{235}U$	$^{207}Pb/^{206}Pb$
83-2										
NM(-2)	683	192	3,050	0.10242	0.0979	0.2673	3.775	1,527	1,587	1,668
M(-2)c	778	215	2,600	0.10217	0.1019	0.2609	3.675	1,494	1,566	1,664
M(-2) f	812	229	1,780	0.10187	0.1068	0.2628	3.691	1,504	1,569	1,659
M(-1)	812	213	2,740	0.10174	0.1060	0.2476	3.473	1,426	1,521	1,656
M(0)	855	216	2,590	0.10164	0.1065	0.2382	3.338	1,377	1,490	1,654
M(1)	893	213	2,220	0.10140	0.1076	0.2239	3.131	1,303	1,440	1,650
83-3										
NM(-2)	806	220	2,230	0.10157	0.1640	0.2453	3.435	1,414	1,512	1,653
M(-2)	861	216	2,070	0.10119	0.1152	0.2335	3.258	1,353	1,471	1,646
M(0)	906	199	1,460	0.10027	0.1249	0.2008	2.775	1,179	1,349	1,629
M(2)	1,064	189	1,020	0.09826	0.1543	0.1561	2.114	935	1,154	1,591

Notes: 1. NM = nonmagnetic, M = magnetic, numbers in parentheses indicate side tilt used on Frantz separator at 1.5 amp power; mesh sizes for sieved samples are f = -200; c = +200; AA = air abraded sample; HP = hand-picked sample.
2. Total U and Pb, corrected for analytical blank.
3. Measured ratio, uncorrected for blank or non-radiogenic Pb; see text for corrections.
4. Pb corrected for blank and non-radiogenic Pb; U corrected for blank (see text).
5. Ages given in Ma; see text for decay constants used.

Ages of Archean rocks north of the Cheyenne belt

Data from the pink quartzofeldspathic orthogneiss (sample 81-40) yield an age of 2,683 ± 6 Ma (Fig. 3); whereas, analyses from the Spring Lake Granodiorite yield an age of 2,710 ± 10 Ma. The Spring Lake Granodiorite intruded all pre-plutonic Archean units, and the age of 2,710 Ma is considered a minimum age for deposition of sediments and volcanics of the Phantom Lake Metamorphic Suite. The 25 to 30 m.y. difference in age between the quartzofeldspathic orthogneiss and the Spring Lake Granodiorite may indicate separate magmatic events, although field relationships between these two units are difficult to interpret due to limited exposure.

The Baggot Rocks Granite (sample 82-5; Fig. 1) has an Early Proterozoic age of 2,429 ± 4 Ma (Fig. 3).

Ages of Proterozoic rocks north of the Cheyenne belt

Analyses of two samples from the Deep Lake Group help define age limits for its deposition. A morphologically homogeneous suite of zircon was separated from metamorphosed quartz-pebble conglomerate (sample 82-26) from the basal unit of the Deep Lake Group, the Magnolia Formation, in the central Sierra Madre and yields an apparent age of 2,451 ± 9 Ma (Fig. 4). Although detrital populations may be derived from more than one source, the morphological homogeneity of this population suggests a single source. The zircons are euhedral, slightly pitted, translucent to semiopaque, and contain opaque inclusions. The near-euhedral morphologies indicate a short residence time in the depositional process, and the age is very similar to that of the nearby Baggot Rocks Granite. This age represents a maximum age limit for deposition of the Magnolia Formation.

Sample 82-30 is from a pegmatitic metagabbro plug that intrudes the Cascade Quartzite of the Deep Lake Group in the Sierra Madre. Analysis of four zircon fractions yields an age of 2,092 ± 9 Ma (Fig. 4). This result establishes a minimum age for deposition of the Cascade Quartzite and, perhaps, for the entire Deep Lake Group. The emplacement of numerous mafic intrusions into the Deep Lake Group may be related to rifting of an intracratonic basin in which rocks of the Deep Lake Group were deposited (Karlstrom and others, 1981). If so, the age of 2,092 Ma may date the rifting event, and deposition of the Deep Lake Group is constrained between 2,450 and 2,100 Ma.

Ages of rocks south of the Cheyenne belt

Gneisses and migmatites. Two samples from the Big Creek Gneiss in the eastern Sierra Madre were dated (Fig. 2). Tonalitic hornblende-biotite gneiss (sample 82-3) with compositional layering and weak foliation of tabular biotite and hornblende yields an apparent age of 1,684 ± 5 Ma (Fig. 5). However, this chord has a negative lower intercept, either due to random scatter in the closely grouped, nearly concordant analyses or higher proportions of older xenocrystic zircons in the more mag-

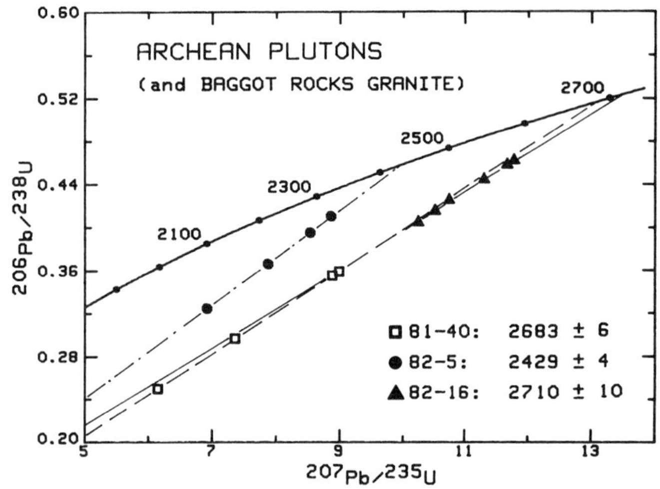

Figure 3. Concordia plot of zircon data from samples of Archean plutons. These data indicate at least two ages of intrusion into the gneissic basement. Note that sample 82-5, the Baggot Rocks Granite, actually defines an earliest Proterozoic age. Uncertainties for all ages are given at the 2-sigma confidence level.

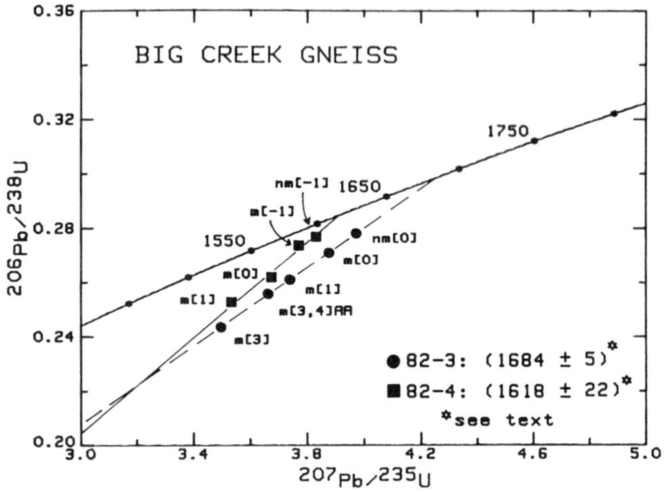

Figure 5. Concordia plot of zircon data from the Big Creek Gneiss, southern Sierra Madre. Note that both chords have negative lower intercepts (Table 1). Their upper intercept ages are not considered primary (see text).

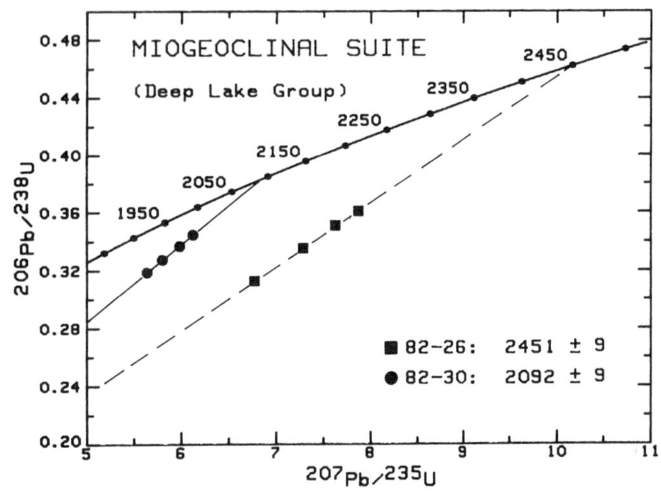

Figure 4. Concordia plot of zircon data for samples from the Deep Lake Group and a cross-cutting mafic plug. The two ages constrain deposition of the Deep Lake Group to between 2,450 and 2,100 Ma.

netic fractions. The presence of cores in igneous zircons may indicate additional radiogenic Pb incorporated within the sample, which will yield ages that are too old (e.g., Aftalion and van Breeman, 1980; Pin, 1981; Todt and Busch, 1981). Petrographic examination of the zircons revealed subhedral to anhedral, transparent overgrowths enclosing dark, metamict, round cores. Closer inspection of zircon populations from the individual magnetic fractions reveals that the more magnetic fractions (m[3] and m[3,4]AA) tend to contain more zircon with cores, and larger cores per zircon, than do the lesser magnetic (more concordant) fractions (nm[0] and m[0]). Thus, the negative lower intercept might be explained by the inherited radiogenic Pb from higher volumes of core material in the more magnetic fractions, as this would result in counterclockwise rotation of the chord. In any case, the age is certainly not a primary age for the protolith of the gneiss.

The most discordant fraction, m[3], has a $^{207}Pb/^{206}Pb$ age of 1,699 Ma, the greatest of all the fractions in sample 82-3. This age is a minimum age for the core material. It is likely that the protolith of the gneiss is older than 1699 Ma. Air abrasion on a mixed fraction (m[3,4]AA) was attempted, but failed to provide a significantly different Pb-Pb age. However, the air abrasion technique was applied for only 15 to 30 minutes and may not have eroded the overgrowths to any significant degree.

A hornblende-pyroxene gneiss phase of the Big Creek Gneiss from the southeastern Sierra Madre (sample 82-4) yields an age of 1,618 ± 22 Ma (Fig. 5). However, these analyses are also characterized by a negative lower intercept. The more magnetic fractions (m[0] and m[1]) contain more core material, whereas the least magnetic fractions (nm[−1] and m[−1]) probably contain less than 5 percent core material. The significance of a negative lower intercept is probably similar to that for sample 82-3. $^{207}Pb/^{206}Pb$ ages for the most concordant fractions of sample 82-4 are 1,629 and 1,621 Ma, respectively. The youngest $^{207}Pb/^{206}Pb$ age, 1,621 Ma, is considered, therefore, a maximum for the true age of overgrowth crystallization. The average age for these two fractions, about 1,625 ± 4 Ma, is very similar to numerous Rb-Sr and K-Ar mineral ages, discussed later in this chapter. The other more magnetic fractions of sample 82-4 have $^{207}Pb/^{206}Pb$ ages of 1,655 and 1,649 Ma, respectively, and are interpreted as hybrid ages due to the presence of xenocrystic cores in these fractions.

Green Mountain Formation. The determination of a primary age for volcanic rocks of the Green Mountain Formation

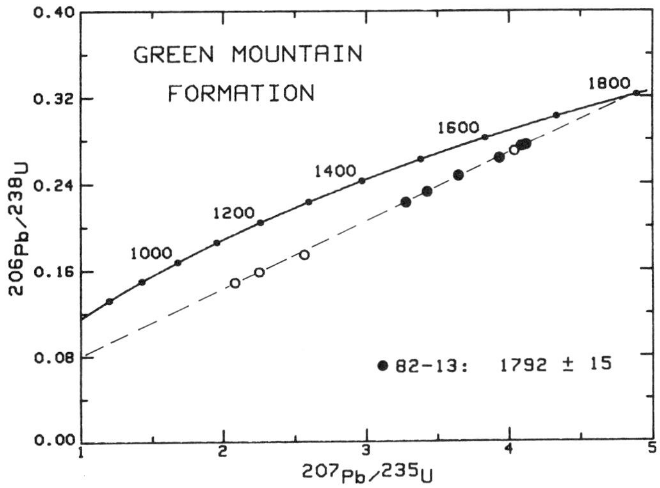

Figure 6. Concordia plot of zircon data from five samples of the Green Mountain Formation, south-central Sierra Madre. Single analyses for samples 81-18, 82-15, 82-18, and 82-28 (open circles) represent total zircon populations. The age given is based on regression for zircon fractions from sample 82-13 only, and the upper intercept age of 1,792 Ma is considered close to the true crystallization age.

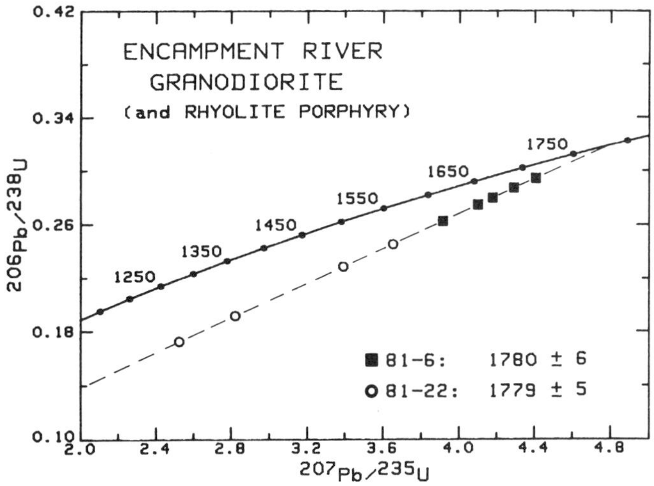

Figure 7. Concordia plot of zircon data from the Encampment River Granodiorite (sample 81-22) and a rhyolite porphyry dike (sample 81-6), which crosscuts the Green Mountain Formation in the area of Fletcher Park, central Sierra Madre. Both samples give essentially identical chords (Table 1).

was hampered by the paucity of zircons in most samples, even though compositions most likely to yield adequate quantities (andesites, dacites, and rhyolites) were sampled extensively.

Only one sample (82-13; Fig. 2), a quartz-feldspar-mica gneiss (metadacite porphyry) from the east side of Green Mountain, yielded enough zircons (about 75 mg) for reliable analysis of multiple fractions. This rock is apparently a flow interbedded with other metavolcanic units stratigraphically near the bottom of the metavolcanic section in the Green Mountain area (T. Schmidt, oral communication, 1983). Zircon analyses yield an age of 1,792 ± 15 Ma (Fig. 6). The zircons are near-euhedral, strongly zoned, and lack xenocrystic cores. The age is accepted as the best estimate of a primary age for metavolcanic rocks. Analyses from four other metavolcanic samples (81-18, 82-15, 82-18, and 82-28; Fig. 2)—although most are quite discordant—are consistent with this result (Fig. 6, open circles). These analyses represent total zircon yields from their respective samples.

Plutonic rocks. Samples of the Encampment River Granodiorite (sample 81-22) and a rhyolite porphyry dike (sample 81-6) that crosscuts the Green Mountain Formation yield essentially identical upper intercept ages of 1,780 ± 6 Ma and 1,779 ± 5 Ma (Fig. 7). Zircons from both samples are euhedral, transparent, and zoned. The Keystone Quartz Diorite (1,781 ± 7 Ma; sample 82-21; Fig. 8) and the granite at Horse Creek (1,777 ± 4 Ma; sample 82-20) in the Medicine Bow Mountains (Fig. 1) yield similar ages. These ages are analytically indistinguishable from the 1,790 Ma age inferred for the Green Mountain Formation. These plutons are interpreted as syngenetic with the volcanics, a view consistent with geochemistry of the same rocks (Condie and Shadel, 1984). Although this interpretation is not proven, major magmatic activity clearly occurred at 1790 to 1,780 Ma.

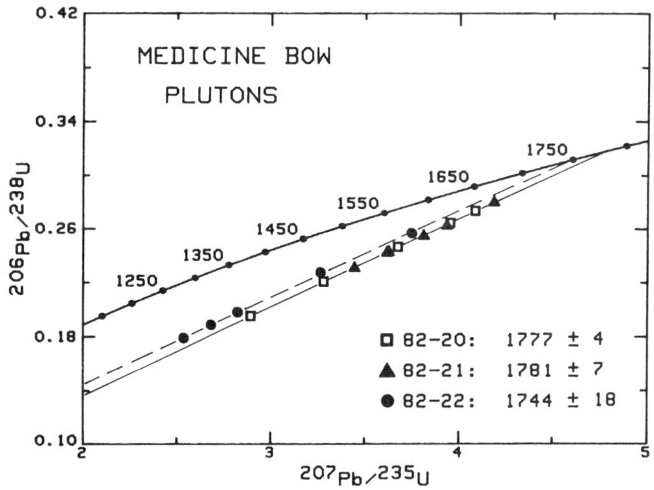

Figure 8. Concordia plot of zircon data from the Keystone Quartz Diorite (sample 82-21), the granite at Horse Creek (sample 82-20), and a red granite (sample 82-22) from the central Medicine Bow Mountains, Wyoming. Note that these ages are very similar to plutonic ages in the Sierra Madre (Figs. 7 and 9).

Three dated samples from the Sierra Madre Granite represent different textural and compositional varieties within the batholith. A massive red granite (sample 82-6) yields an age of 1,763 ± 6 Ma (Fig. 9). A microcline-bearing augen gneiss (sample 82-34) from the west side of Hog Park (Merry, 1963; Divis, 1976) yields an age of 1,749 ± 8 Ma. Gneissic microcline-hornblende granite (sample 81-24) from the east side of Hog Park yields an age of 1,744 ± 14 Ma. A similar age of 1,744 ± 18 Ma

(Fig. 8) was obtained for a red granite (sample 82-22) in the Medicine Bow Mountains (Fig. 1).

Zircons from gneissic granite and pegmatite of the Buffalo Pass pluton (sample 82-12A, 12B) in the Park Range (Fig. 1) yield ages of 1,746 ± 6 and 1,765 ± 8 Ma, respectively (Fig. 10), indicating that the Sierra Madre Granite and other similar granites in the southeastern Wyoming and northern Colorado are about 15 to 30 m.y. younger than the Green Mountain Formation and related plutons. The lack of foliation in the Sierra Madre Granite and its cross-cutting relationship with the Cheyenne belt suggest that the Sierra Madre Granite post-dates the belt, so that the granite's age of emplacement is interpreted as a minimum age for shearing along the Cheyenne belt.

Other major plutons in northern Colorado yield ages that define two later pulses of magmatism after intrusion of the granitic plutons previously mentioned. The first pulse involved emplacement mainly of tonalitic to granitic plutons between 1,740 and 1,705 Ma. In the Park Range, foliated quartz diorite (sample 82-11; Fig. 1) yields an age of 1,735 ± 8 Ma (Fig. 11). Foliated quartz monzonite from the Rawah batholith (sample 82-17) in the northern Front Range (Fig. 1) yields an age of 1,720 ± 8 Ma. Granodiorite (sample 83-1) of the Boulder Creek batholith in the Front Range (Fig. 1) yields an age of 1714 ± 6 Ma. The second pulse occurred between 1,680 and 1,665 Ma with the emplacement of granodiorite to granite composition plutons throughout central Colorado. Two samples of quartz monzonite from the Boulder Creek batholith (samples 83-2 and 83-3) yielded essentially the same age: 1,672 ± 8 and 1,670 ± 4 Ma, respectively (Fig. 12). Similar ages have also been reported for plutons in the Wet Mountains farther to the south (Bickford and others, 1984; Bickford and Boardman, 1984; Reed and others, 1987).

The age for the quartz monzonite of the Boulder Creek batholith is about 40 m.y. younger than the age for a granodiorite phase. Stern and others (1971) reported that nearly all zircon populations from samples of the Boulder Creek batholith contain core material, possibly from digested country rock. Although zircons from sample 83-1 appear "cleaner" than those from the quartz monzonite, inheritance of an older zircon component in zircons from sample 83-1 cannot be ruled out. Nonetheless, the age discrepancy leaves in doubt the exact meaning of "Boulder Creek age."

A still later magmatic pulse is represented by a white quartz monzonite in the Sierra Madre (Fig. 2). This rock (sample 82-8) crosscuts the Sierra Madre Granite and yields an age of 1,627 ± 4 Ma (Fig. 13). Some of the zircon fractions contain minor amounts of inherited cores. It is possible that sample 82-8 contains a small amount of inherited radiogenic Pb. The age of 1,627 Ma is tentatively accepted as a maximum estimate for the age of crystallization. This age is very similar to the youngest ages for

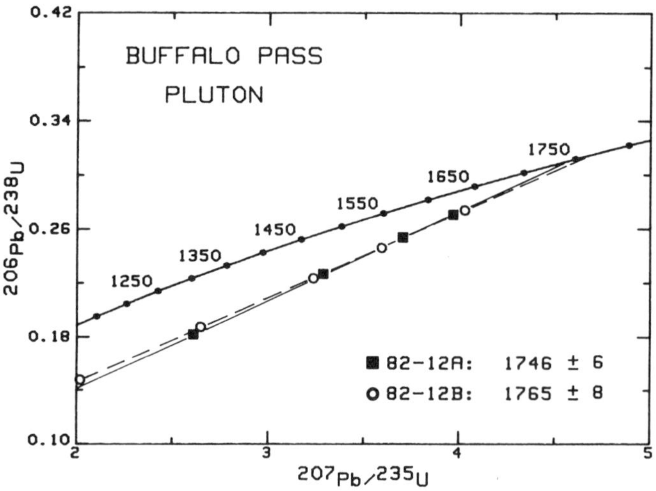

Figure 10. Concordia plot of zircon data from granitic gneiss and pegmatite of the Buffalo Pass pluton in the central Park Range, Colorado. Note that these results are also similar to those from the Sierra Madre Granite (Fig. 9).

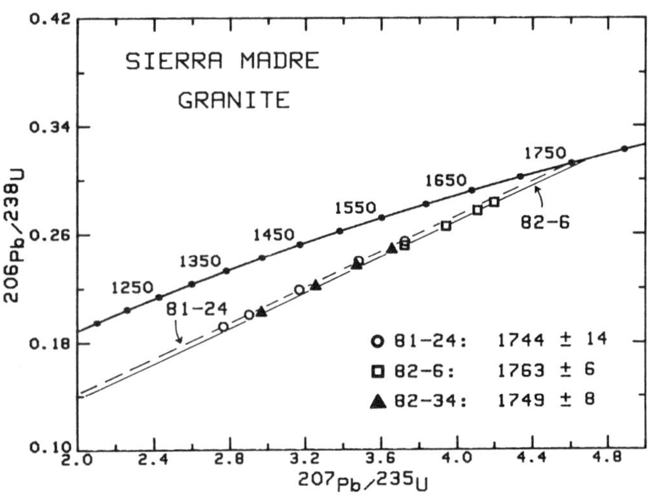

Figure 9. Concordia plot of zircon data from three samples of the Sierra Madre granite, Wyoming. These ages overlap within the 2-sigma error limits and suggest a mean age of about 1,753 Ma.

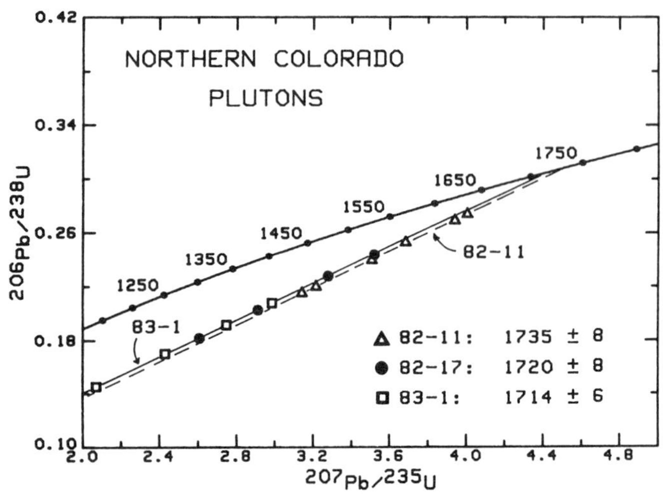

Figure 11. Concordia plot of zircon data from samples of large "Boulder Creek age" plutons in northern Colorado (Fig. 1).

zircons from the Big Creek Gneiss (sample 82-4). Other samples of the white quartz monzonite, which contain xenocrystic zircon and yield older apparent ages, are reported in Premo (1984).

Ages of approximately 1,625 Ma from different southeastern Sierra Madre samples suggest that a local thermal event occurred at that time. This thermal event, with upper amphibolite-grade temperatures and pressures as discussed above, could have produced overgrowths on zircons in the Big Creek Gneiss and caused partial melting to form migmatites, pegmatites, and possible anatectic units such as the white quartz monzonite. The survival of some inherited zircon cores in zircons of the white quartz monzonite would be favored by the peraluminous chemistry (Divis, 1976) of the anatectic quartz monzonite melt (Watson, 1979; Watson and Harrison, 1982).

A summary of the U-Pb zircon ages for plutonic units and one volcanic unit south of the Cheyenne belt is shown in Figure 14. Based on the sampling in this study, there exists a younging of plutonic ages from southeastern Wyoming into northern Colorado with possibly five different pulses of magmatism at approximately 1,780; 1,750; 1,720; 1,670; and 1,625 Ma. The four oldest magmatic events are illustrated as shaded areas in Figure 14. The oldest event (field A) is tightly constrained to approximately 1,780 Ma and is characterized by emplacement of small mafic to intermediate composition plutons just south of the Cheyenne belt. The next event (field B), loosely constrained at 1,750 Ma, is characterized by massive plutons of granite mainly emplaced just south of the Cheyenne belt, although smaller bodies are found farther to the south. Field C, very poorly constrained at approximately 1,720 Ma (only three analyses), includes only the large quartz monzonitic batholiths of northern Colorado. It is possible that fields A, B, and C actually represent one continuous event; however, we have tentatively separated them based on their lithologic and geographic differences. The fourth event (field D), at approximately 1,670 Ma, is not well defined (two analyses) within this study area (see discussion). The last event, at approximately 1,625 Ma, is not well defined and may be restricted to the southern Sierra Madre and possibly Medicine Bow Mountains.

DISCUSSION

The U-Pb zircon geochronology presented here provides time constraints for some of the Early Proterozoic events in the tectonic models of Hills and Houston (1979), Karlstrom and others (1981), Van Schmus and Bickford (1981), and Condie (1982). The chronology is generally consistent with field evidence and previous radiometric ages on rocks north of the Cheyenne belt (Karlstrom and others, 1981). However, previously reported Rb-Sr whole-rock isochron ages for rocks south of the Cheyenne belt (Divis, 1976, 1977) tend to be younger than our U-Pb zircon ages reported for the same units. This discrepancy is probably due to the mobility of Rb and Sr during subsequent metamorphism of these rocks (Divis, 1976).

Tonalitic plutonism during the Late Archean

The zircon geochronology shows that tonalitic plutons were emplaced into an older gneissic terrane about 2,700 Ma. Our U-Pb ages are as much as 40 m.y. older than U-Pb ages obtained by other workers for granitic gneiss exposed in the Wyoming Province. The granite of Long Creek Mountain in the Granite Mountains, about 150 km north of the Sierra Madre, yielded an age of 2,640 ± 20 Ma (Ludwig and Stuckless, 1978), which is similar to an age of 2,642 ± 13 Ma (Naylor and others, 1970) and 2,630 ± 20 Ma (Stuckless and others, 1985) for the granodiorite of Louis Lake in the Wind River Range, about 100 km west of the Granite Mountains. However, a similar $^{207}Pb/^{206}Pb$ minimum age of 2,670 ± 55 Ma was reported for a granite from the Granite Mountains by Stuckless and Nkomo (1978).

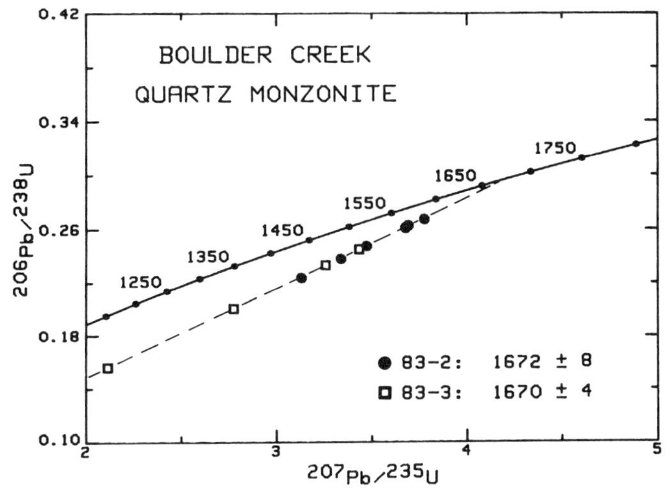

Figure 12. Concordia plot of zircon data from two samples of quartz monzonite of the Boulder Creek batholith. Note that these ages are about 40 m.y. younger than the age for a granodiorite of the same batholith (Fig. 11).

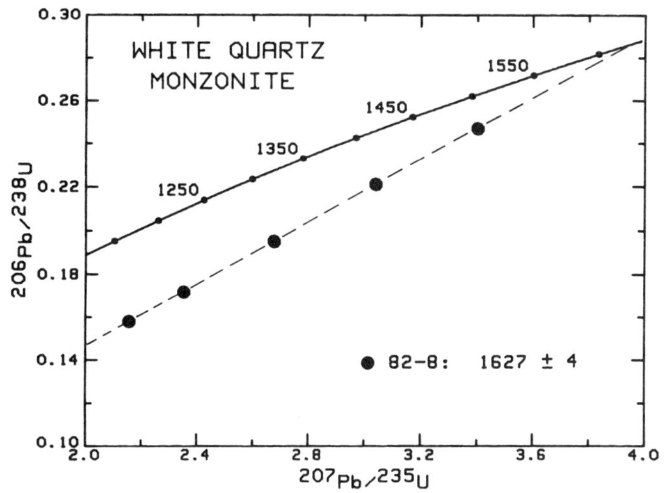

Figure 13. Concordia plot of zircon data for a white quartz monzonite, south-central Sierra Madre, Wyoming. The observation of some core material in fractions of this sample suggests that the age is a maximum estimate of the true age for the quartz monzonite (see text for explanation).

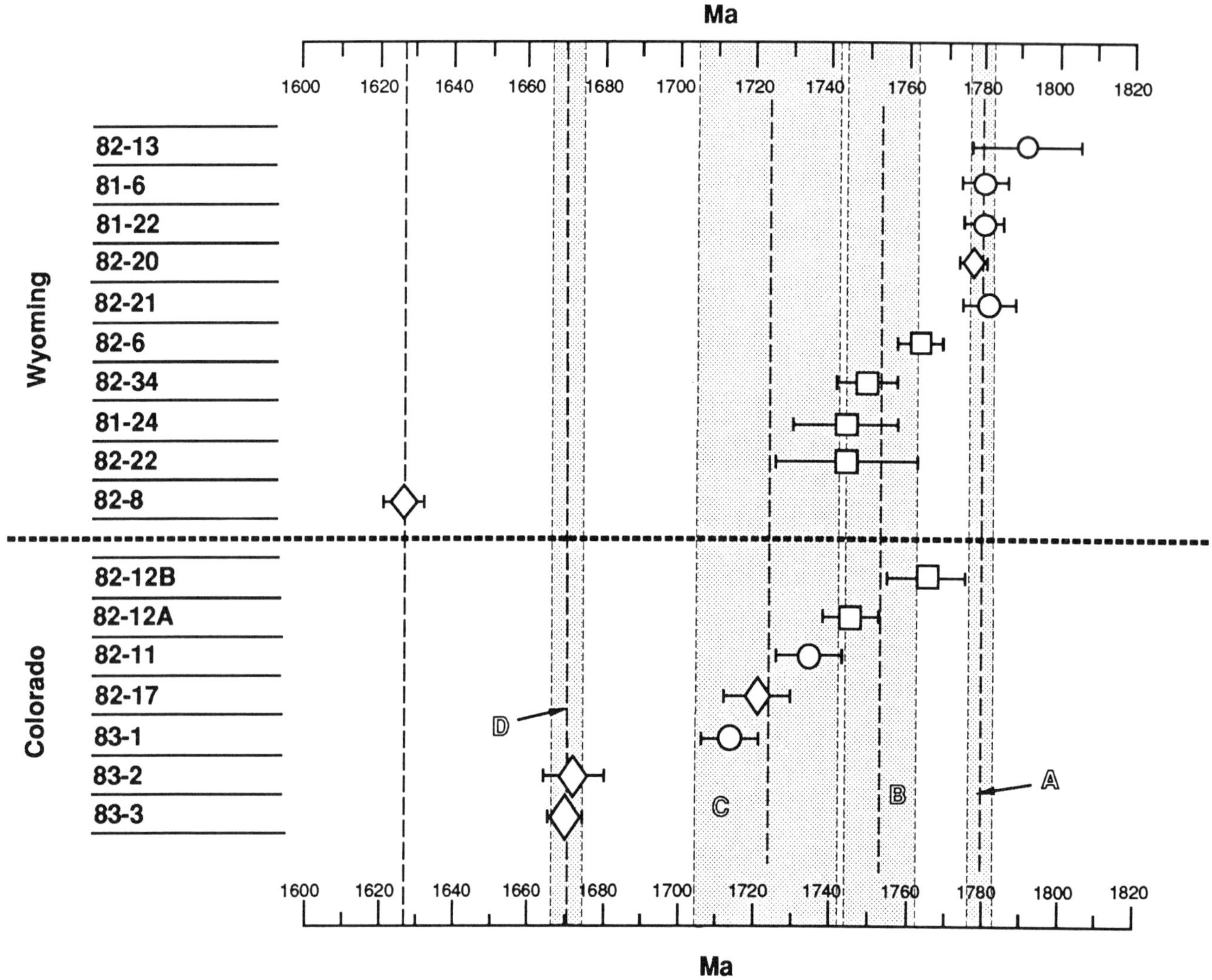

Figure 14. A summary of the ages of igneous rock samples south of the Cheyenne belt illustrates the decreasing age pattern from southern Wyoming into northern Colorado with possibly discrete plutonic episodes or pulses at approximately 1,780; 1,750; 1,720; and 1,670 Ma. Open circles represent samples of quartz diorite to granodiorite, open diamonds represent quartz monzonite samples, and open squares are granites. The heavily dashed vertical lines mark the mean ages of samples from approximately the same region and similar lithology. The shaded areas represent the 95 percent confidence limits for the mean ages. Although another magmatic episode is observed at about 1,625 Ma, it is apparently a minor event and restricted to the southern Sierra Madre and possibly the Medicine Bow Mountains.

Earliest Proterozoic granitic plutonism

From the Teton Range in northwestern Wyoming to the Laramie Range in southeastern Wyoming, intrusion of high-K granitic plutons occurred between 2,550 and 2,350 Ma (Hills and others, 1968; Peterman and Hildreth, 1978; Ludwig and Stuckless, 1978). Our age of 2,429 ± 4 Ma for the Baggot Rocks Granite at its type locality in the western Medicine Bow Mountains compares with a U-Pb age of 2,365 ± 50 Ma reported by Hills and others (1968) for the Baggot Rocks Granite at the same locality. Numerous Rb-Sr whole-rock ages for similar granites fall between 2,550 and 2,400 Ma (Hills and others, 1968; Divis, 1976; Reed and Zartman, 1973; Hills and Armstrong, 1974), but they are not precise enough to define individual events.

Duration of epicratonic sedimentation

The U-Pb age of 2,450 Ma for detrital zircons from the Magnolia Formation, the basal unit of the Deep Lake Group, is comparable to the age reported for the Baggot Rocks Granite. The Magnolia Formation lies upon an eroded surface of Archean rocks and was probably deposited in response to rifting or basin-forming tectonics (Karlstrom and others, 1981). The nearly euhedral zircon and poorly sorted granitic mineral assemblages (quartz, microcline, albite-oligoclase), which characterize parts of the Magnolia Formation, suggest a local source. The source for the conglomerate in the Magnolia Formation may be the Early Proterozoic high-K granites exposed in the southern Wyoming Province. The time of earliest Deep Lake Group sedimentation is unknown, but it certainly could not have occurred before 2,450 Ma, and was probably somewhat later, in view of the time required for uplift and erosion of the source area.

The age of about 2,100 Ma for the metagabbro intrusive into the Cascade Quartzite establishes a minimum age for the quartzite and perhaps for the entire Deep Lake Group. If the Deep Lake Group was deposited in an intracratonic basin, then the numerous mafic intrusions may have been emplaced during rifting about 2,100 Ma (Karlstrom and others, 1981), assuming that most or all of the mafic intrusions are about the same age as the rock we dated. Our dates suggest that the Deep Lake Group was deposited sometime between 2,450 and 2,100 Ma, an age bracket also reported for deposition of the Boxelder Formation in the Black Hills, South Dakota (Redden, 1980).

Duration of passive margin sedimentation

The Libby Creek Group, which overlies the Deep Lake Group, was apparently deposited on a passive margin, which may have developed as a result of a rifting at about 2,100 Ma (Karlstrom and others, 1981, 1983). Carbonates, submarine volcanics, and shales of the upper Libby Creek Group were probably deposited well offshore along the continental margin (Karlstrom and others, 1981). The end of deposition of the upper Libby Creek Group is constrained at 1,780 Ma, the earliest age estimate for shearing of rocks of the upper Libby Creek Group along the Cheyenne belt. If the tectonic model of Hills and Houston (1979) and Karlstrom and others (1981) is correct, the passive continental margin would have remained undeformed for about 300 m.y.

Timing of arc volcanism and plutonism

Metavolcanic and metasedimentary rocks south of the Cheyenne belt probably represent part of a deformed oceanic arc complex (Hills and Houston, 1979; Houston and others, 1979; Karlstrom and others, 1981). The most reliable age obtained for volcanic rocks south of the Cheyenne belt is 1,792 $\pm$ 15 Ma for a metavolcanic unit in the Green Mountain Formation. The age is similar to results from Sm-Nd isotopic studies of various metamorphic rock types from the Colorado Front Range, which indicate that the Proterozoic crust south of the Cheyenne belt was derived from the mantle about 1,800 $\pm$ 90 Ma and was relatively uncontaminated by pre-existing continental crust (DePaolo, 1981; Nelson and DePaolo, 1984, 1985).

Small, pre- to syn-tectonic mafic and intermediate plutons were emplaced into metavolcanic and metasedimentary rocks about 1,780 Ma (Fig. 14). These include the Encampment River Granodiorite in the Sierra Madre, and the Keystone Quartz Diorite and the granite at Horse Creek in the Medicine Bow Mountains. Snyder (1980b) reported a zircon $^{207}Pb/^{206}Pb$ age of 1,781 Ma for a dioritic phase of the gabbro of Elkhorn Mountain in the southwestern Sierra Madre (Figs. 1 and 2), which intruded the Green Mountain Formation in the Hog Park area. The same zircon separate was recently used to refine the age to 1,774 $\pm$ 2 Ma (Pallister and Aleinikoff, 1987). This age represents a minimum age for major volcanic activity in the early arc complex and a maximum age for shearing along the Cheyenne belt in the Sierra Madre.

Pre- to post-tectonic plutonism and deformation along the Cheyenne belt

Emplacement of the pre- to post-tectonic Sierra Madre Granite between 1,765 and 1,740 Ma (Figs. 9 and 14) establishes the time of cessation of major shearing along this section of the Cheyenne belt and constrains the duration of movement to less than 40 m.y. and perhaps less than 20 m.y. If the models presented by Hills and Houston (1979) and Karlstrom and others (1981) are correct, then the same time span is the duration of deformation related to suturing of the southern Proterozoic terrane to the Archean craton. Recent Sm-Nd isotopic studies on a similar granite from the Medicine Bow Mountains yielded a model age around 2.0 Ga (Bennett and DePaolo, 1987). The difference of this age from the zircon crystallization ages we found suggests that the granites exposed just south of the Cheyenne belt were at least partially derived from melting of pre-existing crustal material. The source could be the Early Proterozoic marginal sediments subducted and melted beneath the accreting oceanic island arc.

Zircon ages for some of the major pre- to post-tectonic plutons in northern Colorado suggest two later pulses of magmatism (Fig. 14): (1) Tonalitic to granitic plutons emplaced between 1,740 and 1,705 Ma. The majority of these bodies are batholiths that show textural features consistent with emplacement during a major deformational event. These include the Rawah batholith, the Boulder Creek batholith, and the batholith in the southern Park Range. (2) Dominantly granitic plutons emplaced between 1,680 and 1,665 Ma. These are commonly smaller plutons and stocks exposed mainly in central and southern Colorado (Reed and others, 1987).

Post-tectonic plutonism

The intrusion of the white quartz monzonite and high-grade metamorphism of the Big Creek Gneiss in the Sierra Madre occurred about 1,625 Ma. This same age is reflected in numerous Rb-Sr and K-Ar whole-rock and mineral ages (Hills and others, 1968; Hills and Houston, 1979) in southeastern Wyoming, including some rocks north of the Cheyenne belt. The source of heat required to reset these mineral ages is not known, but it is possible the ages reflect the temperature at the depth of burial of these rocks or may reflect the extent of the thermo-tectonism of a later arc collision event from the south (Condie, 1982, 1984;

Silver, 1984). Regardless, upper amphibolite- to lower granulite-facies metamorphism about 1,625 Ma is implied from analyses of least-cored zircon fractions from the Big Creek Gneiss in the Sierra Madre of southern Wyoming.

CONCLUSIONS

The U-Pb zircon data from this study define eight periods of magmatism in Precambrian rocks of the Sierra Madre and adjacent ranges of southeastern Wyoming and northern Colorado. In addition, the data shows that Proterozoic plutonism progressed from north to south, occurring in perhaps five distinct pulses at approximately 1,780; 1,750; 1,720; 1,670; and 1,625 Ma. The last magmatic episode at 1,625 Ma is apparently minor and restricted to the southern Sierra Madre and possibly the Medicine Bow Mountains. Our U-Pb zircon dates are consistent with the chronology of tectonic events proposed by Karlstrom and others (1981). The new geochronology constrains the possible duration of tectonic events as follows:

(1) Deposition of epicratonic sedimentary rocks north of the Cheyenne belt must have occurred between 2,450 and 1,780 Ma. Estimated ages of deposition for the major sedimentary packages are: the Deep Lake Group, 2,450 to 2,100 Ma; and the Libby Creek Group, no later than 1,780 Ma.

(2) Southeast-dipping subduction and island-arc formation must have been initiated before about 1,780 Ma.

(3) The Proterozoic terrane just south of the Cheyenne belt was accreted to the Archean Wyoming craton approximately 1,780 to 1,750 Ma.

(4) No major movement has occurred along the Cheyenne belt in the central Sierra Madre since about 1,750 Ma.

(5) Extensive pre- to post-tectonic plutonism occurred south of the Cheyenne belt between 1,765 and 1,665 Ma.

(6) Heating of some rocks south of the Cheyenne belt occurred at about 1,625 Ma, resulting in local high-grade metamorphism and possibly anatexis of Early Proterozoic crust.

ACKNOWLEDGMENTS

The analytical work and field collecting were supported by National Science Foundation Grants EAR 79-19544, 81-18234, and 83-03110 to W. R. Van Schmus. We wish to acknowledge the help provided during the field work by R. Houston, K. Condie, C. Shadel, C. Nelson, R. Lawrence, and T. Schmidt. Interaction in the laboratory with S. Bowring, R. Shuster, W. Hoppe, M. Bickford, and G. Girty aided considerably during the analytical work. Also, we would like to thank Z. Peterman, J. Reed, Jr., G. Snyder, J. Grambling, and W. Olszewski for their valuable reviews of the manuscript.

REFERENCES CITED

Aftalion, M., and van Breeman, O., 1980, U-Pb zircon, monazite, and Rb-Sr whole rock systematics of granitic gneiss and psammitic to semi-pelitic host gneiss from Glenfinnian, north-western Scotland: Contributions to Mineralogy and Petrology, v. 72, p. 87–98.

Allmendinger, R. W., Brewer, J. A., Brown, L. D., Kaufman, S., Oliver, J. E., and Houston, R. S., 1982, COCORP profiling across the Rocky Mountain Front Range in southern Wyoming; Part 2, Precambrian basement structure and its influence on Laramide deformation: Geological Society of America Bulletin, v. 93, p. 1253–1263.

Barovich, K. M., 1986, Age constraints on Early Proterozoic deformation in the northern Front Range, Colorado [M.S. thesis]: Boulder, University of Colorado, 49 p.

Bennett, V. C., and DePaolo, D. J., 1987, Proterozoic crustal history of the western United States as determined by neodymium isotopic mapping: Geological Society of America Bulletin, v. 99, p. 674–685.

Bickford, M. E. and Boardman, S. J., 1984, A Proterozoic volcano-plutonic terrane, Gunnison and Salida areas, Colorado: Journal of Geology, v. 92, p. 657–666.

Bickford, M. E., Cullers, R. L., and Van Schmus, W. R., 1984, U-Pb zircon chronology of Early and Middle Proterozoic igneous events in the Gunnison, Salida, and Wet Mountains areas, Colorado: Geological Society of America Abstracts with Programs, v. 16, p. 215.

Braddock, W. A., 1966, Precambrian geology on the east flank of the Front Range near Fort Collins, Colorado, *in* Geological Society of America Abstracts for 1965: Geological Society of America Special Paper 87, p. 277.

Chappell, B. W., and White, A.J.R., 1974, Two contrasting granite types: Pacific Geology, v. 8, p. 173–174.

Condie, K. C., 1982, Plate-tectonic model for Proterozoic continental accretion in the southwestern United States: Geology, v. 10, p. 37–42.

—— , 1984, Early Proterozoic supracrustal associations in the southwest; An Update: Geological Society of America Abstracts with Programs, v. 16, p. 218.

Condie, K. C., and Shadel, C. A., 1984, An early Proterozoic volcanic arc succession in southeastern Wyoming: Canadian Journal of Earth Sciences, v. 21, p. 415–427.

DePaolo, D. J., 1981, Neodymium isotopes in the Colorado Front Range and crust-mantle evolution in the Proterozoic: Nature, v. 291, p. 193–196.

Divis, A. F., 1976, Geology and geochemistry of the Sierra Madre Range, Wyoming: Colorado School of Mines Quarterly, v. 71, no. 3, 127 p.

—— , 1977, Isotopic studies on a Precambrian geochronologic boundary, Sierra Madre Mountains, Wyoming: Geological Society of America Bulletin, v. 88, p. 96–100.

Duebendorfer, E. M., and Houston, R. S., 1986, Kinematic history of the Cheyenne belt, Medicine Bow Mountains, southeastern Wyoming: Geology, v. 14, p. 171–174.

Gable, D. J., 1969, Geologic map of the Nederland Quadrangle, Boulder and Gilpin Counties, Colorado: U.S. Geological Survey Map GQ–833, scale 1:24,000.

—— , 1972, Geologic map of the Tungsten Quadrangle, Boulder, Gilpin, and Jefferson Counties, Colorado: U.S. Geological Survey Map GQ–978, scale 1:24,000.

Graff, P. J., 1978, Geology of the lower part of the Early Proterozoic Snowy Range Supergroup, Sierra Madre, Wyoming, with chapters on Proterozoic regional tectonics and uraniferous quartz-pebble conglomerates [Ph.D. thesis]: Laramie, University of Wyoming, 85 p.

—— , 1979, A review of the stratigraphy and uranium potential of Early Proterozoic (Precambrian X) metasediments in the Sierra Madre, Wyoming: Contributions to Geology, v. 17, no. 2, p. 1047–1069.

Hedge, C. E., Peterman, Z. E., and Braddock, W. A., 1967, Age of the major Precambrian regional metamorphism in the northern Front Range, Colorado: Geological Society of America Bulletin, v. 78, p. 551–558.

Hills, F. A., and Armstrong, R. L., 1974, Geochronology of Precambrian rocks in the Laramie Range and implications for the tectonic framework of Precambrian southern Wyoming: Precambrian Research, v. 1, p. 213–225.

Hills, F. A., and Houston, R. S., 1979, Early Proterozoic tectonics of the central Rocky Mountains, North America: Contributions to Geology, v. 17, p. 89–109.

Hills, F. A., Gast, P. W., Houston, R. S., and Swainbank, I. G., 1968, Precambrian geochronology of the Medicine Bow Mountains, southeastern Wyoming: Geological Society of America Bulletin, v. 79, p. 1757–1784.

Houston, R. S., and Ebbett, B. E., 1977, Geologic map of the Sierra Madre and western Medicine Bow Mountains, southeastern Wyoming: U.S. Geological Survey Map MF–827, scale 1:125,000.

Houston, R. S. and others, 1968, A regional study of rocks of Precambrian age in that part of the Medicine Bow Mountains lying in southeastern Wyoming, with a chapter on the relationship between Precambrian and Laramide structure: Wyoming Geological Survey Memoir 1, 167 p.

Houston, R. S., Graff, P. J., Karlstrom, K. E., and Root, F. K., 1977, Preliminary report on radioactive conglomerates of Middle Precambrian age in the Sierra Madre and Medicine Bow Mountains of southeastern Wyoming: U.S. Geological Survey Open-File Report 77–584, 31 p.

Houston, R. S., Karlstrom, K. E., Hills, F. A., and Smithson, S. B., 1979, The Cheyenne belt; A major Precambrian crustal boundary in the western United States: Geological Society of America Abstracts with Programs, v. 11, p. 446.

Karlstrom, K. E., and Houston, R. S., 1984, The Cheyenne belt; Analysis of a Proterozoic suture in southern Wyoming: Precambrian Research, v. 25, p. 415–446.

Karlstrom, K. E., Houston, R. S., Flurkey, A. J., Coolidge, C. M., Kratochvil, A. L., and Sever, O. K., 1981, A summary of the geology and uranium potential of Precambrian conglomerates in southeastern Wyoming: Special DOE Report GJBX-139, v. 1, 541 p.

Karlstrom, K. E., Flurkey, A. J., and Houston, R. S., 1983, Stratigraphy and depositional setting of the Proterozoic Snowy Pass Supergroup, southeastern Wyoming; Record of an Early Proterozoic Atlantic-type cratonic margin: Geological Society of America Bulletin, v. 94, p. 1257–1274.

Krogh, T. E., 1973, A low contamination method for the hydrothermal decomposition of zircon and extraction of U-Pb for isotopic age determinations: Geochimica et Cosmochimica Acta, v. 37, p. 485–494.

Lackey, L. L., 1965, Petrography of metavolcanic and igneous rocks of Precambrian age in the Huston Park area, Sierra Madre, Wyoming [M.A. thesis]: Laramie, University of Wyoming, 78 p.

Love, J. D., and Christiansen, A. C., 1985, Geologic map of Wyoming: U.S. Geological Survey, scale 1:500,000.

Ludwig, K. R., 1980, Calculation of uncertainties of U-Pb data: Earth and Planetary Letters, v. 46, p. 212–220.

Ludwig, K. R., and Stuckless, J. S., 1978, Uranium-lead isotope systematics and apparent ages of zircons and other minerals in Precambrian granitic rocks, Granite Mountains, Wyoming: Contributions to Mineralogy and Petrology, v. 65, p. 243–254.

McCallum, M. E., and Hedge, C. E., 1976, Rb-Sr ages of granitic rocks in the Rawah batholith, Medicine Bow Mountains, northern Colorado: Isochron West, v. 17, no. 12, p. 33–37.

McCallum, M. E., Hartmann, L. A., Hedge, C. E., and Filson, R. H., 1975, The Rawah batholith; A Boulder Creek pluton in northern Colorado: Geological Society of America Abstracts with Programs, v. 7, p. 627–628.

Merry, R. D., 1963, Precambrian geology, shear zones, and associated mineral deposits of the Hog Park area, Carbon County, Wyoming [M.S. thesis]: Laramie, University of Wyoming, 73 p.

Naylor, R. S., Steiger, R. H., and Wasserburg, G. J., 1970, U-Th-Pb and Rb-Sr systematics in 2,700 m.y. old plutons from the southern Wind River Range, Wyoming: Geochimica et Cosmochimica Acta, v. 34, p. 1133–1159.

Nelson, B. K., and DePaolo, D. J., 1984, Origin of Precambrian metavolcanic rocks from New Mexico, Colorado, and Wyoming, and the isotopic evolution of Proterozoic mantle: Geological Society of America Abstracts with Programs, v. 16, p. 249.

—— , 1985, Rapid production of continental crust 1.7 to 1.9 b.y. ago; Nd isotopic evidence from the basement of the North American mid-continent: Geological Society of America Bulletin, v. 96, p. 746–754.

Pallister, J. S., and Aleinikoff, J. N., 1987, Gabbroic plutons south of the Cheyenne belt; Underpinnings of an Early Proterozoic continental-margin arc: Geological Society of America Abstracts with Programs, v. 19, p. 325.

Peterman, Z. E., and Hildreth, R. A., 1978, Reconnaissance geology and geochronology of the Precambrian of the Granite Mountains, Wyoming: U.S. Geological Survey Professional Paper 1055, 22 p.

Peterman, Z. E., Hedge, C. E., and Braddock, W. A., 1968, Age of Precambrian events in the northeastern Front Range, Colorado: Journal of Geophysical Research, v. 73, p. 2277–2296.

Premo, W. R., 1984, U-Pb zircon geochronology of some Precambrian rocks in the Sierra Madre Range, Wyoming [M.S. thesis]: Lawrence, University of Kansas, 106 p.

Pin, C., 1981, Old inherited zircons in two synkinematic variscan granitoids; The "granite du Pinet" and the "Orthogneiss de Marvejols" (southern French Massif Central): Neues Jharbuch für Mineralogie Abhandlungen, v. 142, p. 27–48.

Redden, J. A., 1980, Geology and uranium resources in Precambrian conglomerate of the Nemo area, Black Hills, South Dakota: U.S. Department of Energy Open-File GJBX–127(80), 147 p.

Reed, J. C., Jr., and Zartman, R. E., 1973, Geochronology of Precambrian rocks of the Teton Range, Wyoming: Geological Society of America Bulletin, v. 84, p. 561–582.

Reed, J. C., Jr., Bickford, M. E., Premo, W. R., Aleinikoff, J. N., and Pallister, J. S., 1987, Evolution of the Early Proterozoic Colorado province; Constraints from U-Pb geochronology: Geology, v. 15, p. 861–865.

Ridgley, N. H., 1971, Precambrian rocks in the Blackhall Mountain area, Carbon County, Wyoming [M.S. thesis]: Laramie, University of Wyoming, 50 p.

Silver, L. T., 1984, Observations on the Precambrian evolution of northern New Mexico and adjacent regions: Geological Society of America Abstracts with Programs, v. 16, p. 256.

Snyder, G. L., 1980a, Geologic map of the northernmost Gore Range and southernmost northern Park Range, Grand, Jackson, and Routt Counties, Colorado: U.S. Geological Survey Miscellaneous Investigations Map I–1114, scale 1:48,000.

—— , 1980b, Geologic map of the northern-most Park Range and southern-most Sierra Madre, Jackson and Routt Counties, Colorado: U.S. Geological Survey Miscellaneous Investigations Map I–1113, scale 1:48,500.

—— , 1980c, Geologic map of the central part of the northern Park Range, Jackson and Routt Counties, Colorado: U.S. Geological Survey Miscellaneous Investigations Map I–1112, scale 1:48,000.

Snyder, G. L., and Hedge, C. E., 1978, Intrusive rocks northeast of Steamboat Springs, Park Range, Colorado: U.S. Geological Survey Professional paper 1041, 42 p.

Stacey, J. S. and Kramers, J. D., 1975, Approximation of terrestrial lead isotope evolution by a two-stage model: Earth and Planetary Science Letters, v. 26, p. 207–221.

Steiger, R. H. and Jager, E., 1977, Subcommission on geochronology; Convention on the use of decay constants in geo- and cosmochronology: Earth and Planetary Science Letters, v. 36, p. 359–362.

Stern, T. W., Phair, G., and Newell, M. F., 1971, Boulder Creek batholith, Colorado; Part 2, Isotopic age of emplacement and morphology of zircon: Geological Society of America Bulletin, v. 82, p. 1615–1634.

Stuckless, J. S., and Nkomo, I. T., 1978, Uranium-lead isotope systematics in uraniferous alkali-rich granites from the Granite Mountains, Wyoming; Implications for uranium source rocks: Economic Geology, v. 73, p. 427–441.

Stuckless, J. S., Hedge, C. E., Worl, R. G., Simmons, K. R., Nkomo, I. T., and Wenner, D. B., 1985, Isotopic studies of the Late Archean plutonic rocks of the Wind River Range, Wyoming: Geological Society of America Bulletin, v. 96, p. 850–860.

Swift, P. N., 1982, Precambrian metavolcanic rocks and associated volcanogenic mineral deposits of the southwestern Sierra Madre, Wyoming [M.A. thesis]: Laramie, University of Wyoming, 61 p.

Todt, W. A., and Busch, W., 1981, U-Pb investigations on zircons from pre-Variscan gneisses; 1, A study from the Schwarzwald, West Germany: Geochimica et Cosmochimica Acta, v. 45, p. 1789–1801.

Tweto, O., 1976, Geologic map of the Craig 1° by 2° Quadrangle, northwestern Colorado: U.S. Geological Survey Miscellaneous Investigations Series Map I-972, scale 1:250,000.

——, 1979, Geologic map of Colorado: U.S. Geological Survey, scale 1:500,000.

Van Schmus, W. R., and Bickford, M. E., 1981, Proterozoic chronology and evolution of the midcontinent region, North America, *in* Kroner, A., ed., Precambrian plate tectonics: Amsterdam, Elsevier, p. 261–290.

Watson, E. B., 1979, Zircon saturation in felsic liquids; Experimental results and applications to trace element geochemistry: Contributions to Mineralogy and Petrology, v. 70, p. 407–419.

Watson, E. B., and Harrison, T. M., 1982, Zircon saturation in crustal melts; Implications for trace element and Pb redistribution during anatexis: EOS Transactions of the American Geophysical Union, v. 63, no. 45, p. 1142.

Manuscript Accepted by the Society September 27, 1988

Printed in U.S.A.

Geological Society of America
Special Paper 235
1989

U-Pb geochronology of the Proterozoic volcano-plutonic terrane in the Gunnison and Salida areas, Colorado

M. E. Bickford and R. D. Shuster*
Department of Geology, University of Kansas, Lawrence, Kansas 66045
S. J. Boardman
Department of Geology, Carleton College, Northfield, Minnesota 55057

ABSTRACT

Early Proterozoic supracrustal rocks near Gunnison and Salida, Colorado, include sequences of tholeiitic metabasalt, metarhyolite to metadacite, and interbedded volcaniclastic turbidite. These rocks were intruded by synchronous gabbroic sheets, complexly folded and metamorphosed in upper greenschist to upper amphibolite facies, and intruded by plutons ranging from quartz diorite to granite. U-Pb ages of zircons show that an early period of volcanism in the Gunnison area occurred from 1,770 to 1,760 Ma and was followed by emplacement of plutons from 1,755 to 1,750 Ma. A younger sequence of volcanic rocks was formed in both the Gunnison and Salida areas between 1,740 and 1,730 Ma. In the Gunnison area these rocks were intruded by major plutons from 1,725 to 1,714 Ma. Near Howard, Colorado, southeast of Salida, metarhyolite yielding ages of 1,713 and 1,668 Ma is believed to be part of the younger sequence. Late, post-tectonic granite plutons were emplaced in both areas from 1,700 to 1,670 Ma. The age data, petrography, and geochemistry of these rocks indicates that they are part of a broad belt of juvenile, arc-related terranes, exposed from southern California across Arizona, New Mexico, Colorado, and known in the subsurface as far east as western Missouri, that was accreted to the southern edge of the continent during the Early Proterozoic.

INTRODUCTION AND GEOLOGIC SETTING

A terrane of Early Proterozoic volcanic, plutonic, and interbedded metasedimentary rocks is exposed across the southwestern United States and northern Mexico (Fig. 1). These rocks have been studied carefully in southern California and Arizona by Silver (1968), Condie and Demalas (1985), Condie and others (1985), Ludwig (1974), Conway (1976), and Conway and Silver (1987), whereas in New Mexico they have been studied by Barker and others (1976), Condie (1980), Condie and Budding (1979), Condie and McCrink (1982), Robertson and Condie (this volume), and numerous others. Condie (1986) has provided an excellent summary of the occurrence of these rocks and of their chemical characteristics. Silver (1968) recognized that two belts, or terranes, of volcanic and related plutonic rocks occur in Arizona and extend into New Mexico. The more northerly of these includes volcanic rocks, formed 1,740 to 1,780 m.y. ago, that were intruded by plutons 1,730 to 1,750 m.y. ago. The more southerly terrane includes volcanic rocks, formed 1,680 to 1,700 m.y. ago, that were intruded by plutons formed 1,630 to 1,670 m.y. ago.

Similar rocks in the Gunnison and Salida areas of Colorado form the basis for this study. These rocks (Figs. 1 and 2) have been mapped by personnel of the U.S. Geological Survey (Hedlund, 1974; Hedlund and Olson 1974, 1975; Olson, 1974, 1976a,b; Olson and Steven, 1976); excellent theses have been done there by Afifi (1981), Tobin (1982), Bennett (1984), Shonk (Shonk and Geissman, 1985), and Vance (1984). The Proterozoic rocks north of Salida have been mapped and extensively studied by Boardman (1976).

*Present address: Department of Geology and Geography, University of Nebraska–Omaha, Omaha, Nebraska 68182.

Bickford, M. E., Shuster, R. D., and Boardman, S. J., 1989, U-Pb geochronology of the Proterozoic volcano-plutonic terrane in the Gunnison and Salida areas, Colorado, *in* Grambling, J. A., and Tewksbury, B. J., Proterozoic geology of the southern Rocky Mountains: Boulder, Colorado, Geological Society of America Special Paper 235.

The only previous age determinations from this region were done in the early 1960s by Rb-Sr methods (Wetherill and Bickford, 1965). A whole-rock isochron defined an age of 1,650 Ma for plutonic rocks, but, because at that time the region had not been geologically mapped, the samples analyzed were not all from the same pluton. Mineral isochrons from three whole-rock samples yielded ages ranging from 1,350 to 1,420 Ma. Additionally, age data by both Rb-Sr and U-Pb (zircon) methods have been reported from the Needle Mountains area, some 100 km to the southwest of the Gunnison area (Fig. 1). There a major metavolcanic unit, the Twilight Gneiss, has yielded a Rb-Sr whole-rock isochron age of 1,780 Ma (Barker and others, 1969), whereas plutons have yielded ages of 1,720 to 1,680 Ma by both U-Pb (zircon; Silver and Barker, 1968) and Rb-Sr whole-rock methods (Bickford and others, 1969).

The purpose of this study was to obtain accurate ages for many rock bodies in the Gunnison and Salida areas to see whether they are extensions of the older or younger terranes recognized by Silver, or whether they were formed during yet another period of igneous activity in the Proterozoic. In this paper we report the details of the geochronological measurements and interpretations of the resulting ages. A second goal of our work was to obtain major- and trace-element chemical data and to combine these with petrographic and field study to understand the petrogenesis and tectonic setting of these rocks. Some aspects of that work are in the theses of Tobin (1982) and Bennett (1984) and have also been presented by Bickford and Boardman (1984), Bennett and others (1984), and Boardman and Bickford (1984). Detailed papers on the geochemical and petrological studies are in preparation, and this work will be discussed only briefly here.

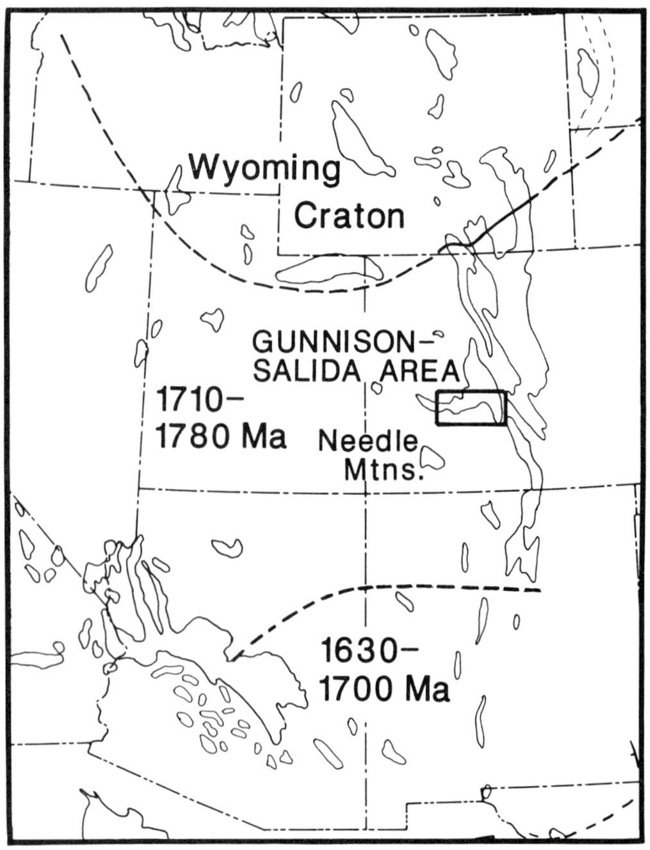

Figure 1. Sketch map showing Archean and Proterozoic age provinces in the southwestern United States.

Gunnison area

Proterozoic rocks exposed in the eastern and northeastern part of the Gunnison area are referred to here as the Cochetopa Succession after their excellent exposures in Cochetopa Canyon (Fig. 2). They include a bimodal sequence of metavolcanic rocks that were originally tholeiitic basalt to basaltic andesite and rhyolite to dacite, and a thick sequence of interbedded metasedimentary rocks, originally quartz and feldspar wackes, displaying features indicating deposition as turbidites. These rocks were intruded by gabbroic sills, multiply folded and metamorphosed in upper greenschist to lower amphibolite facies. Despite deformation and metamorphism, primary features in these rocks are well preserved; the metabasalts show well-developed pillows with vesiculated rinds; many of the metarhyolites have relict eutaxitic textures and show such features as lithic fragments and compacted pumice fragments; and the metasedimentary rocks preserve graded bedding, flame structures, sole marks, and similar features. A number of plutons, both foliated and non-foliated and ranging in composition from quartz diorite and tonalite to granite, have intruded the supracrustal sequence.

Rocks exposed in the southwestern part of the study area in the vicinity of Powderhorn, Colorado (Fig. 2), are significantly different from those exposed in the eastern and northeastern part. These rocks, the Dubois Greenstone of Hedlund and Olson (1975), are here called the Dubois Succession; they include a bimodal volcanic suite that is compositionally similar to that in the Cochetopa Succession, but the interbedded metasedimentary rocks are mostly more mafic, including chloritic schists and abundant Fe-rich metachert bodies. These rocks were also intruded by bodies of gabbro and then later intruded by major dioritic to granitic plutons. Deformation, including pervasive shearing, is much more prevalent than farther east, and preservation of primary extrusive and depositional features is poorer; nevertheless, pillowed flows and pillow breccias are common, and gabbroic plugs and sills show relict ophitic texture. As will be shown, these rocks predate the Cochetopa Succession by about 30 m.y.

Salida and Howard areas

The sequence of Proterozoic rocks exposed in the Salida area (Fig. 2) is strikingly similar to that in the Gunnison area. Metamorphism is generally higher grade, ranging from lower amphibolite in the south to upper amphibolite in the north, but penetrative deformation is generally less severe than in the Gunnison area and primary features are commonly spectacularly preserved (Boardman, 1976, 1986). The metavolcanic rocks are

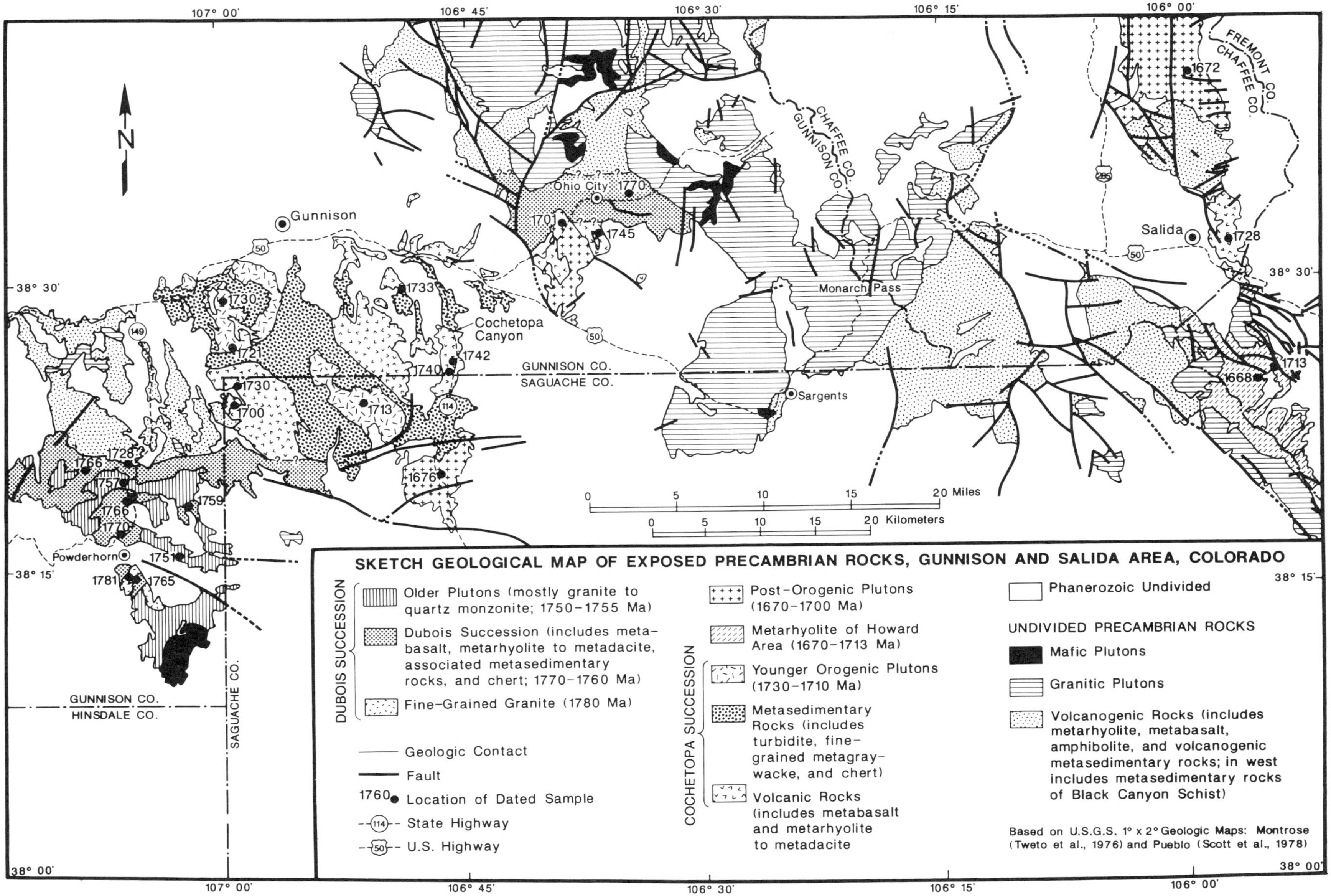

Figure 2. Sketch geologic map of exposed Proterozoic rocks, Gunnison and Salida areas, Colorado. Localities and ages of dated samples indicated by dots. Map modified from Montrose and Pueblo 1° by 2° geologic maps of U.S. Geological Survey.

distinctly bimodal and have been intruded by thick gabbroic sills. In addition to fine- to medium-grained wackes, metasedimentary rocks include massive mafic volcaniclastic units up to 150 m thick containing angular to subrounded volcanic rock clasts and crudely to well-bedded and cross-bedded coarse-grained volcaniclastic rocks with calcareous matrix; these rocks suggest near-source deposition in relatively shallow water (Bickford and Boardman, 1984; Boardman, 1986). A major granitic pluton intrudes this sequence at the northern end of the outcrop belt (Fig. 2).

Proterozoic rocks in the Howard area (Fig. 2), about 14 km southeast of Salida, are more deformed and probably at somewhat higher metamorphic grade than those in the Salida area. This area has been mapped only at reconnaissance scale (Taylor and others, 1975), and details are not available. We observed and collected strongly recrystallized metarhyolites with clearly recognizable features such as lithic fragments, crushed pumice fragments, and primary phenocrysts. Other rocks in the area include abundant amphibolite and micaceous schists.

Geochemistry

Boardman and Condie (1986) have presented detailed studies of the geochemistry of rocks in the Salida area. A paper presenting geochemical studies in the Gunnison area is in preparation by Bickford and several of his students. Only a brief summary will be presented here. The first extensive geochemical study of rocks in the Gunnison area was made by Condie and Nuter (1981), who analyzed metavolcanic rocks, mostly collected from the Cochetopa Succession, for major and trace elements, including the rare earth elements (REE). They concluded that: (1) two suites of metabasalts are represented but no geographic pattern to their distribution is apparent, and (2) on the basis of the trace element data, the most likely tectonic setting for the development of these rocks was a back-arc region formed upon at least somewhat older continental crust.

Bennett (1984) studied the major- and trace-element chemistry of mafic and felsic metavolcanic rocks and interbedded metasedimentary rocks of the Cochetopa Succession in the northern part of the Iris 7½-minute Quadrangle (Olson, 1976a; this paper Fig. 2). Bennett found that the mafic rocks range in composition from basalt to basaltic andesite, but most are typical tholeiites with trace-element abundances indicating formation either in an ocean floor or continental rift setting (e.g., Pearce and Cann, 1973; Floyd and Winchester, 1975; Holm, 1982). Felsic metavolcanic rocks formed as flows, ash-flows, and subaqueous debris flows (Afifi, 1981) and are compositionally dacite to rhyolite, whereas most of the sedimentary rocks are intermediate in composition. According to Bennett (1984), the fractionation trends of the complete suite on trace-element diagrams of Pearce and Norry (1979) fit island-arc settings better than within-plate settings. Tobin (1982) studied the major- and trace-element chemistry and petrography of an annular plutonic complex about 5 km southwest of Gunnison (Fig. 2). He found that rocks in the complex ranged from quartz diorite, through tonalite and granodiorite, to granite. These rocks have both major- and trace-element characteristics indicating calcalkaline chemistry.

Boardman and Condie (1986) have determined major- and trace-element characteristics of the metavolcanic rocks in the Salida area. The suite is bimodal, consisting of tholeiitic basalt and related thick gabbroic sills and felsic rocks of both ash-flow and subaqueous debris-flow origin. The chemistry of these rocks is strikingly similar to those in the Gunnison area.

Bickford and Boardman (1984) summarized the occurrence, age, and chemical characteristics of rocks in both the Gunnison and Salida areas, observing that although petrologic and chemical characteristics are consistent with origin in a back-arc setting, as proposed by Condie and Nuter (1981), there is no evidence of significantly older continental crust any closer than southern Wyoming, some 300 km to the north. Bickford and Boardman (1984) also pointed out that bimodal volcanic suites and MORB-like basaltic rocks are known from a number of primitive modern island arcs, and suggested that this setting might be more consistent with the highly fragmental nature of the supracrustal rocks in both the Gunnison and Salida areas and with the thick sequences of turbidite in the Gunnison area. The Sm-Nd data of Nelson and DePaolo (1984) are important here, for they show that the mantle separation age of some rocks in the Gunnison area is 1,800 ± 100 Ma; as will be shown below, this is essentially the crystallization age of the rocks and precludes derivation from significantly older crust. Nelson and DePaolo also presented some initial $^{87}Sr/^{86}Sr$ data for basaltic rocks from this area. As expected, the values are primitive (about 0.7030); unfortunately, Nelson and DePaolo (1984) did not determine the initial $^{87}Sr/^{86}Sr$ values for the felsic metavolcanic rocks and related plutons.

GEOCHRONOLOGY

Analytical methods

The analyses of zircon fractions for U and Pb concentrations and Pb isotopic compositions were done in the Isotope Geochemistry Laboratory at the University of Kansas by standard techniques of isotope dilution analysis and mass spectrometry. In most cases the zircon fractions were separated on the basis of their magnetic susceptibility. For same samples (Table 1) we also analyzed different size fractions in an attempt to identify possible metamorphic zircon populations, but these yielded data points that were essentially colinear with the magnetic splits. Dissolution and ion-exchange techniques are slightly modified from those of Krogh (1973). Mass spectrometry was done on a 23-cm-radius Nier-type instrument using single rhenium filaments on which samples were loaded with silica gel and phosphoric acid (Cameron and others, 1969). Analytical blanks measured during the course of these analyses were 1.0 ng or less total Pb and less than 0.5 ng U. Decay constants used are $\lambda^{238}U = 0.155125 \times 10^{-9}$ yrs^{-1} and $\lambda^{235}U = 0.98458 \times 10^{-9}$ yrs^{-1} (Steiger and Jager, 1977).

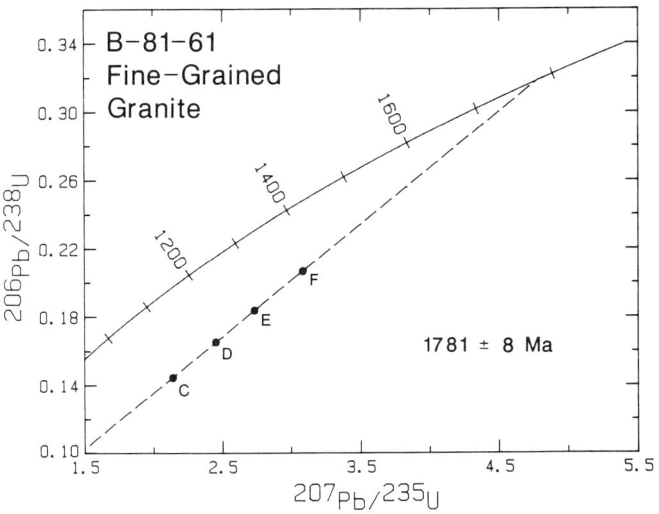

Figure 3. Concordia plot for sample B-81-61. Model 2 solution based on 4 zircon fractions, probability of fit 15.2 percent, MSWD = 1.88. Intercepts at 1781 ± 8 and 41 ± 19 Ma.

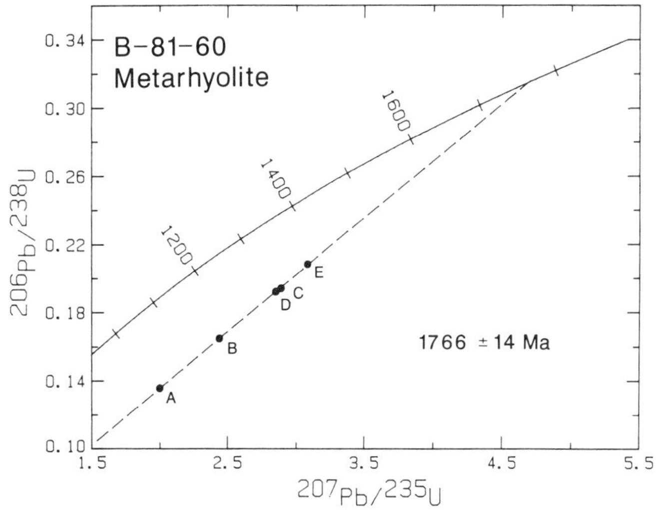

Figure 5. Concordia plot for sample B-81-60. Model 2 solution based on 5 zircon fractions, probability of fit 0 percent, MSWD = 6.45. Intercepts at 1,766 ± 14 and 25 ± 34 Ma.

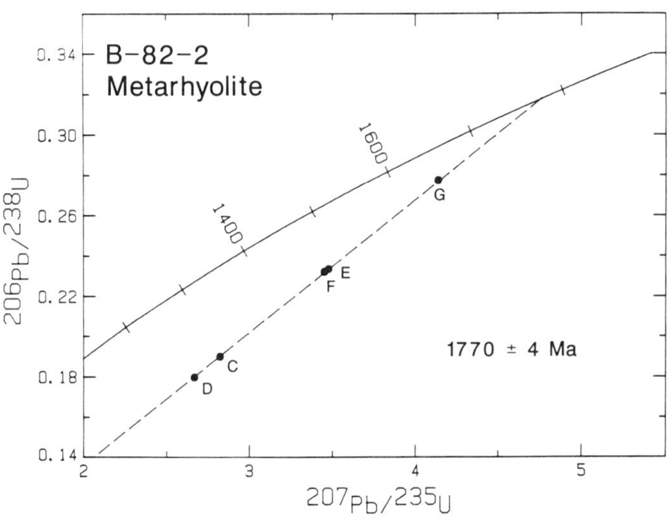

Figure 4. Concordia plot for sample B-82-2. Model 1 solution based on 5 zircon fractions, probability of fit 57.8 percent, MSWD = 0.66. Intercepts at 1,770 ± 4 and 23 ± 14 Ma.

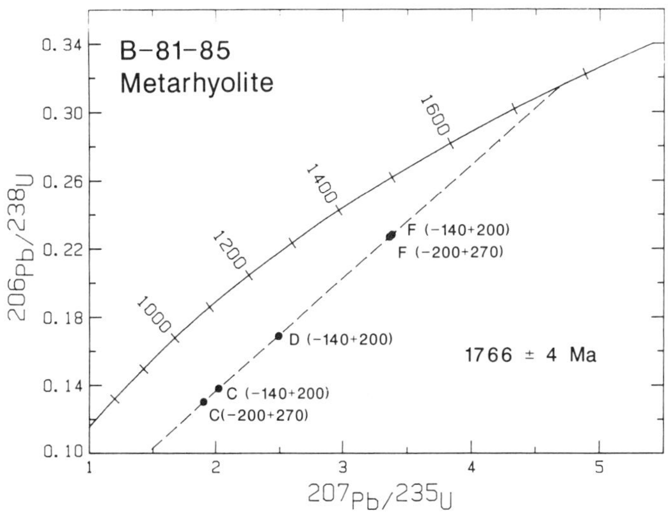

Figure 6. Concordia plot for sample B-81-85. Model 1 solution based on 5 zircon fractions, probability of fit 33.2 percent, MSWD = 1.14. Intercepts at 1,766 ± 4 and 42 ± 7 Ma.

Uncertainties of measured $^{207}Pb/^{206}Pb$ ratios are estimated to be 0.1 percent (2σ); relative uncertainty of $^{204}Pb/^{206}Pb$ ratios varied with the magnitude of the ratio, but absolute uncertainty is estimated as ± 0.00002, or about 2 percent for a sample with $^{204}Pb/^{206}Pb$ of 0.00100. The U-Pb ratios are considered accurate to ± 1.0 percent (2σ). After correction for all sources of uncertainty, radiogenic $^{207}Pb/^{206}Pb$ ratios are considered accurate to ± 0.25 percent (2σ).

Ages were calculated using the regression and error analysis methods of Ludwig (1980, 1983). The calculated $^{207}Pb/^{235}U$ and $^{206}Pb/^{238}U$ ratios were assigned uncertainties of 1.0 percent with a correlation coefficient of 0.98. If the probability of fit from the regression was 30 percent or better, ages were calculated with Ludwig's Model 1 solution; the Model 2 solution was used if the probability of fit was less than 30 percent. Analytical data are given in Table 1, and ages (± 2σ) are given in Table 2 and Figures 3 through 27.

Results

The ages determined are presented below, beginning with the oldest rocks studied. It will be seen that two sequences of volcanism followed by emplacement of plutons are represented; the older sequence, the Dubois Succession as defined earlier, occurs mostly in the southwestern part of the Gunnison area in the vicinity of Powderhorn, Colorado, and is also represented by a sample from the north-central part of the area near Ohio City (Fig. 2), whereas the younger sequence, the Cochetopa Succession, mostly occurs in the eastern and central part of the Gunnison area and in the Salida area. Sample locations are given in Figure 2.

Gunnison area; Dubois Succession. The sample that yielded the greatest zircon age is B-81-61, a fine-grained granite from the Powderhorn area. The data (Fig. 3) yield an age of 1,781 ± 8 Ma. Interpretation of this result is difficult, for the geologic maps of the Powderhorn (Hedlund and Olson, 1975) and Rudolph Hill (Olson, 1974) Quadrangles indicate that this granite intrudes metarhyolite and the Powderhorn Granite. Our dated sample of the metarhyolite (SOC-2, see below) yielded an age of 1,765 ± 44 Ma, whereas the Powderhorn Granite (B-81-71, see below) yielded 1,751 ± 6 Ma. Thus the field data and the geochronologic data are in conflict. We have not attempted to remap this area in detail, but a reconnaissance (M. Knoper, personal communication, 1986) indicated that the mapped field relations are correct. The most likely explanation for the discrepancy is that the zircons from the fine-grained granite have an inherited older zircon component. However, we have not observed cores within the zircons microscopically.

Five metarhyolite samples are represented by samples B-82-2, B-81-60, B-81-85, B-81-82, and SOC-2. These rocks are all porphyries carrying both quartz and feldspar phenocrysts in a fine-grained felsic matrix. All are deformed, especially B-81-85, which has a schistose matrix containing metamorphic biotite and muscovite, but there is no question of their volcanic origin. Each

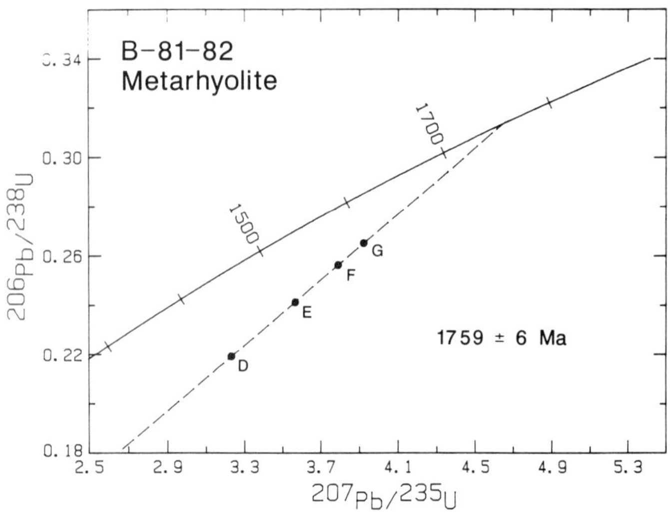

Figure 7. Concordia plot for sample B-81-82. Model 1 solution based on 4 zircon fractions, probability of fit 82.5 percent, MSWD = 0.19. Intercepts at 1,759 ± 6 and 48 ± 36 Ma.

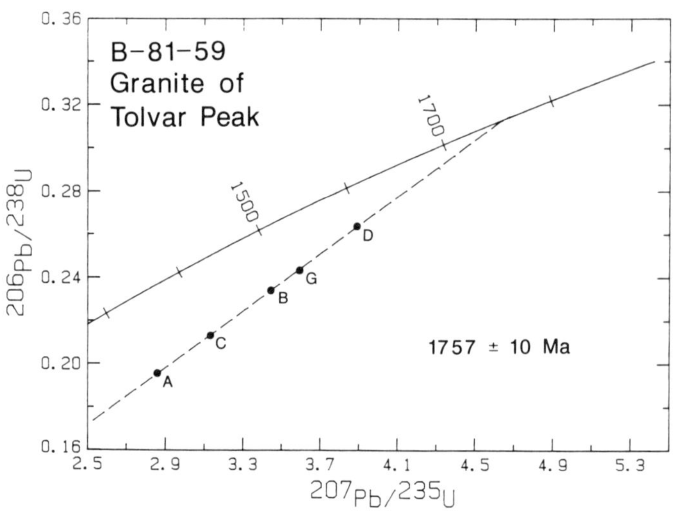

Figure 9. Concordia plot for sample B-81-59. Model 2 solution based on 4 zircon fractions, probability of fit 1.9 percent, MSWD = 3.99. Intercepts at 1,757 ± 10 and 72 ± 46 Ma.

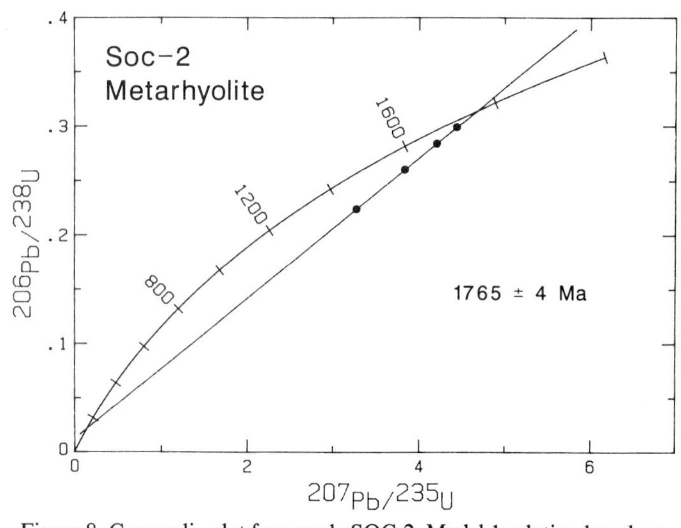

Figure 8. Concordia plot for sample SOC-2. Model 1 solution based on 4 zircon fractions, probability of fit 58 percent, MSWD = 0.538. Intercepts at 1,765 ± 4 and 135 ± 25 Ma.

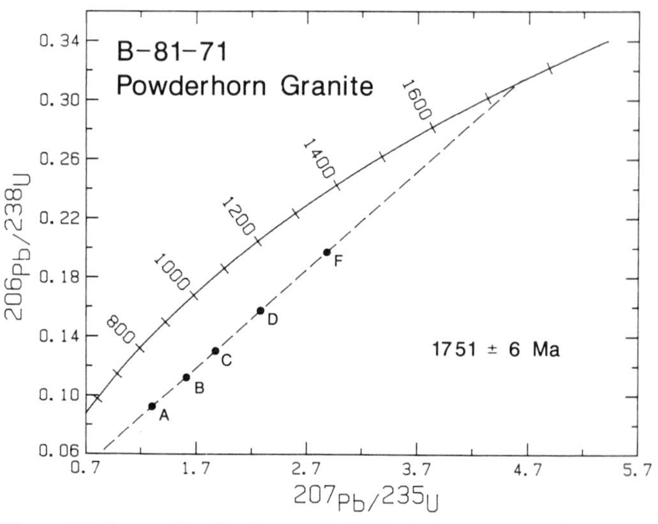

Figure 10. Concordia plot for sample B-81-71. Model 2 solution based on 5 zircon fractions, probability of fit 2.7 percent, MSWD = 3.06. Intercepts at 1,751 ± 6 and 61 ± 8 Ma.

sample yielded euhedral, relatively clear zircons showing the growth zonation typical of igneous crystallization. These samples yield the following ages: *B-82-2,* 1,770 ± 4 Ma (Fig. 4); *B-81-60,* 1,766 ± 7 Ma (Fig. 5); *B-81-85,* 1,766 ± 4 Ma (Fig. 6); *B-81-82,* 1,759 ± 6 Ma (Fig. 7); and *SOC-2,* 1,765 ± 4 Ma (Fig. 8). Because the metarhyolite units sampled are intercalated with metabasalt and metasedimentary rocks in this region, we conclude that 1,765 Ma is a good estimate of the mean time of the volcanic activity and related sedimentation that produced these rocks.

Two plutons that intrude the volcanic sequence were also dated. These are the Granite of Tolvar Peak (B-81-59) and the Powderhorn Granite (B-81-71). Sample B-81-59 yields an age of 1,757 ± 10 Ma (Fig. 9), whereas sample B-81-71 yields an age of 1,751 ± 6 Ma (Fig. 10). Both of these plutons are granitic in composition and are moderately to strongly foliated. The outcrop pattern of the Granite of Tolvar Peak is elongate, with numerous convolutions along its contacts with metavolcanic rocks, suggesting that it may have been a large, sill-like mass that was emplaced in the volcanic pile and subsequently folded along with the supracrustal rocks. The outcrop pattern of the Powderhorn Granite, in contrast, suggests that it intruded the supracrustal rocks as a cross-cutting pluton.

The area near Ohio City in the north-central part of the study area (Fig. 2) has not been mapped geologically. However, reconnaissance studies by Michael Knoper of the New Mexico Institute of Mining and Technology have shown that a sequence of metarhyolite and metabasalt similar to that elsewhere in the

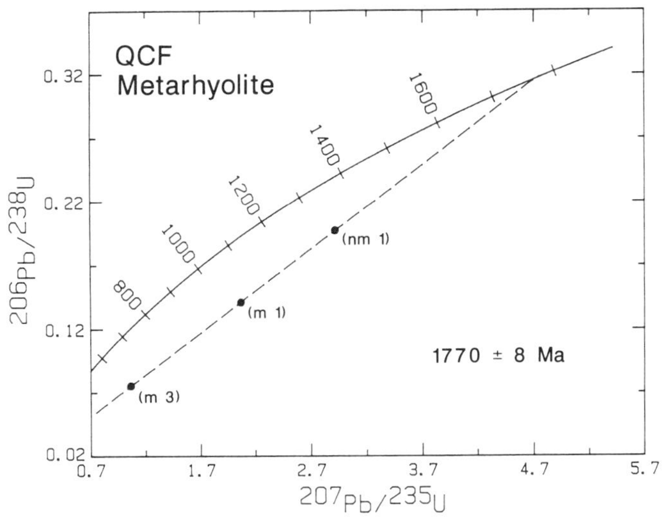

Figure 11. Concordia plot for sample QCF. Model 2 solution based on 3 zircon fractions, probability of fit 3.5 percent, MSWD = 4.43. Intercepts at 1,770 ± 8 and 59 ± 11 Ma.

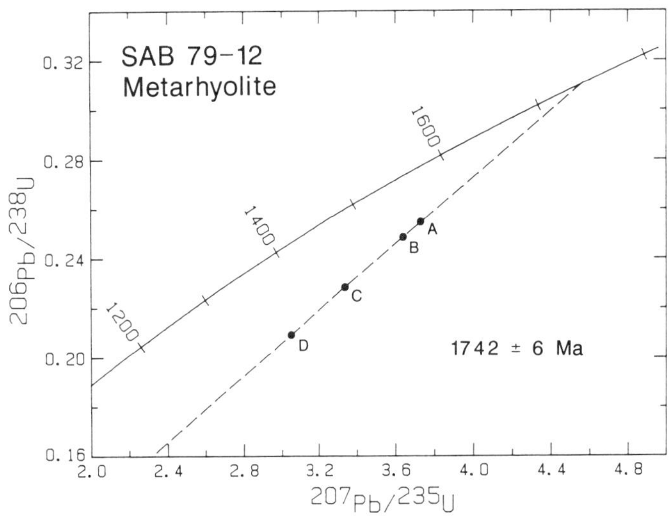

Figure 13. Concordia plot for sample SAB-79-12. Model 1 solution based on 4 zircon fractions, probability of fit 71.0 percent, MSWD = 0.346. Intercepts at 1,742 ± 6 and 64 ± 33 Ma.

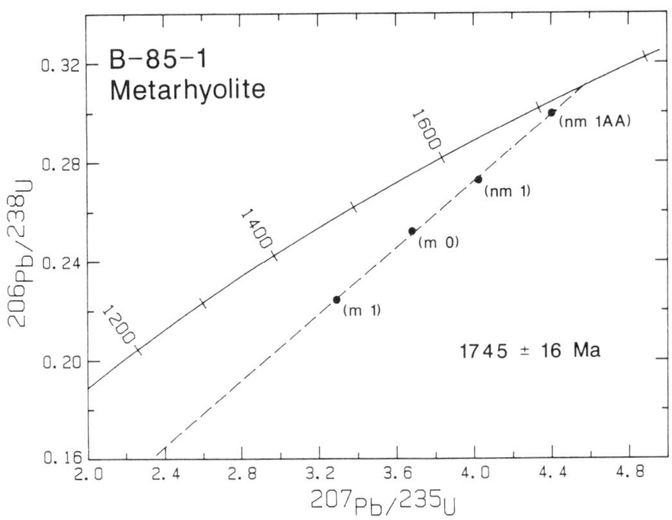

Figure 12. Concordia plot for sample B-85-1. Model 2 solution based on 4 zircon fractions, probability of fit 0 percent, MSWD = 25.9. Intercepts at 1,745 ± 16 and 50 ± 140 Ma.

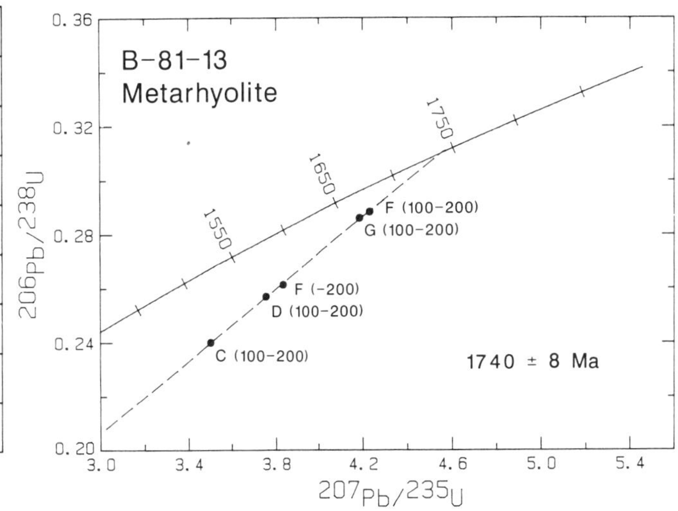

Figure 14. Concordia plot for sample B-81-13. Model 2 solution based on 5 zircon fractions, probability of fit 1.1 percent, MSWD = 3.72. Intercepts at 1,740 ± 8 and 66 ± 74 Ma.

Gunnison area is exposed there. At least one sample belongs to the older volcanic-plutonic sequence of rocks. Sample QCF is metarhyolite that is interbedded with metabasalt; it yielded zircons that give an age of 1,770 ± 8 Ma (Fig. 11). A second metarhyolite sample from this area, B-85-1, yields an age of 1,745 ± 16 Ma; it seems most likely that this sample is part of the younger sequence, to be discussed below.

Gunnison area: Cochetopa Succession. The younger sequence of volcanic and plutonic rocks is similar to the older sequence except that it is not as pervasively sheared. These rocks are mostly exposed in the central and eastern part of the Gunnison area, although sample SOC-2 was collected in the northwestern part of the area (Fig. 2).

Several felsic volcanic rocks whose compositions are in the rhyolite to dacite range were dated. These samples yield the following ages: *B-85-1,* 1,745 ± 16 Ma (Fig. 12); *SAB-79-12,* 1,742 ± 6 Ma (Fig. 13); *B-81-13,* 1,740 ± 8 Ma (Fig. 14); *B-83-1,* 1,733 ± 11 Ma (Fig. 15); *B-81-43,* 1,730 ± 7 Ma (Fig. 16); and *SOC-1,* 1,728 ± 16 Ma (Fig. 17).

Five plutons that intrude the younger volcanic sequence were dated. Two of these are from the annular intrusive complex about 5 km southwest of Gunnison (Fig. 2). Quartz diorite from its center (sample T-81-10; Tobin, 1982) yields an age of 1,730 ± 6 Ma (Fig. 18), whereas granite from the outer ring (Granite of South Beaver Creek, sample B-81-40; Hedlund and Olson, 1974; Tobin, 1982) yields an age of 1,721 ± 7 Ma (Fig. 19). The tonalite of Gold Basin, located about 10 km south-southeast of Gunnison (Fig. 2; sample B-81-24; Olson, 1976a) yields an age of

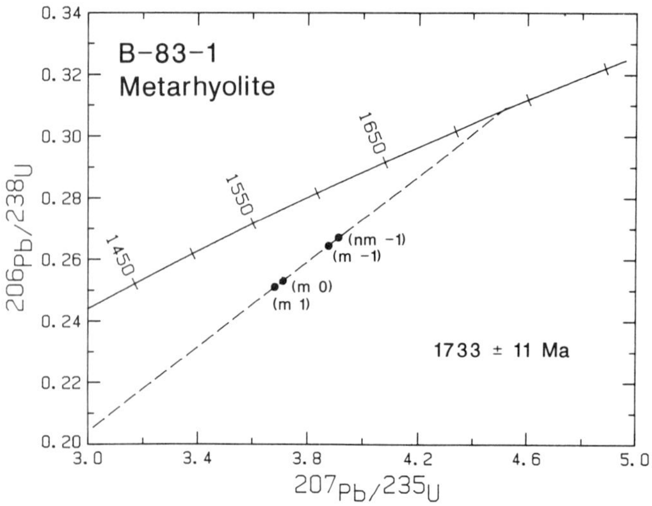

Figure 15. Concordia plot for sample B-83-1. Model 1 solution based on 4 zircon fractions, probability of fit 85.3 percent, MSWD = 0.16. Intercepts at 1,733 ± 11 and –33 ± 110 Ma.

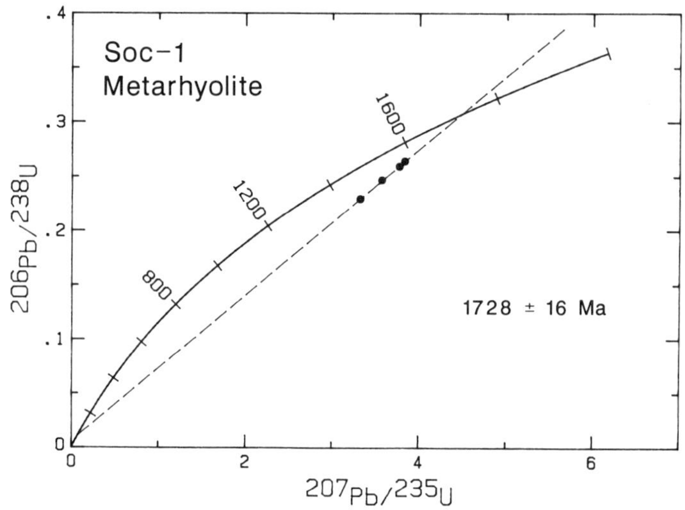

Figure 17. Concordia plot for sample SOC-1. Model 2 solution based on 4 zircon fractions, probability of fit 0.61 percent, MSWD = 5.09. Intercepts at 1,728 ± 16 and 71 ± 120 Ma.

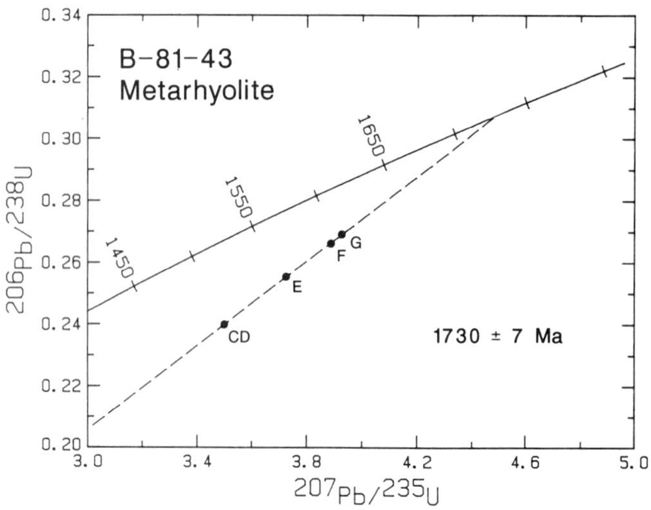

Figure 16. Concordia plot for sample B-81-43. Model 1 solution based on 4 zircon fractions, probability of fit 62.3 percent, MSWD = 0.47. Intercepts at 1,730 ± 7 and 12 ± 62 Ma.

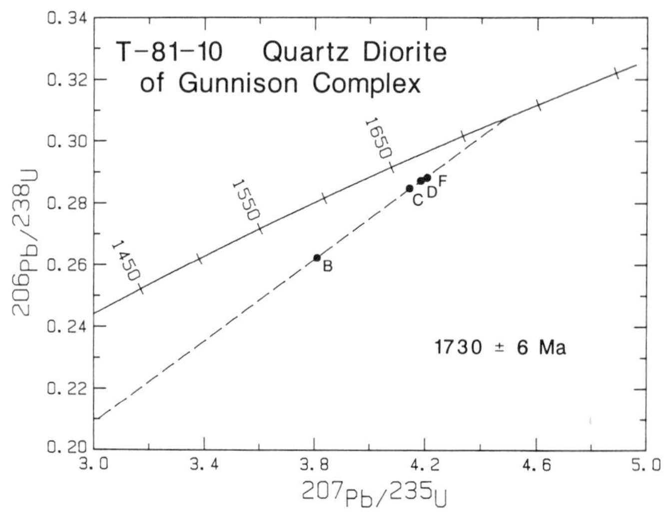

Figure 18. Concordia plot for sample T-81-10. Model 2 solution based on 4 zircon fractions, probability of fit 23.2 percent, MSWD = 1.46. Intercepts at 1,730 ± 6 and 121 ± 100 Ma.

1,713 ± 10 Ma (Fig. 20). These plutons have ages between 1,730 and 1,714 Ma and are probably associated with the volcanism. Two other samples—a granite located south of Gunnison in the extreme northwest corner of Saguache County (Fig. 2; sample B-81-42), and the Granite of Wood Gulch (Olson, 1976b; Fig. 2; sample B-81-70)—yield ages of about 1,700 Ma (Figs. 21 and 22); their relation to the volcanism is equivocal until definitive geochemical data are in hand. The Granite of Cochetopa Canyon (Olson and Steven, 1976; Fig. 2; sample B-81-68) yields an age of 1,676 ± 5 Ma (Fig. 23) and is apparently a post-tectonic granite.

Salida and Howard areas. Four samples have been dated from the Salida and Howard areas. These include a metadacite (sample Z-3) and a granite (sample SAL-1) from the Salida area and two metarhyolites (samples HO-1 and HO-2) from the Howard area (Fig. 2). The results are as follows: *Z-3,* 1,728 ± 6 Ma (Fig. 24); *SAL-1,* 1,672 ± 5 Ma (Fig. 25); *HO-1,* 1,668 ± 18 Ma (Fig. 26); and *HO-2,* 1,713 ± 14 Ma (Fig. 27).

DISCUSSION

A major conclusion of the work presented here is that the volcanic and plutonic rocks in the Gunnison and Salida areas are broadly coeval with the older volcano-plutonic belt defined by Silver (1968). The range of ages determined for volcanic rocks, 1,770 to 1,730 Ma, clearly falls within the range of the older belt studied by Silver in Arizona and New Mexico. Thus, that terrane continues northeasterly across the continent at least as far as central Colorado. Study of basement rocks recovered from deep

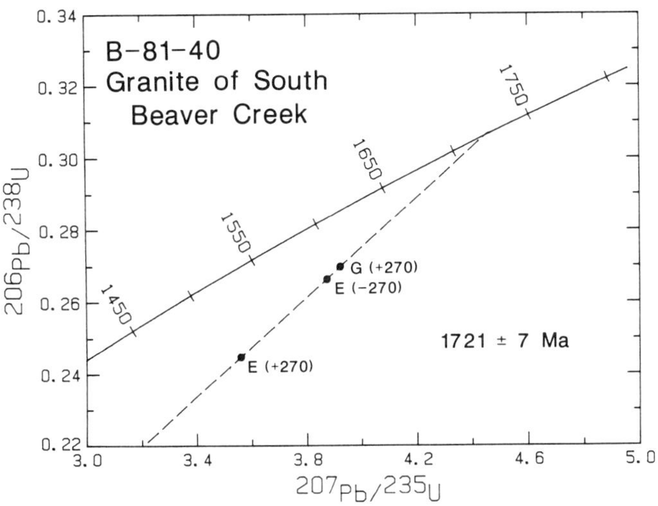

Figure 19. Concordia plot for sample B-81-40. Model Model 1 solution based on 3 zircon fractions, probability of fit 88 percent, MSWD = 0.02. Intercepts at 1,721 ± 7 and –11 ± 77 Ma.

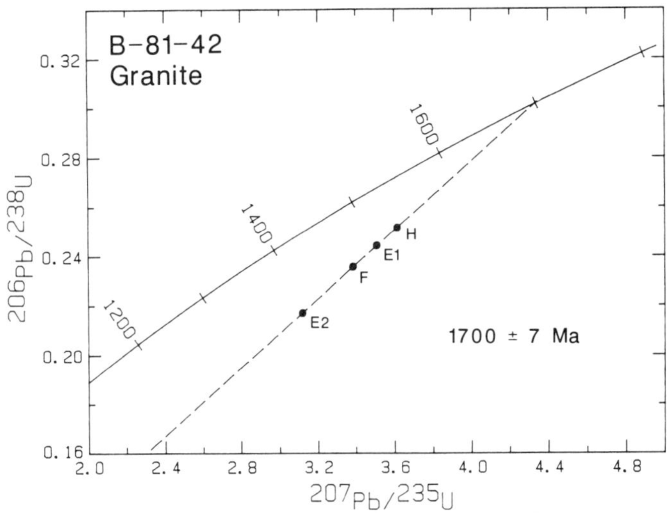

Figure 21. Concordia plot for sample B-81-42. Model 1 solution based on 4 zircon fractions, probability of fit 91.1 percent, MSWD = 0.09. Intercepts at 1,700 ± 7 and 1 ± 48 Ma.

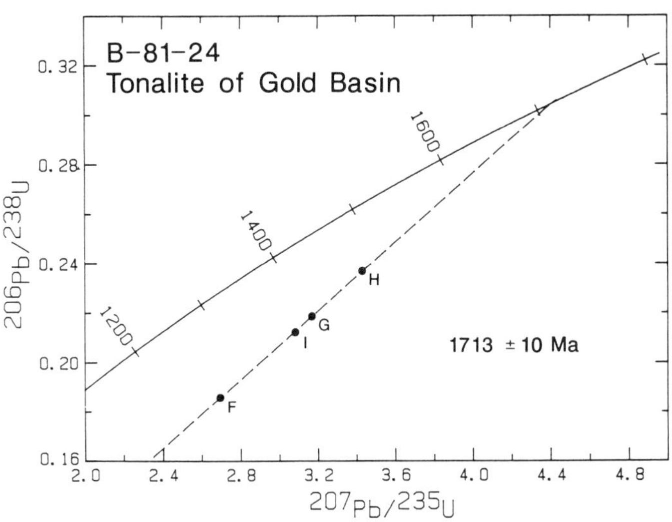

Figure 20. Concordia plot for sample B-81-24. Model 2 solution based on 4 zircon fractions, probability of fit 10.5 percent, MSWD = 2.26. Intercepts at 1,713 ± 10 and –14 ± 44 Ma.

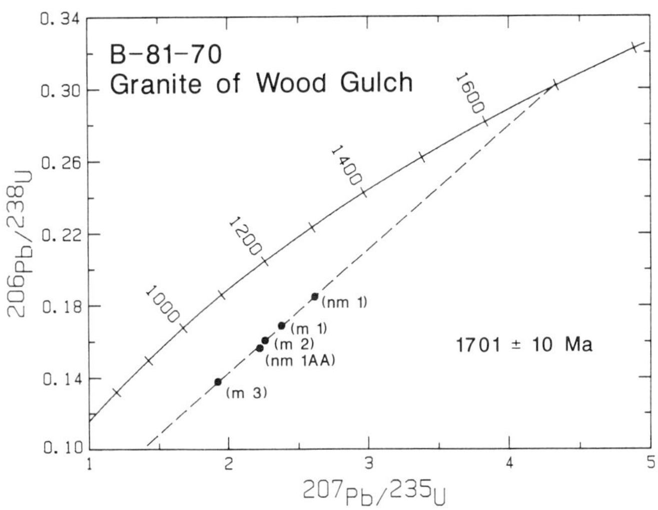

Figure 22. Concordia plot for sample B-81-70. Model 2 solution based on 4 zircon fractions, probability of fit 30 percent, MSWD = 1.21. Intercepts at 1,701 ± 10 and 84 ± 20 Ma.

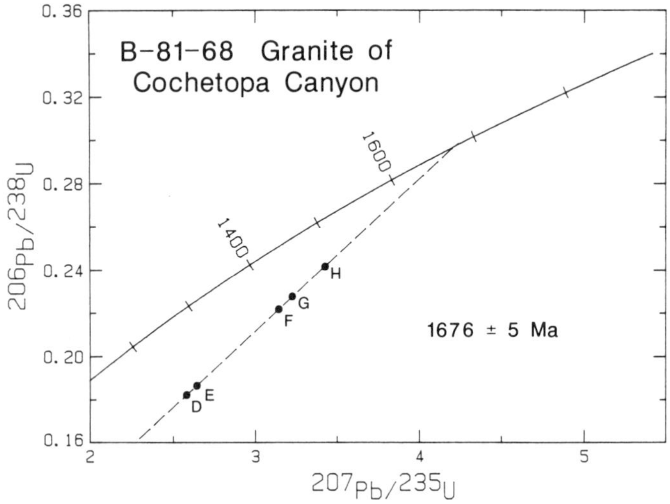

Figure 23. Concordia plot for sample B-81-68. Model 1 solution based on 5 zircon fractions, probability of fit 78.6 percent, MSWD = 0.35. Intercepts at 1,676 ± 5 and 4 ± 19 Ma.

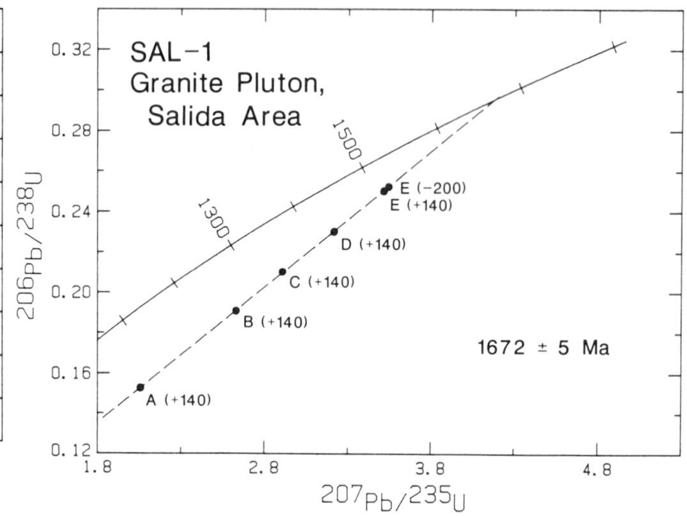

Figure 25. Concordia plot for sample SAL-1. Model 2 solution based on 6 zircon fractions, probability of fit 8.0 percent, MSWD = 2.09. Intercepts at 1,672 ± 5 and 164 ± 18 Ma.

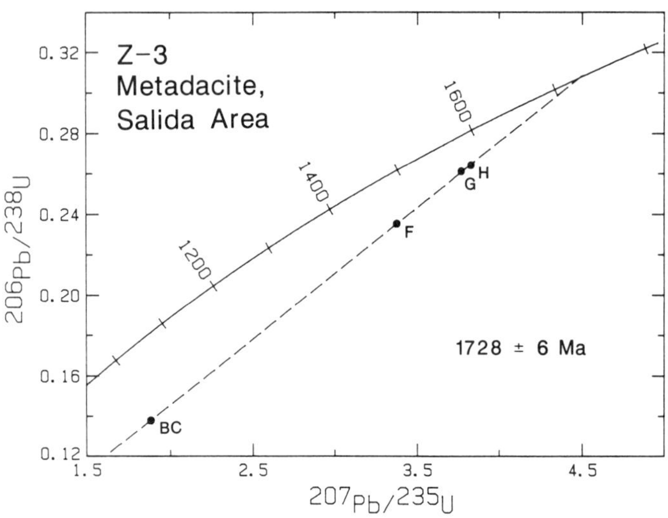

Figure 24. Concordia plot for sample Z-3. Model 2 solution based on 4 zircon fractions, probability of fit 4.3 percent, MSWD = 3.16. Intercepts at 1,728 ± 6 and 175 ± 22 Ma.

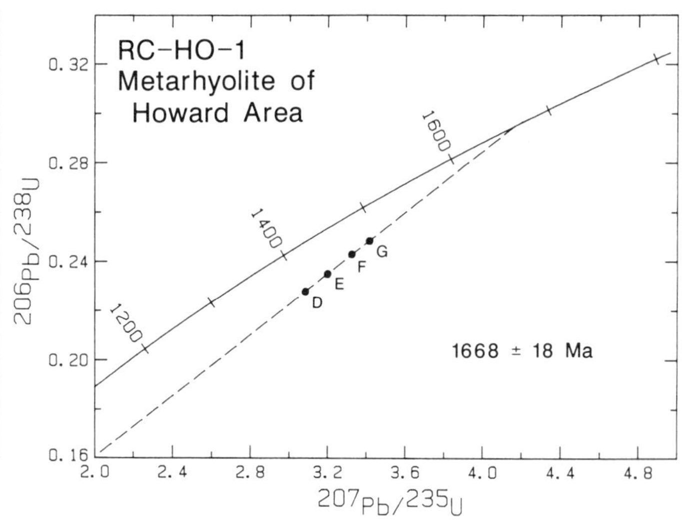

Figure 26. Concordia plot for sample RC-HO-1. Model 1 solution based on 4 zircon fractions, probability of fit 84 percent, MSWD = 0.18. Intercepts at 1,668 ± 18 and 431 ± 83 Ma.

drilling across eastern Colorado, northern Kansas, and southern Nebraska indicates that rocks of similar age (1,720 to about 1,800 Ma) and type (granite to granodiorite) extend much farther to the east beneath the Phanerozoic rocks of the plains (Van Schmus and others, 1987; Bickford and others, 1986). It seems clear that a major terrane of volcanogenic, and presumably arc-related, rocks was accreted to the southern edge of the growing continent during the Early Proterozoic (Bickford and others, 1986; Condie, 1986).

Although petrologically similar, the Dubois Succession is some 30 m.y. older than the Cochetopa Succession. Both sequences include thick accumulations of volcanogenic metasedimentary rocks intercalated with a bimodal volcanic assemblage. These rocks accumulated 1,770 to 1,760 m.y. ago in the Dubois Succession and were intruded by granodioritic to granitic plutons (Granite of Tolvar Peak and Powderhorn Granite) about 1,755 Ma. Volcanism that formed the Cochetopa Succession occurred between 1,740 and 1,730 Ma and was followed by intrusion of quartz diorite, tonalite, and granodiorite (annular complex of the Gunnison area and Tonalite of Gold Basin) from 1,725 to 1,714 Ma.

Structural and stratigraphic relations between the Dubois and Cochetopa Successions in the Gunnison area are almost completely unknown. The boundaries drawn in Figure 2 are

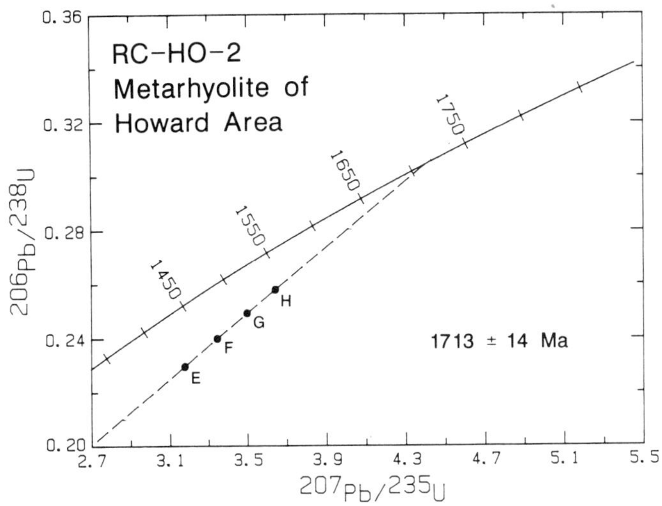

Figure 27. Concordia plot for sample RC-HO-2. Model 1 solution based on 4 zircon fractions, probability of fit 93 percent, MSWD = 0.072. Intercepts at 1,713 ± 14 and 419 ± 63 Ma.

based on the distribution of ages, but not even the location of the contact between the older and younger sequence is known. Needless to say, the nature of the contact remains completely obscure. The level of mapping and the quality and accessibility of outcrop available in this area are better than in most other regions of Proterozoic rocks in Colorado. A program of careful structural mapping and analysis could greatly elucidate relations between these volcanogenic terranes.

Geochemical study of these successions now in progress (A. Astatke, personal communication, 1986; M. Knoper, personal communication, 1986) indicates that, on the basis of trace-element abundances, the mafic volcanic rocks of the Dubois Succession are more primitive than those of the Cochetopa Succession. A tentative interpretation is that the Dubois Succession represents an accreted arc, whereas the Cochetopa Succession was formed in an adjacent or superposed back-arc setting. Although this interpretation is consistent with the chronological and geochemical data at hand, a fuller understanding must await completion of detailed geochemical study and the structural analysis discussed above.

The volcanic sequence in the Salida area is correlated with the Cochetopa Succession on the basis of the age of 1,728 ± 3 Ma determined from a metadacite in the Salida section. It seems likely that the volcanic rocks to the southeast in the Howard area (Fig. 2) are also part of this sequence, but the ages obtained for two metarhyolites there are equivocal; one sample yielded 1,713 ± 14 Ma and could clearly be part of the Cochetopa Succession, but the other yielded 1,668 ± 18 Ma and may be part of a still younger volcanic sequence.

Little is currently known about the younger plutons that intrude the Cochetopa Succession and its equivalent in the Salida area. Plutons with ages of 1,700 and about 1,670 Ma are known to the east in the Wet Mountains and the southern Front Range (Bickford and others, this volume), to the north in the Sierra Madre Range of south-central Wyoming (Premo and Van Schmus, this volume), and in the buried Proterozoic of the mid-continent as far east as western Kansas (Van Schmus and others, 1987; Bickford and others, 1986). These rocks may be isolated plutons formed inboard of an orogenic belt that formed at this time to the south in New Mexico and Arizona (Bickford and others, 1986; Condie, 1986) or they may have formed during terminal collision of the collage of volcanic-arc and back-arc terranes with the Archean craton to the north.

ACKNOWLEDGMENTS

Financial support for this work was provided by NSF Grant EAR 8025257. The authors are grateful for discussions with K. C. Condie and W. R. Van Schmus. The manuscript benefited greatly from careful reviews by T. E. Krogh, J. Aleinikoff, and J. Grambling.

TABLE 1. ANALYTICAL DATA FOR ZIRCON FRACTIONS

Fraction	Concentrations		Pb Isotopic Composition			Radiogenic Ratios			Ages		
	U (ppm)	Pb (ppm)	$^{204}Pb/^{206}Pb$	$^{207}Pb/^{206}Pb$	$^{208}Pb/^{206}Pb$	$^{206}Pb/^{238}U$	$^{207}Pb/^{235}U$	$^{207}Pb/^{206}Pb$	$^{206}Pb/^{238}U$	$^{207}Pb/^{235}U$	$^{207}Pb/^{206}Pb$
GUNNISON AREA											
DuBois Succession											
B-81-61-F	466	109	0.00197	0.13509	0.13843	0.20668	3.0852	0.10826	1211.1	1429.0	1770.2
B-81-61-E	515	113	0.00285	0.14679	0.17321	0.18372	2.7346	0.10795	1087.3	1337.9	1765.2
B-81-61-D	581	119	0.00357	0.15617	0.20140	0.16536	2.4524	0.10756	986.5	1258.1	1758.5
B-81-61-C	781	144	0.00412	0.16362	0.22254	0.14450	2.1419	0.10750	870.1	1162.4	1757.3
B-82-2-G	222	65	0.00024	0.11153	0.11091	0.27747	4.1396	0.10820	1578.6	1662.2	1769.5
B-82-2-F	346	85	0.00030	0.11195	0.11321	0.23209	3.4519	0.10787	1345.5	1516.3	1763.6
B-82-2-E	385	95	0.00021	0.11095	0.11263	0.23340	3.4775	0.10806	1352.3	1522.1	1767.1
B-82-2-D	523	101	0.00043	0.11356	0.12354	0.17971	2.6694	0.10773	1065.4	1320.0	1761.3
B-82-2-C	406	81	0.00015	0.10986	0.11553	0.18999	2.8243	0.10781	1121.3	1362.0	1762.8
B-81-60-E	653	146	0.00078	0.11790	0.11209	0.20847	3.0842	0.10730	1220.7	1428.8	1754.0
B-81-60-D	700	147	0.00104	0.12168	0.12467	0.19230	2.8505	0.10751	1133.8	1368.9	1757.7
B-81-60-C	782	166	0.00101	0.12158	0.12541	0.19428	2.8886	0.10783	1144.5	1378.9	1763.1
B-81-60-B	772	143	0.00144	0.12695	0.14862	0.16487	2.4403	0.10735	938.8	1254.6	1754.9
B-81-60-A	1130	180	0.00204	0.13457	0.18333	0.13571	1.9987	0.10682	820.3	1115.1	1745.9
B-81-85-F (-200-270)	575	134	0.00020	0.11026	0.08023	0.22682	3.3612	0.10748	1317.8	1495.4	1757.0
B-81-85-F (-140-200)	579	136	0.00020	0.11019	0.07999	0.22811	3.3802	0.10747	1324.6	1499.8	1757.0
B-81-85-D (-140-200)	798	142	0.00050	0.11387	0.10141	0.16870	2.4901	0.10705	1004.9	1269.2	1749.9
B-81-85-C (-140-200)	940	146	0.00149	0.12665	0.14962	0.13802	2.0240	0.10636	833.4	1123.6	1738.0
B-81-85-C (-200-270)	1058	148	0.00059	0.11417	0.11994	0.13022	1.9045	0.10607	789.1	1082.7	1733.1
SOC-2 (nm 1 AA)	393	118	0.00002	0.10794	0.06311	0.29896	4.4385	0.10768	1686.1	1719.6	1760.6
SOC-2 (nm 1)	427	123	0.00007	0.10848	0.06709	0.28365	4.2067	0.10756	1609.7	1675.3	1758.7
SOC-2 (m 2)	512	136	0.00013	0.10880	0.07518	0.26004	3.8356	0.10698	1490.0	1600.3	1748.7
SOC-2 (m 1)	729	169	0.00027	0.10994	0.09091	0.22375	3.2770	0.10622	1301.7	1475.6	1735.4
B-81-82-G	320	87	0.00006	0.10816	0.08026	0.26509	3.9211	0.10728	1515.8	1618.0	1753.7
B-81-82-F	384	101	0.00008	0.10835	0.08364	0.25622	3.7885	0.10724	1470.5	1590.3	1753.0
B-81-82-E	403	100	0.00007	0.10820	0.08608	0.24119	3.5645	0.10718	1392.9	1541.7	1752.2
B-81-82-D	477	108	0.00012	0.10861	0.09039	0.21919	3.2310	0.10691	1277.6	1464.6	1747.3
B-81-59-G	361	91	0.00026	0.11063	0.09119	0.24337	3.59212	0.10715	1404.2	1547.8	1749.6
B-81-59-D	335	93	0.00043	0.11280	0.10147	0.26379	3.88828	0.10690	1509.2	1611.3	1747.3
B-81-59-C	459	103	0.00036	0.11150	0.10377	0.21322	3.13157	0.10652	1246.0	1440.5	1740.6
B-81-59-B	390	96	0.00036	0.11151	0.10084	0.23416	3.44356	0.10666	1356.3	1514.4	1743.0
B-81-59-A	317	66	0.00040	0.11141	0.11038	0.19564	2.85907	0.10599	1151.9	1371.2	1731.6
B-81-71-F	911	188	0.00061	0.11444	0.09273	0.19706	2.88259	0.10609	1159.5	1377.4	1733.2
B-81-71-D	1198	196	0.00052	0.11230	0.08678	0.15743	2.28433	0.10523	942.5	1207.5	1718.4
B-81-71-C	1078	147	0.00064	0.11307	0.09308	0.13024	1.87310	0.10430	789.3	1071.6	1702.0
B-81-71-B	1546	184	0.00074	0.11414	0.10092	0.11234	1.61153	0.10404	686.3	974.7	1697.5
B-81-71-A	1769	178	0.00110	0.11770	0.12234	0.09233	1.30629	0.10261	569.3	848.5	1671.9
QCF-nm 1	5-5	137	0.00533	0.17968	0.27955	0.19738	2.91353	0.10706	1161.2	1385.4	1750.0
QCF-m 1	342	80	0.00840	0.22080	0.44152	0.14125	2.06779	0.10617	851.7	1138.2	1734.8
QCF- m 3	861	117	0.01020	0.24198	0.50048	0.07547	1.06414	0.10226	469.0	735.9	1665.6

TABLE 1. ANALYTICAL DATA FOR ZIRCON FRACTIONS (Continued)

Fraction	Concentrations U (ppm)	Concentrations Pb (ppm)	Pb Isotopic Composition $^{204}Pb/^{206}Pb$	Pb Isotopic Composition $^{207}Pb/^{206}Pb$	Pb Isotopic Composition $^{208}Pb/^{206}Pb$	Radiogenic Ratios $^{206}Pb/^{238}U$	Radiogenic Ratios $^{207}Pb/^{235}U$	Radiogenic Ratios $^{207}Pb/^{206}Pb$	Ages $^{206}Pb/^{238}U$	Ages $^{207}Pb/^{235}U$	Ages $^{207}Pb/^{206}Pb$
Cochetopa Succession											
B-85-1 (nm-1-AA)	272	84	0.00010	0.10795	0.08379	0.29956	4.40178	0.10657	1689.1	1712.7	1741.6
B-85-1 (nm-1)	294	83	0.00029	0.11105	0.09097	0.27265	4.02341	0.10703	1554.2	1638.9	1749.4
B-85-1 (m0)	364	95	0.00037	0.11092	0.09633	0.25210	3.67902	0.10584	1449.3	1566.8	1729.1
B-85-1 (m1)	426	101	0.00051	0.11470	0.10908	0.22440	3.24097	0.10637	1305.1	1478.9	1738.0
SAB-79-12A	297	82	0.00076	0.11647	0.12070	0.25476	3.72802	0.10613	1463.0	1577.4	1733.8
SAB-79-12B	310	83	0.00071	0.11591	0.12414	0.24852	3.63781	0.10616	1430.8	1557.8	1734.6
SAB-79-12C	350	89	0.00120	0.12239	0.14843	0.22845	3.33776	0.10597	1326.4	1489.9	1731.1
SAB-79-12D	383	92	0.00150	0.12606	0.16754	0.20920	3.04706	0.10564	1224.5	1419.5	1725.6
B-81-13-F (100-200)	270	84	0.00086	0.11811	0.12522	0.28835	4.23067	0.10641	1633.3	1680.0	1738.8
B-81-13-G (100-200)	266	83	0.00103	0.12020	0.13005	0.28606	4.18561	0.10612	1621.8	1671.2	1733.8
B-81-13-F (-200)	316	94	0.00158	0.12793	0.15299	0.26131	3.83192	0.10636	1496.5	1599.5	1738.0
B-81-13-D (100-200)	288	84	0.00153	0.12678	0.15728	0.25696	3.75304	0.10593	1474.2	1582.8	1730.6
B-81-13-C (100-200)	314	99	0.00421	0.16333	0.25775	0.24008	3.50253	0.10581	1387.1	1527.8	1728.3
B-83-1 (nm-1)	216	61	0.00055	0.11368	0.10843	0.26716	3.91132	0.10618	1526.4	1616.0	1734.8
B-83-1 (m-1)	209	58	0.00048	0.11284	0.10644	0.26448	3.87526	0.10627	1512.7	1608.5	1736.5
B-83-1 (m0)	230	62	0.00064	0.11497	0.11414	0.25316	3.71026	0.10629	1454.7	1573.6	1736.9
B-83-1 (m1)	244	65	0.00060	0.11450	0.11366	0.25177	3.68030	0.10627	1444.5	1567.1	1736.5
B-81-43-G	220	61	0.00013	0.10756	0.08726	0.26917	3.92549	0.10577	1536.6	1619.0	1727.9
B-81-43-F	221	62	0.00035	0.11074	0.09594	0.26617	3.88683	0.10591	1521.4	1610.9	1730.0
B-81-43-E	229	61	0.00040	0.11122	0.09856	0.25535	3.72465	0.10579	1466.0	1576.7	1728.3
B-81-43-CD	254	64	0.00043	0.11165	0.10101	0.23988	3.49878	0.10578	1386.1	1526.9	1728.0
SOC-1 (nm1 AA)	228	62	0.00023	0.10857	0.09187	0.26375	3.83143	0.10536	1509.0	1599.4	1720.5
SOC-1 (m-1)	245	66	0.00037	0.11062	0.10018	0.25896	3.76927	0.10557	1484.5	1586.2	1724.1
SOC-1 (m0)	274	75	0.00136	0.12360	0.13391	0.24642	3.56694	0.10498	1420.0	1542.2	1713.9
SOC-1 (m1)	301	74	0.00070	0.11473	0.11614	0.22894	3.31869	0.10514	1328.9	1485.5	1716.6
T-81-10-F	152	61	0.00608	0.18882	0.27321	0.28826	4.20538	0.10508	1632.8	1675.1	1728.3
T-81-10-D	205	60	0.00016	0.10774	0.07775	0.28726	4.18274	0.10560	1627.8	1670.6	1724.8
T-81-10-C	237	69	0.00010	0.10695	0.07352	0.28480	4.14316	0.10551	1615.5	1662.9	1723.2
T-81-10-B	208	55	0.00013	0.10706	0.06826	0.26233	3.80834	0.10529	1501.8	1594.5	1719.3
B-81-40-G (+270)	415	116	0.00036	0.11036	0.09669	0.26980	3.92237	0.10544	1539.8	1618.3	1722.0
B-81-40-E (-270)	392	109	0.00035	0.11032	0.10039	0.26634	3.87314	0.10547	1522.2	1608.1	1722.5
B-81-40-E (+270)	616	159	0.00047	0.11188	0.10474	0.24474	3.55970	0.10549	1411.3	1540.6	1722.7

TABLE 1. ANALYTICAL DATA FOR ZIRCON FRACTIONS (Continued)

Fraction	Concentrations		Pb Isotopic Composition			Radiogenic Ratios			Ages		
	U (ppm)	Pb (ppm)	$^{204}Pb/^{206}Pb$	$^{207}Pb/^{206}Pb$	$^{208}Pb/^{206}Pb$	$^{206}Pb/^{238}U$	$^{207}Pb/^{235}U$	$^{207}Pb/^{206}Pb$	$^{206}Pb/^{238}U$	$^{207}Pb/^{235}U$	$^{207}Pb/^{206}Pb$
B-81-24-H	324	79	0.00032	0.10936	0.08604	0.23693	3.42949	0.10498	1370.7	1511.2	1714.1
B-81-24-G	354	79	0.00040	0.11054	0.08432	0.21869	3.16779	0.10506	1275.0	1449.3	1715.3
B-81-24-I	317	69	0.00064	0.11409	0.09920	0.21227	3.08201	0.10530	1240.9	1428.2	1719.6
B-81-24-F	407	79	0.00047	0.11149	0.09563	0.18586	2.69357	0.10511	1098.9	1326.7	1716.1
Younger Plutons											
B-81-42-N	227	61	0.00083	0.11561	0.15402	0.25152	3.61336	0.10419	1446.3	1552.5	1700.0
B-81-42-E1	251	68	0.00106	0.11867	0.15989	0.24427	3.50726	0.10413	1408.9	1528.9	1699.2
B-81-42-F	290	79	0.00193	0.13060	0.19455	0.23570	3.38449	0.10414	1364.3	1500.8	1699.2
B-81-42-E2	311	76	0.00124	0.12118	0.16731	0.21704	3.11729	0.10417	1266.2	1437.0	1699.6
B-81-70 (nm1AA)	815	134	0.00091	0.11550	0.09105	0.15639	2.22173	0.10304	936.7	1183.9	1679.7
B-81-70 (nm1)	883	173	0.00112	0.11805	0.09753	0.18478	2.61639	0.10270	1093.0	1305.2	1673.4
B-81-70 (m1)	995	183	0.000156	0.12349	0.11491	0.16882	2.37617	0.10208	1005.6	1235.5	1662.5
B-81-70 (m2)	1073	190	0.00169	0.12516	0.12115	0.16067	2.25921	0.10198	960.5	1199.7	1660.5
B-81-70	1382	218	0.00231	0.13288	0.14707	0.13786	1.92456	0.10125	832.6	1089.7	1647.3
B-81-68-N	133	37	0.00166	0.12559	0.19899	0.24169	3.42837	0.10288	1395.5	1510.9	1676.6
B-81-68-G	158	40	0.00085	0.11444	0.17143	0.22778	3.22700	0.10275	1322.8	1463.7	1674.2
B-81-68-F	169	43	0.00167	0.12565	0.20473	0.22184	3.14381	0.10278	1291.6	1443.5	1675.0
B-81-68-E	212	49	0.00284	0.14171	0.25185	0.18651	2.64428	0.10283	1102.4	1313.1	1675.8
B-81-68-D	186	40	0.00201	0.13035	0.21448	0.18219	2.58097	0.10275	1078.9	1295.3	1674.2
SALIDA AREA											
Z-3-H (-100)	318	87	0.00019	0.10774	0.10424	0.26425	3.82966	0.10511	1511.6	1599.0	1716.3
Z-3-G (-100)	309	84	0.00019	0.10723	0.10670	0.26118	3.76825	0.10464	1495.9	1586.0	1708.1
Z-3-F (-100)	410	101	0.00024	0.10729	0.11603	0.23527	3.37567	0.10406	1362.1	1498.8	1697.8
Z-3-BC	848	126	0.00045	0.10525	0.13862	0.13790	1.88289	0.09903	832.8	1075.1	1605.8
SAL-1-E (-200)	427	123	0.00230	0.13327	0.13991	0.25245	3.53976	0.10169	1451.1	1536.1	1655.3
SAL-1-E (+140)	390	106	0.00148	0.12207	0.11495	0.25028	3.51035	0.10172	1439.9	1529.5	1655.9
SAL-1-D (+140)	410	101	0.00125	0.11838	0.10788	0.23016	3.21255	0.10123	1335.3	1460.2	1646.9
SAL-1-C (+140)	535	121	0.00123	0.11721	0.10767	0.21021	2.90579	0.10026	1229.9	1383.4	1628.9
SAL-1-B (+140)	605	122	0.00092	0.11260	0.09944	0.19080	2.62888	0.09993	1125.7	1308.8	1622.7
SAL-1-A (+140)	781	124	0.00080	0.10893	0.08906	0.15264	2.05919	0.09784	915.8	1135.3	1583.2
HOWARD AREA											
RC-HO-1-G	329	82	0.00012	0.10134	0.06403	0.24839	3.41518	0.09972	1430.2	1507.9	1618.8
RC-HO-1-F	307	75	0.00011	0.10070	0.06345	0.24297	3.32419	0.09923	1402.1	1486.7	1609.8
RC-HO-1-E	396	94	0.00016	0.10095	0.06662	0.23497	3.19833	0.09872	1360.5	1456.8	1600.0
RC-HO-1-D	419	96	0.00018	0.10066	0.07096	0.22782	3.08287	0.09815	1323.0	1428.4	1589.1
RC-HO-2-H	255	67	0.00022	0.10540	0.08523	0.25801	3.64313	0.10241	1479.6	1559.0	1668.2
RC-HO-2-G	276	71	0.00032	0.10618	0.09082	0.24935	3.49867	0.10176	1435.1	1526.9	1656.3
RC-HO-2-F	316	78	0.00022	0.10419	0.08916	0.23998	3.34699	0.10115	1386.6	1492.6	1645.4
RC-HO-2-E	352	84	0.00032	0.10479	0.09337	0.22967	3.17740	0.10034	1332.8	1451.7	1630.2

TABLE 1. ANALYTICAL DATA FOR ZIRCON FRACTIONS (Continued)

Fraction	Concentrations		Pb Isotopic Composition			Radiogenic Ratios			Ages		
	U (ppm)	Pb (ppm)	$^{204}Pb/^{206}Pb$	$^{207}Pb/^{206}Pb$	$^{208}Pb/^{206}Pb$	$^{206}Pb/^{238}U$	$^{207}Pb/^{235}U$	$^{207}Pb/^{206}Pb$	$^{206}Pb/^{238}U$	$^{207}Pb/^{235}U$	$^{207}Pb/^{206}Pb$

Notes: For most samples designation A, B, C, ... indicates decreasing magnetic susceptibility. For other samples, NM or M followed by a number refers to zircon fractions that are magnetic or nonmagnetic on the Frantz separator at the side-tilt indicated by the number. Numbers in parentheses for some samples indicate size-fractions. All ratios are atomic ratios.

TABLE 2. U–PB AGES OF ZIRCONS

Sample	Rock Type	Number of Zircon Fractions	Age (m.y. ± 2σ)
		Gunnison Area	
B-61	fine-grained granite	4	1781 ± 8
B-2	rhyolite	5	1770 ± 4
B-60	rhyolite	5	1766 ± 14
QCF	rhyolite	3	1770 ± 8
B-85	rhyolite	5	1766 ± 4
SOC-2	rhyolite	4	1765 ± 4
B-82	rhyolite	4	1759 ± 6
B-59	granite of Tolvar Peak	5	1757 ± 10
B-71	Powderhorn granite	5	1751 ± 6
B-85-1	rhyolite	4	1745 ± 16
SAB-12	rhyolite	4	1742 ± 6
B-13	rhyolite	5	1740 ± 8
B-1	rhyolite	4	1733 ± 11
B-43	rhyolite	4	1730 ± 7
SOC-1	rhyolite	4	1728 ± 16
Gunnison Complex			
T-10	quartz diorite	4	1730 ± 6
B-40	granite of South Beaver Creek	3	1721 ± 7
B-24	tonalite of Gold Basin	4	1713 ± 10
B-42	granite	4	1700 ± 7
B-70	granite of Wood Gulch	4	1701 ± 10
B-68	granite of Cochetopa Canyon	5	1676 ± 5
		Salida Area	
Z-3	dacite	4	1728 ± 6
SAL-1	granite	6	1672 ± 5
		Howard Area	
HO-1	rhyolite	4	1668 ± 18
HO-2	rhyolite	4	1713 ± 14

REFERENCES CITED

Afifi, A. M., 1981, Precambrian geology of the Iris area, Gunnison and Saguache Counties, Colorado [M.S. thesis]: Golden, Colorado School of Mines, 197 p.

Barker, F., Peterman, Z. E., and Hildreth, R. A., 1969, A rubidium-strontium study of the Twilight Gneiss, West Needle Mountains, Colorado: Contributions to Mineralogy and Petrology, v. 23, p. 271–283.

Barker, F., Arth, J. G., Peterman, Z. E., and Friedman, I., 1976, The 1.7–1.8 b.y.-old trondhjemites of southwestern Colorado and northern New Mexico; Geochemistry and depths of genesis: Geological Society of America Bulletin, v. 87, p. 189–198.

Bennett, G. S., 1984, Geochemistry of bimodal volcanic and volcaniclastic metasedimentary rocks, Cochetopa Canyon area, central Colorado [M.S. thesis]: Lawrence, University of Kansas, 82 p.

Bennett, G. S., Bickford, M. E., and Girty, G. H., 1984, Geochemistry and petrotectonic setting of bimodal volcanic and volcaniclastic rocks, Cochetopa Canyon area, central Colorado: Geological Society of America Abstracts with Programs, v. 16, p. 214.

Bickford, M. E., and Boardman, S. J., 1984, A Proterozoic volcano-plutonic terrane, Gunnison and Salida area, Colorado: Journal of Geology, v. 92, p. 657–666.

Bickford, M. E., Wetherill, G. W., Barker, F., and Lee-Hu, Chin-Nan, 1969,

Precambrian Rb-Sr geochronology in the Needle Mountains, southwestern Colorado: Journal of Geophysical Research, v. 74, p. 1660–1676.

Bickford, M. E., Van Schmus, W. R., and Zietz, I., 1986, Proterozoic history of the midcontinent region of North America: Geology, v. 14, p. 492–496.

Boardman, S. J., 1976, Geology of the Precambrian metamorphic rocks of the Salida area, Chaffee County, Colorado: The Mountain Geologist, v. 13, p. 89–100.

—— , 1986, Early Proterozoic bimodal volcanic rocks in central Colorado, U.S.A; Part 1, Petrography, stratigraphy, and depositional history: Precambrian Research, v. 34, p. 1–36.

Boardman, S. J., and Bickford, M. E., 1984, Stratigraphy and geochemistry of Early Proterozoic bimodal volcanogenic rocks near Salida, Colorado: Geological Society of America Abstracts with Programs, v. 16, p. 215.

Boardman, S. J., and Condie, K. C., 1986, Early Proterozoic bimodal volcanic rocks in central Colorado, U.S.A.; Part 2, Geochemistry, petrogenesis, and tectonic setting: Precambrian Geology, v. 34, p. 37–68.

Cameron, A. E., Smith, D. H., and Walker, R. L., 1969, Mass spectrometry of nanogram-size samples of lead: Analytical Chemistry, v. 41, p. 525–526.

Condie, K. C., 1980, The Tijeras greenstone; Evidence for depleted upper mantle beneath New Mexico during the Proterozoic: Journal of Geology, v. 88, p. 603–609.

—— , 1986, Geochemistry and tectonic setting of Early Proterozoic supracrustal rocks in the southwestern United States: Journal of Geology, v. 94, p. 845–864.

Condie, K. C., and Budding, A. J., 1979, Geology and geochemistry of Precambrian rocks, central and south-central New Mexico: New Mexico Bureau of Mines and Mineral Resources Memoir 35, 58 p.

Condie, K. C., and Demalas, J. P., 1985, The Pinal Schist; An Early Proterozoic quartz wacke association in southeastern Arizona: Precambrian Research, v. 27, p. 337–356.

Condie, K. C., and McCrink, T. P., 1982, Geochemistry of Proterozoic volcanic and granitic rocks from the Gold Hill–Wheeler Peak area, northern New Mexico: Precambrian Research, v. 19, p. 141–166.

Condie, K. C., and Nuter, J. A., 1981, Geochemistry of the Dubois greenstone succession; An Early Proterozoic bimodal volcanic association in west-central Colorado: Precambrian Research, v. 15, p. 131–155.

Condie, K. C., Bowling, G. P., and Vance, R. K., 1985, Geochemistry and origin of Early Proterozoic supracrustal rocks, Dos Cabezas Mountains, southeastern Arizona: Geological Society of America Bulletin, v. 96, p. 655–662.

Conway, C. M., 1976, Petrology, structure, and evolution of a Precambrian volcanic and plutonic complex, Tonto Basin, Gila County, Arizona [Ph.D. thesis]: Pasadena, California Institute of Technology, 460 p.

Conway, C. M., and Silver, L. T., 1988, Early Proterozoic rocks (1710–1615 Ma) in central to southeastern Arizona, *in* Reynolds, S. J., and Jenney, J. P., eds., The Geology of Arizona: Arizona Geological Society Digest, v. 17 (in press).

Floyd, P. A., and Winchester, J. A., 1975, Magma type and tectonic setting discrimination using immobile elements: Earth and Planetary Science Letters, v. 27, p. 211–218.

Hedlund, D. C., 1974, Geologic map of the Big Mesa Quadrangle, Gunnison County, Colorado: U.S. Geological Survey map GQ–1153, scale 1:24,000.

Hedlund, D. C., and Olson, J. C., 1974, Geologic map of the Iris NW Quadrangle, Gunnison and Saguache Counties, Colorado: U.S. Geological Survey Map GQ–1134, scale 1:24,000.

—— , 1975, Geologic map of the Powderhorn Quadrangle, Gunnison and Saguache Counties, Colorado: U.S. Geological Survey Map GQ–1178, scale 1:24,000.

Holm, P. E., 1982, Non-recognition of continental tholeiites using the Ti-Y-Zr diagram: Contributions to Mineralogy and Petrology, v. 79, p. 308–310.

Krogh, T. E., 1973, A low contamination method for the hydrothermal decomposition of zircon and extraction of U-Pb for isotopic age determinations: Geochimica et Cosmochimica Acta, v. 37, p. 485–494.

Ludwig, K. R., 1974, Precambrian geology of the central Matzatzal Mountains, Arizona (Part 1) and lead isotope heterogeneity in Precambrian igneous feldspars (Part 2) [Ph.D. thesis]: Pasadena, California Institute of Technology, 363 p.

—— , 1980, Calculation of uncertainties of U-Pb isotopic age data: Earth and Planetary Science Letters, v. 46, p. 212–220.

—— , 1983, Plotting and regression programs for isotope geochemists, for use with HP–86/87 microcomputers: U.S. Geological Survey Open-File Report 83–849, 94 p.

Nelson, B. K., and DePaolo, D. J., 1984, 1,700-Myr greenstone volcanic successions in southwestern North America and isotopic evolution of the Proterozoic mantle: Nature, v. 312, p. 143–146.

Olson, J. C., 1974, Geologic map of the Rudolph Hill Quadrangle, Gunnison, Hinsdale, and Saguache Counties, Colorado: U.S. Geological Survey Map GQ–1177, scale 1:24,000.

—— , 1976a, Geologic map of the Iris Quadrangle, Gunnison and Saguache Counties, Colorado: U.S. Geological Survey Map GQ–1286, scale 1:24,000.

—— , 1976b, Geologic map of the Houston Gulch Quadrangle, Gunnison and Saguache Counties, Colorado: U.S. Geological Survey Map GQ–1287, scale 1:24,000.

Olson, J. C., and Steven, T. A., 1976, Geologic map of the Sawtooth Mountain Quadrangle, Saguache County, Colorado: U.S. Geological Survey Map MF–733, scale 1:24,000.

Pearce, J. A., and Cann, J. R., 1973, Tectonic settings of basic volcanic rocks determined using trace element analysis: Earth and Planetary Science Letters, v. 19, p. 290–300.

Pearce, J. A., and Norry, M. J., 1979, Petrogenetic implications of Ti, Zr, Y, and Nb variations in volcanic rocks: Contributions to Mineralogy and Petrology, v. 29, p. 33–47.

Scott, G. R., Taylor, R. B., Epis, R. C., and Wobus, R. A., 1978, Geologic map of the Pueblo 1° by 2° Quadrangle, south-central Colorado: U.S. Geological Survey Miscellaneous Investigations Series Map I–1022, scale 1:250,000.

Shonk, K. N., and Geissman, J. W., 1984, Time-stratigraphic equivalence of the Dubois greenstone and felsic volcanic-sedimentary terrain, Gunnison Uplift, Gunnison and Saguache Counties, Colorado: Geological Society of America Abstracts with Programs, v. 16, p. 255.

Silver, L. T., 1968, Precambrian batholiths of Arizona: Geological Society of America Abstracts for 1968, Special Paper 121, p. 558–559.

Silver, L. T., and Barker, F., 1968, Geochronology of Precambrian rocks of the Needle Mountains, southwestern Colorado; Part 1, U-Pb results: Geological Society of America Abstracts for 1967, Special paper 115, p. 204–205.

Steiger, R. H., and Jaeger, E., 1977, Subcommission on geochronology; Convention on use of decay constants in geo- and cosmo-chronology: Earth and Planetary Science Letters, v. 36, p. 359–362.

Taylor, R. B., Scott, G. R., and Wobus, R. A., 1975, Reconnaissance geologic map of the Howard (15 minute) Quadrangle, central Colorado: U.S. Geological Survey Miscellaneous Investigations Map I–892, scale 1:62,500.

Tobin, R. J., 1982, Petrology and geochronology of a Proterozoic annular complex near Gunnison, Colorado [M.S. thesis]: Lawrence, University of Kansas, 62 p.

Tweto, O., Steven, T. A., Hail, W. J., Jr., and Moench, R. H., 1976, Preliminary geologic map of the Montrose 1° by 2° Quadrangle, southwestern Colorado: U.S. Geological Survey Miscellaneous Field Studies Map MF 761, scale 1:250,000.

Vance, R. K., 1984, Geology and geochemistry of the Gunnison intrusive complex, Gunnison County, Colorado [M.S. thesis]: Lexington, University of Kentucky, 261 p.

Van Schmus, W. R., Bickford, M. E., and Zietz, I., 1987, Early and Middle Proterozoic provinces in the central United States, *in* Kroner, A., ed., Proterozoic lithospheric evolution: American Geophysical Union Geodynamics Series, v. 17, p. 43–68.

Wetherill, G. W., and Bickford, M. E., 1965, Primary and metamorphic Rb-Sr chronology in central Colorado: Journal of Geophysical Research, v. 71, p. 4669–4689.

Manuscript Accepted by the Society September 26, 1988

Printed in U.S.A.

Geological Society of America
Special Paper 235
1989

U-Pb zircon geochronology of Proterozoic and Cambrian plutons in the Wet Mountains and southern Front Range, Colorado

M. E. Bickford
Department of Geology, University of Kansas, Lawrence, Kansas 66045
R. L. Cullers
Department of Geology, Kansas State University, Manhattan, Kansas 66056
R. D. Shuster,* W. R. Premo,* and W. R. Van Schmus
Department of Geology, University of Kansas, Lawrence, Kansas 66045

ABSTRACT

The Wet Mountains–southern Front Range region is underlain by high-grade granitic gneiss, amphibolite, and schist of Early Proterozoic age. These rocks were intruded by granitic to granodioritic plutons during four episodes: one in the Early Proterozoic (1,660 to 1,700 Ma) and three in the Middle Proterozoic (1,485 to 1,440 Ma, 1,370 to 1,360 Ma, and about 1,060 Ma). We also report here a zircon age determination (536 ± 4 Ma) for syenite of the Cambrian McClure Mountain alkaline-mafic complex.

The granitic gneiss was clearly formed before 1,700 Ma. Its protolith was probably pelitic to psammitic sedimentary rocks, in contrast to the volcanogenic rocks of this age farther west in the Gunnison and Salida areas of Colorado. The early Proterozoic plutons emplaced within the granitic gneiss are mostly somewhat younger than those emplaced within volcanogenic rocks to the west, although some are coeval. The middle Proterozoic rocks are representatives of the widespread "anorogenic granite-rhyolite suite" which is known in the St. Francois Mountains of Missouri and the subsurface of the midcontinent.

INTRODUCTION

The southern Front Range of Colorado and its southern continuation, the Wet Mountains, are mostly underlain by lower and middle Proterozoic rocks, although Phanerozoic sedimentary rocks, a Cambrian alkaline complex, and Cenozoic volcanic rocks are also present (Fig. 1). The Proterozoic rocks of this region include high-grade paragneisses and large granitic to granodioritic plutons, in marked contrast with the dominantly volcanogenic Proterozoic rocks to the west in the Salida and Gunnison areas (Bickford and Boardman, 1984; Bickford and others, this volume). A major goal of studying these ancient rocks is to understand the manner in which they were added to the continental crust of North America. Besides field study and petrography, important approaches are: (1) analysis for major- and trace-element abundances as guides to sources and tectonic setting, and (2) geochronological studies to determine absolute ages and establish time relationships.

This chapter presents the results of U-Pb isotopic measurements of zircons separated from plutonic rocks collected in the Wet Mountains and the southern part of the Front Range (Fig. 1); Rb-Sr data for samples of the San Isabel batholith are also given. Another work, presenting the results of detailed petrographic and geochemical studies of these plutons, is in preparation.

GEOLOGIC RELATIONS

The oldest rocks in the southern Front Range and the Wet Mountains (Fig. 1) are sequences of metasedimentary rocks

*Present address: Shuster, Department of Geology and Geography, University of Nebraska–Omaha, Omaha, Nebraska 68182; Premo, U.S. Geological Survey, MS 963, Box 25046, Denver Federal Center, Denver, Colorado 80225.

Bickford, M. E., Cullers, R. L., Shuster, R. D., Premo, W. R., and Van Schmus, W. R., U-Pb geochronology of Proterozoic and Cambrian plutons in the Wet Mountains and southern Front Range, Colorado, *in* Grambling, J. A., and Tewksbury, B. J., Proterozoic geology of the southern Rocky Mountains: Boulder, Colorado, Geological Society of America Special Paper 235.

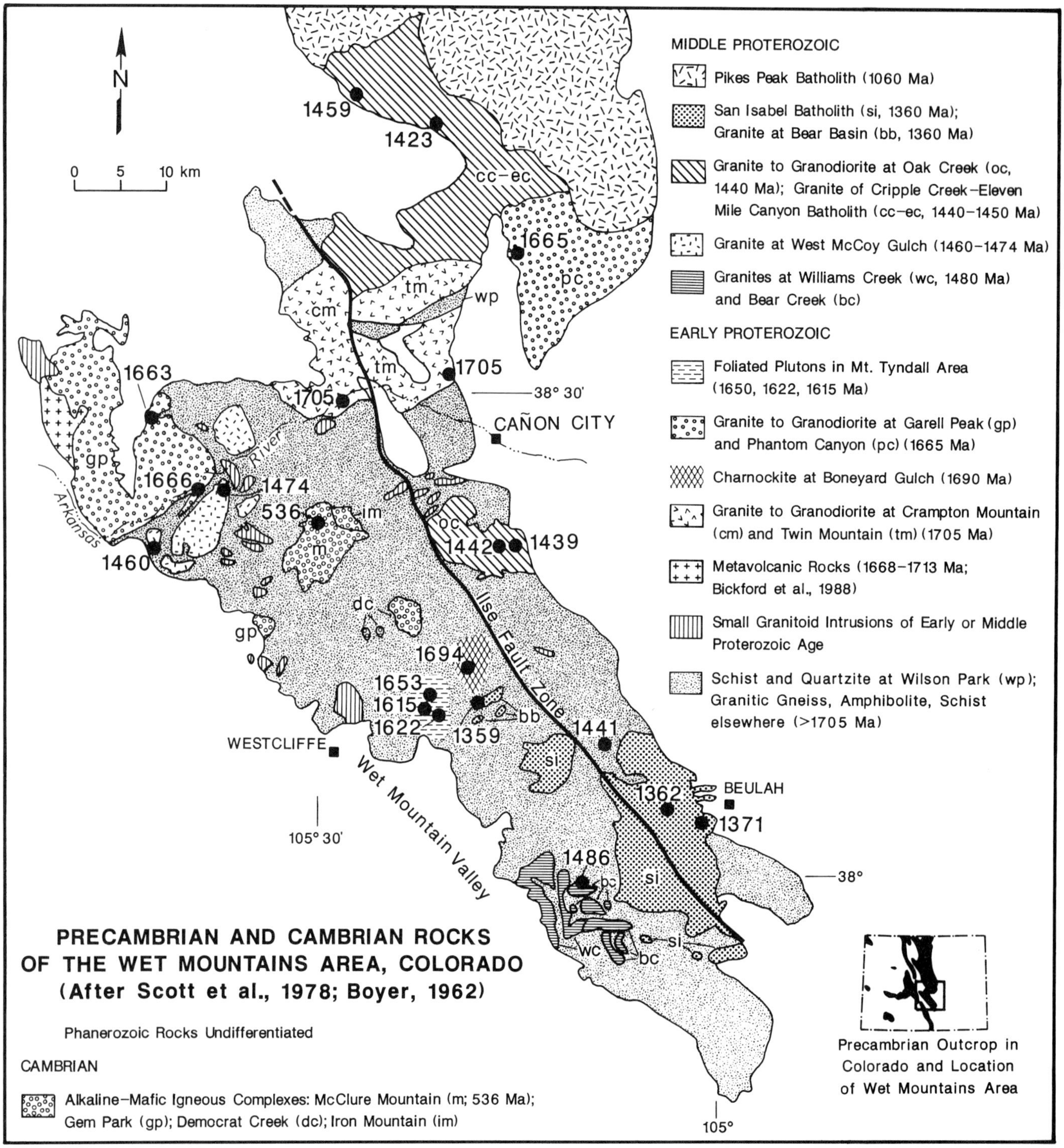

Figure 1. Geologic map of the Wet Mountains and southern Front Range, Colorado (modified after Boyer, 1962, and Scott and others, 1978).

mostly in upper amphibolite facies (Scott and others, 1978; Murray, 1970; Logan, 1966; Christman and others, 1959) but locally in granulite facies (Mt. Tyndall area east of Westcliffe; Fig. 1; Brock and Singewald, 1968). The Wilson Park region (Fig. 1; Knepper, 1972; Reuss, 1974; Vera and Van Schmus, 1974) exposes a folded sequence of quartzite, muscovite-garnet schist, and sillimanite-staurolite gneiss. Most of the rest of the southern Front Range and the Wet Mountains is underlain by quartzofeldspathic gneiss, commonly migmatitic, and amphibolite. The gross structure and mineralogy of these rocks suggest that they are of metasedimentary origin, but little detailed study has been done.

During the Early and Middle Proterozoic these rocks were intruded by at least four generations of plutons ranging in composition from granodiorite to granite. It is these plutons which we have dated and which are the principal subject of this paper. We also report the age of a syenite from the McClure Mountain complex, an assemblage of mafic, ultramafic, and alkaline igneous rocks of Cambrian age (Kilbane and Cullers, 1979).

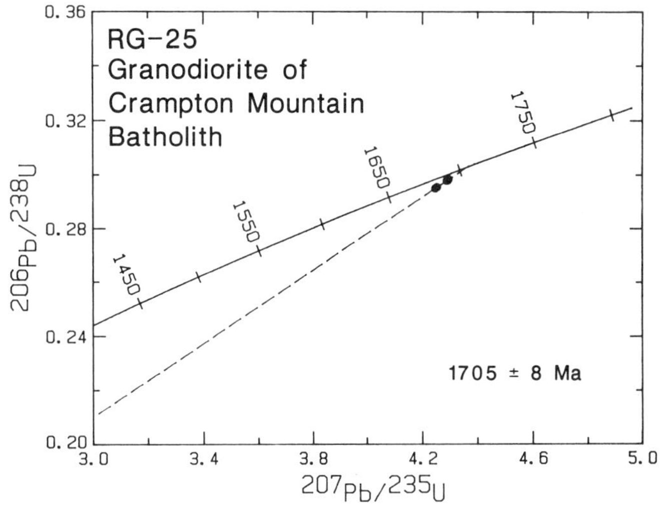

Figure 2. Concordia diagram for sample RG-25, Crampton Mountain batholith. Model 1 solution based on 5 zircon fractions, probability of fit 97 percent, MSWD = 0.0749. Intercepts at 1,705 ± 8 and 70 ± 620 Ma.

GEOCHRONOLOGY

Analytical methods

The analyses of zircon fractions for U and Pb concentrations and Pb isotopic compositions were done in the Isotope Geochemistry Laboratory at the University of Kansas by standard techniques of isotope dilution analysis and mass spectrometry. In all cases the zircon fractions were separated on the basis of their magnetic susceptibility. Dissolution and ion-exchange techniques are slightly modified from those of Krogh (1973). Mass spectrometry was done on a 23-cm-radius Nier-type instrument using single rhenium filaments on which samples were loaded with silica gel and phosphoric acid (Cameron and others, 1969). Analytical blanks measured during the course of these analyses were 1.0 ng or less total Pb and less than 0.5 ng U. Decay constants used are $\lambda^{238}U = 0.155125 \times 10^{-9}$ yrs^{-1} and $\lambda^{235}U = 0.98458 \times 10^{-9}$ yrs^{-1} (Steiger and Jager, 1977).

Uncertainties of measured $^{207}Pb/^{206}Pb$ ratios are estimated to be 0.1 percent (2 σ); relative uncertainty of $^{204}Pb/^{206}Pb$ ratios varied with the magnitude of the ratio, but absolute uncertainty is estimated as $\pm$ 0.00002, or about 2 percent for a sample with $^{204}Pb/^{206}Pb$ of 0.00100. The U-Pb ratios are considered accurate to $\pm$1.0 percent (2σ). After correction for all sources of uncertainty, radiogenic $^{207}Pb/^{206}Pb$ ratios are considered accurate to $\pm$ 0.25 percent (2 σ).

Ages were calculated using the regression and error analysis methods of Ludwig (1980, 1983). The calculated $^{207}Pb/^{235}U$ and $^{206}Pb/^{238}U$ ratios were assigned uncertainties of 1.0 percent with a correlation coefficient of 0.98. If the probability of fit from the regression was 30 percent or better, ages were calculated with Ludwig's Model 1 solution; the Model 2 solution was used if the probability of fit was less than 30 percent. Analytical data are given in Table 1 and ages ($\pm$ 2 σ) in Figure 2 through 21 and Figures 24 and 25.

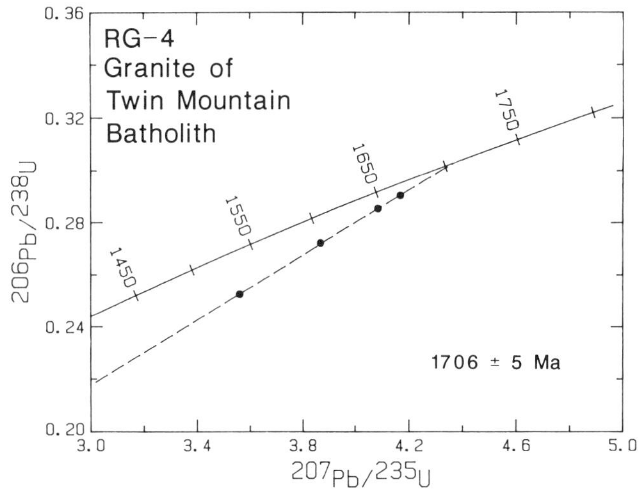

Figure 3. Concordia diagram for sample RG-4, Twin Mountain batholith. Model 1 solution based on 4 zircon fractions, probability of fit 45 percent, MSWD 0.797. Intercepts at 1706 ± 5 and 356 ± 56 Ma.

Some of the samples analyzed in this study yielded zircon suites for which regression gave negative lower intercepts, ranging from –16 to –616 Ma, on concordia diagrams. In each case the zircon suite yielded data points that plotted quite near the upper intercept with concordia and were clustered. We believe that the negative lower intercepts are artifacts of the difficulty of regressing a line through this kind of data, for the zircons do not appear to contain xenocrystic cores, nor do they show any other visible evidence of alteration. We considered simply reporting the average of the $^{207}Pb/^{206}Pb$ ages for these samples; this would have been effectively the same as forcing the regression through the origin and would have yielded a minimum age for the samples. We chose, however, to add an artificial point to the regression

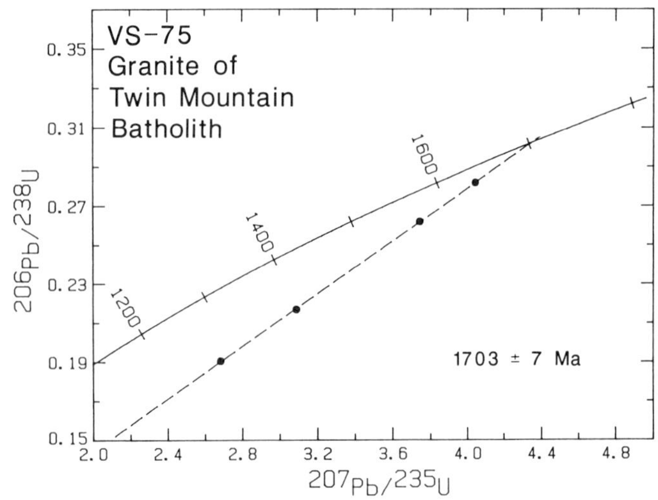

Figure 4. Concordia diagram for sample VS-75, Twin Mountain batholith. Model 2 solution based on 4 zircon fractions, probability of fit 0.41 percent, MSWD = 5.51. Intercepts at 1,703 ± 7 and 112 ± 38 Ma.

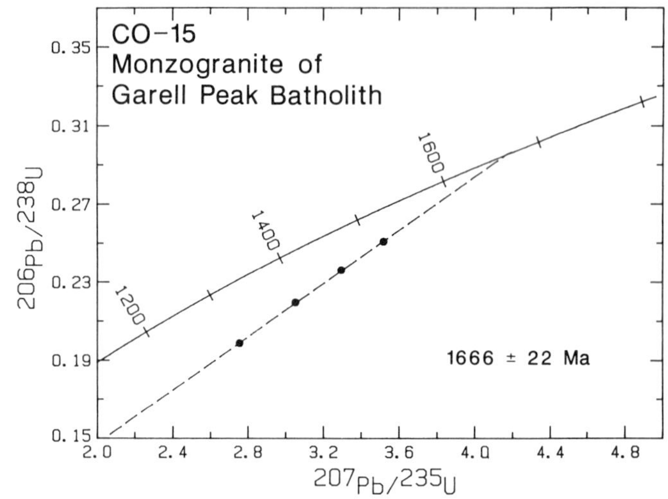

Figure 6. Concordia diagram for sample CO-15, Garell Peak batholith. Model 2 solution based on 4 zircon fractions, probability of fit 4.4 percent, MSWD = 3.13. Intercepts at 1,666 ± 22 and 130 ± 110 Ma.

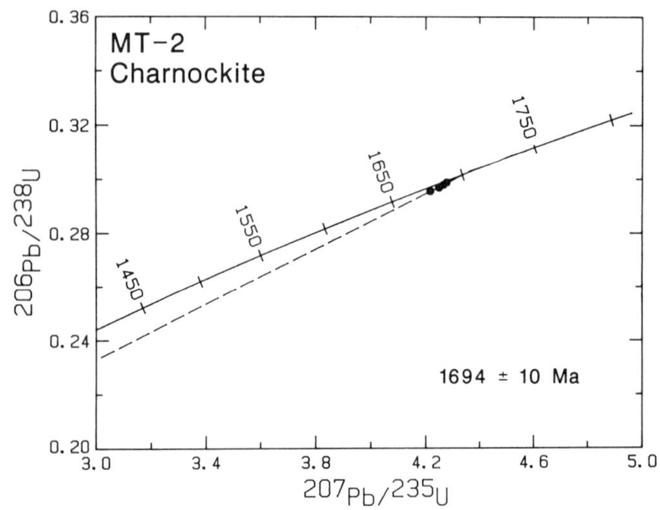

Figure 5. Concordia diagram for sample MT-2, charnockite of Boneyard Gulch. Model 1 solution based on 4 zircon fractions, probability of fit 73 percent, MSWD = 0.309. Intercepts at 1,703 ± 300 and 940 ± 900 Ma. See text for age interpretation.

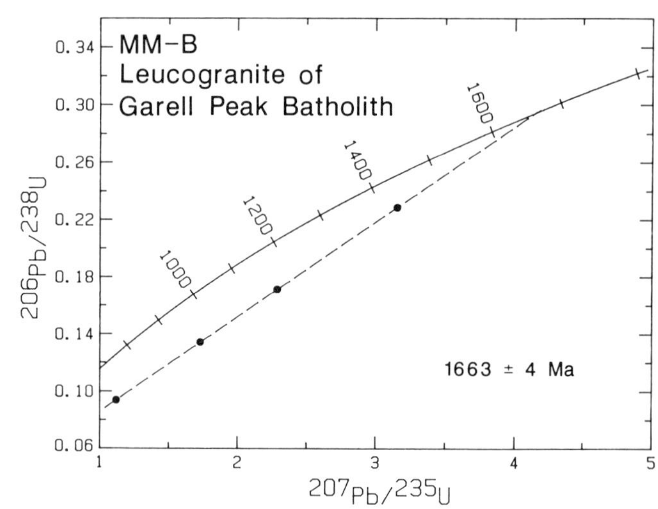

Figure 7. Concordia diagram for sample MM-B, Garell Peak batholith. Model 1 solution based on 4 zircon fractions, probability of fit 55 percent, MSWD = 0.594. Intercepts at 1,663 ± 4 and 232 ± 5 Ma.

procedure to force the chord through a lower intercept of 99 Ma, for the positive lower intercepts in this study scatter approximately about this value. We believe this procedure results in a better estimate of the crystallization age of the rocks. It has the effect of artificially improving the probability of fit and reducing MSWD in some cases, and it reduces the calculated uncertainty in the reported age. The samples for which this was done are MT-9A (Fig. 10), MT-9B (Fig. 11), RV-1C (Fig. 17), VS-9 (Fig. 19), SD-2 (Fig. 21), and MT-3 (Fig. 24). Sample 80-3, syenite of the Cambrian McClure Mountain alkaline complex, also yielded a negative lower intercept; in this case, because of the greatly different geologic history of the sample, we chose to force the regression through the origin.

The results of Rb-Sr whole-rock analyses of samples from the San Isabel batholith are given in Table 2 and Figure 22. Sample locations are shown on Figure 23. Rb and Sr analytical methods, constants used, laboratory blanks, and methods of calculating uncertainties are given by Shuster and Bickford (1985).

The results are presented below, beginning with the oldest rocks studied. Except as noted, the zircons analyzed have the sharp terminations, euhedral form, and commonly, growth zonation that are characteristic of zircons from igneous rocks. Some

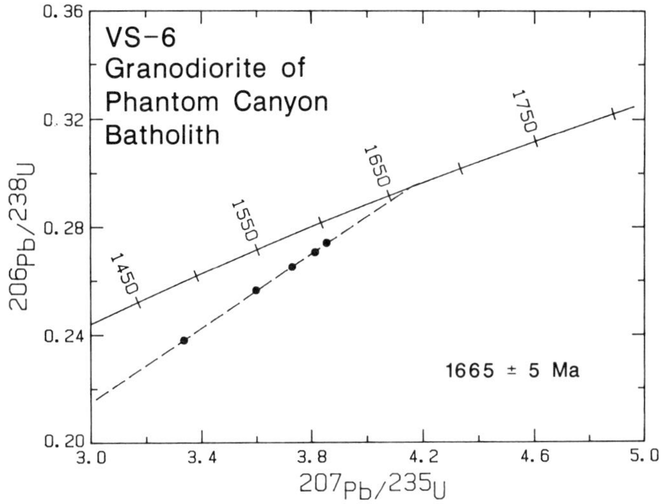

Figure 8. Concordia diagram for sample VS-6, Phantom Canyon batholith. Model 2 solution based on 5 zircon fractions, probability of fit 22 percent, MSWD = 1.46. Intercepts at 1,665 ± 5 and 92 ± 66 Ma.

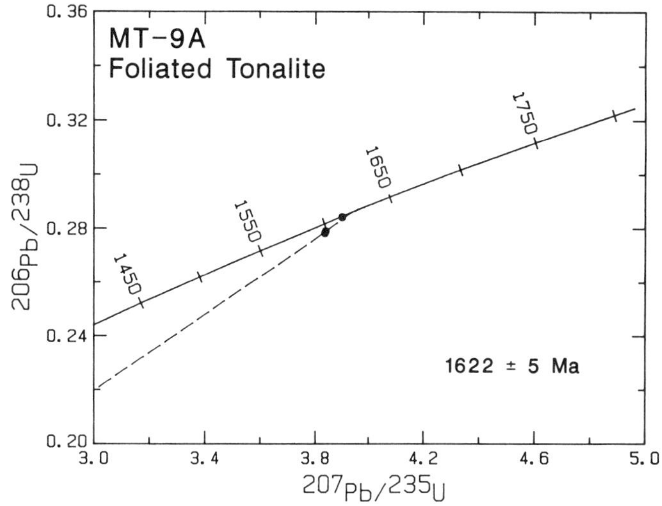

Figure 10. Concordia diagram for sample MT-9A, tonalitic gneiss of Mt. Tyndall area. Model 1 solution based on 3 zircon fractions and an artificial point to force the chord through a lower intercept of 99 Ma, probability of fit 1.3 percent, MSWD = 4.35. See text for explanation. Intercepts at 1,622 ± 5 and 99 ± 14 Ma.

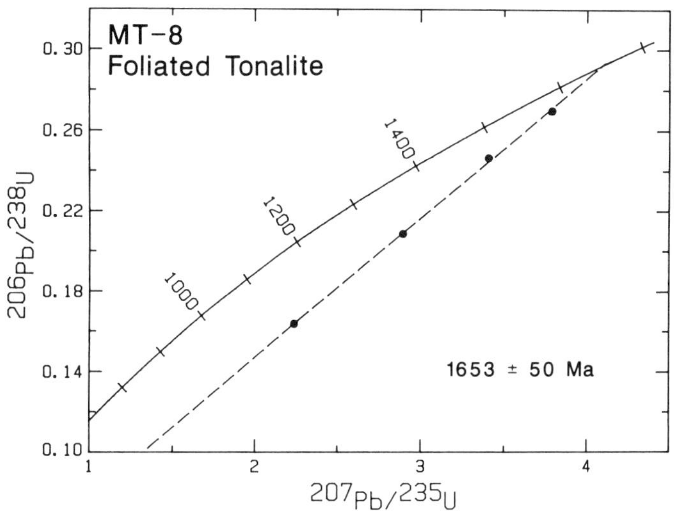

Figure 9. Concordia diagram for sample MT-8, foliated tonalite of Mt. Tyndall area. Model 2 solution based on 4 zircon fractions, probability of fit 0 percent, MSWD = 44.8. Intercepts at 1,653 ± 50 and 100 ± 230 Ma.

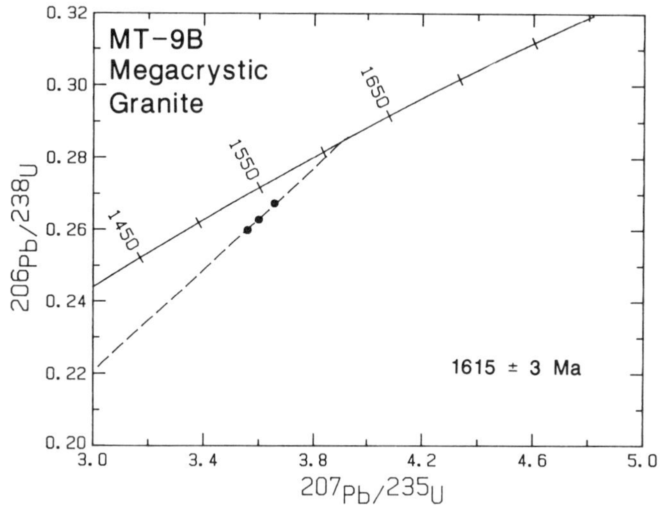

Figure 11. Concordia diagram for sample MT—9B, foliated megacrystic granite of Mt. Tyndall area. Model 2 solution based on 3 zircon fractions and an artificial point to force the chord through 99 Ma, probability of fit 12 percent, MSWD = 2.14. See text for explanation. Intercepts at 1,615 ± 3 and 99 ± 10 Ma.

zircon samples have quite different morphologies; these are discussed below for the appropriate samples.

Early Proterozoic plutons

Sample RG-25 is granodiorite of the Crampton Mountain batholith (Fig. 1). This sample yields an age of 1,705 ± 8 Ma (Fig. 2). Samples RG—4 and VS-75 are granite of the Twin Mountain batholith (Fig. 1) and were collected independently by R. L. Cullers (RG-4) and W. R. Van Schmus (VS-75) from the same outcrop. RG-4 yields an age of 1,706 ± 5 Ma (Fig. 3), whereas VS-75 yields an age of 1,703 ± 7 Ma (Fig. 4). The Twin Mountain and Crampton Mountain batholiths are petrographically similar and contiguous (Fig. 1). These results show that they are also coeval.

Sample MT-2 is a sample of charnockite from the Boneyard Gulch area of the Mt. Tyndall Quadrangle (Brock and Singewald, 1968). This rock, though collected in a region characterized by granulitic gneisses, is massive and unfoliated and presumably represents an intrusive igneous rock or its metamorphosed equivalent. It consists of plagioclase, quartz, hornblende, clinopyroxene, orthopyroxene, and accessory minerals, including euhedral

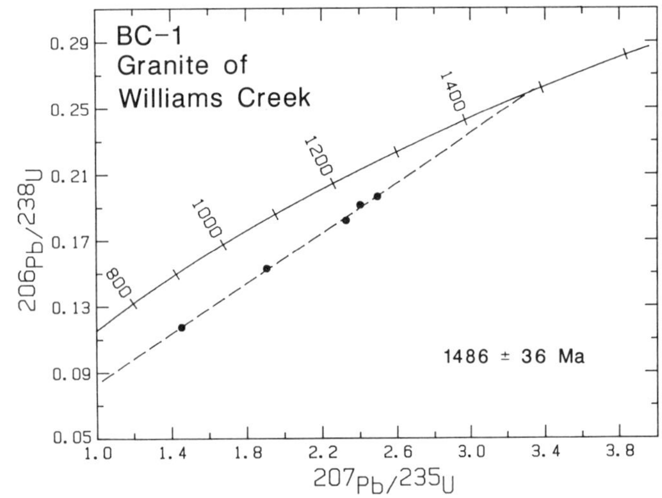

Figure 12. Concordia diagram for sample BC-1, Granite of Williams Creek. Model 2 solution based on 5 zircon fractions, probability of fit 0 percent, MSWD = 73.1. Intercepts at 1,486 ± 36 and 95 ± 97 Ma. See text for interpretation.

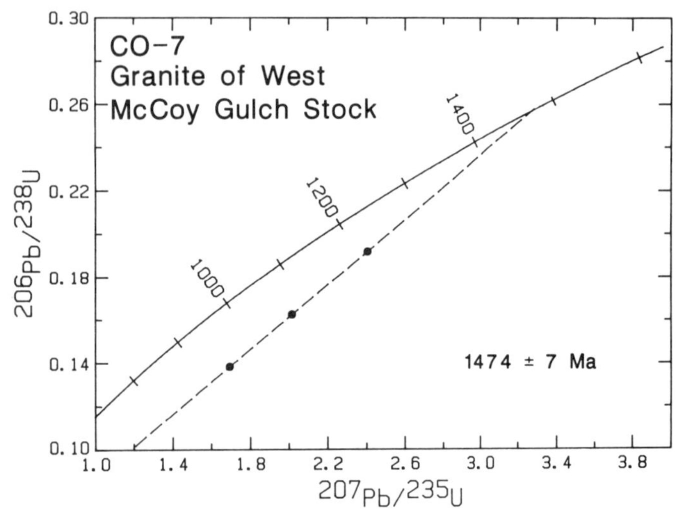

Figure 13. Concordia diagram for sample CO-7, Granite of West McCoy Gulch. Model 1 solution based on 3 zircon fractions, probability of fit 81 percent, MSWD = 0.0561. Intercepts at 1474 ± 7 and 146 ± 17 Ma.

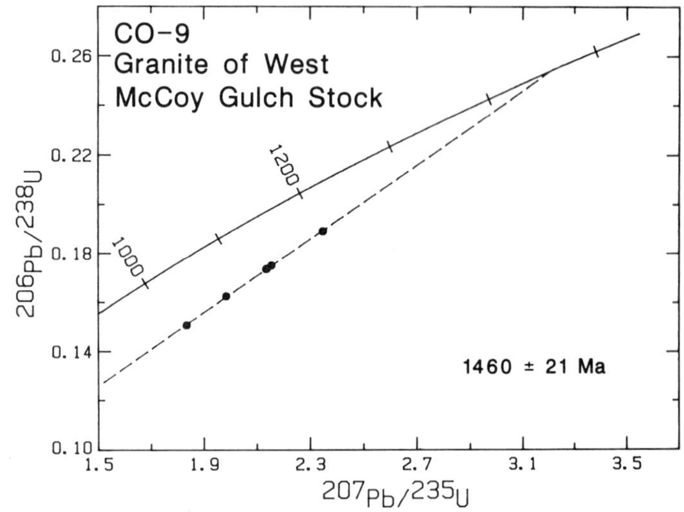

Figure 14. Concordia diagram for sample CO-9, Granite of West McCoy Gulch. Model 2 solution based on 6 zircon fractions, probability of fit 2 percent, MSWD = 2.93. Intercepts at 1,460 ± 21 and 177 ± 57 Ma.

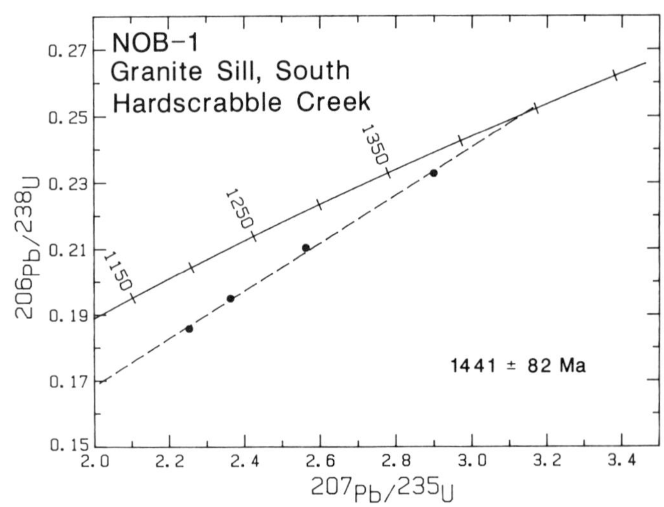

Figure 15. Concordia diagram for sample NOB-1, granitic sill in Hardscrabble Creek area. Model 2 solution based on 4 zircon fractions, probability of fit 0 percent, MSWD = 32.3. Intercepts at 1,441 ± 82 and 311 ± 370 Ma.

zircons. The zircons, however, display neither the sharply terminated habit nor the internal growth zonation that is characteristic of most igneous zircons. Rather, they are clear and stubby with somewhat rounded ends. Four fractions of these zircons yielded almost concordant results (Fig. 5). The regressed data yield an age of 1,703 ± 300 Ma, but these errors are unrealistically high because the analytical points cluster near concordia. The $^{207}Pb/^{206}Pb$ ages of the four fractions analyzed average 1,692 Ma with a standard deviation (1 σ) of 2.9 Ma; we believe that 1,694 ± 10 Ma is a good estimate of the age of this rock. A critical question, however, is whether the zircons analyzed were formed during primary igneous crystallization or were formed during granulite-facies recrystallization. We will not have a definitive answer to this question until these granulites are studied in more detail. Either the charnockite studied was crystallized from magma at about the same time as the batholithic rocks described above or it is an older rock that underwent granulite-facies metamorphism at the time of emplacement of the batholithic rocks.

Sample CO-15 is monzogranite from the coarse-grained facies of the Garell Peak batholith, whereas sample MM-B is from its leucogranite facies (Fig. 1; McCabe and Stone, 1984). CO-15

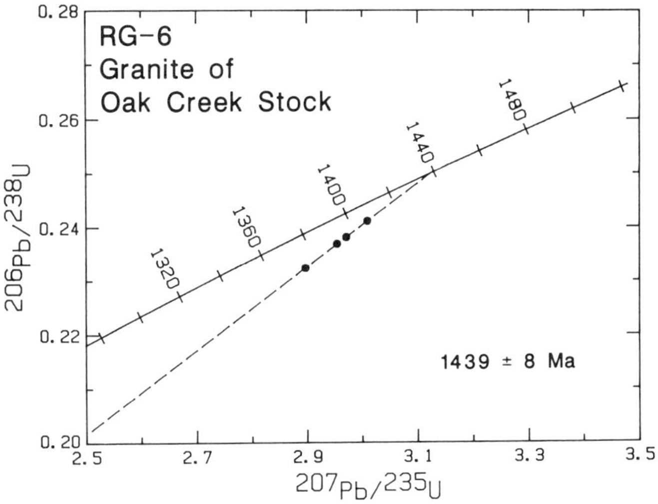

Figure 16. Concordia diagram for sample RG-6, Oak Creek pluton. Model 1 solution based on 4 zircon fractions, probability of fit 97 percent, MSWD = 0.0352. Intercepts at 1,439 ± 8 and 110 ± 220 Ma.

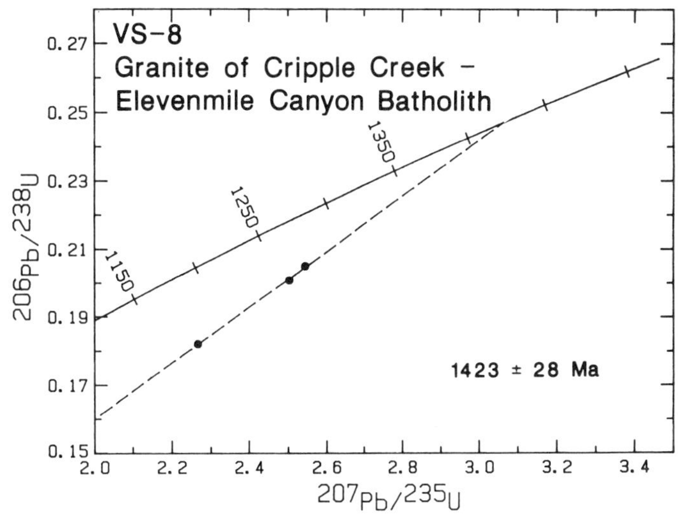

Figure 18. Concordia diagram for sample VS-8, Cripple Creek–Elevenmile Canyon batholith. Model 2 solution based on 3 zircon fractions, probability of fit 0.29 percent, MSWD = 8.86. Intercepts at 1,423 ± 28 and –48 ± 190 Ma.

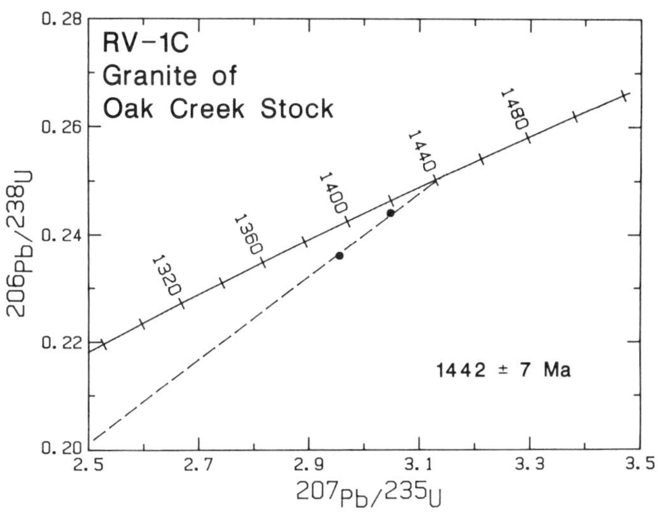

Figure 17. Concordia diagram for sample RV-1C, Oak Creek pluton. Model 2 solution based on 2 zircon fractions and an artificial point to force the chord through 99 Ma, probability of fit 1 percent, MSWD = 6.63. Intercepts at 1,442 ± 7 and 99 ± 17 Ma.

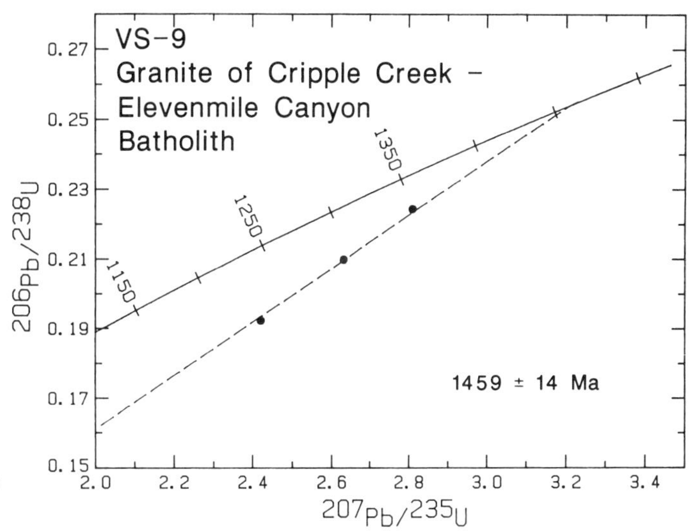

Figure 19. Concordia diagram for sample VS-9, Cripple Creek–Elevenmile Canyon batholith. Model 2 solution based on 3 zircon fractions and an artificial point to force the chord through 99 Ma, probability of fit 0 percent, MSWD = 32.9. Intercepts at 1,459 ± 14 and 97 ± 31 Ma.

yields an age of 1,666 ± 22 Ma (Fig. 6), whereas MM-B yields an age of 1,663 ± 4 Ma (Fig. 7). Sample VS-6 is granodiorite from the Phantom Canyon batholith (Fig. 1). This sample yields an age of 1,665 ± 5 Ma (Fig. 8). These plutons are coeval with similar bodies in the Gunnison and Salida areas (Granite of Cochetopa Canyon and pluton north of Salida; *Bickford and others, this volume*), the Boulder Creek batholith of the northern Front Range (*Premo and Van Schmus, this volume*), the buried basement of western Kansas (Bickford and others, 1986; Van Schmus and others, 1987), and several plutons in New Mexico (Bowring and others, 1984). The age of this ca. 1,670 Ma generation of plutons is analytically distinct from the ca. 1,700 Ma plutons discussed above. It will be of great interest to see whether they are chemically distinct as well.

A group of deformed, foliated, plutonic rocks in the Mt. Tyndall area, east of Westcliffe (Fig. 1), has yielded enigmatic results. Sample MT-8 is a foliated tonalite. The zircons from this sample are anhedral to rounded and are not typical of igneous

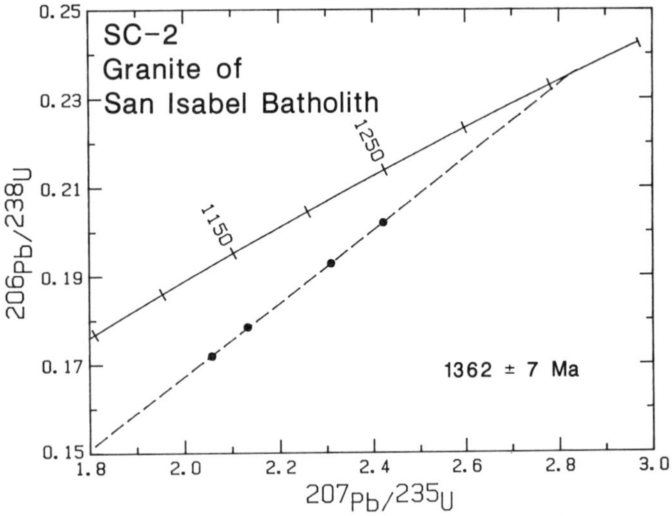

Figure 20. Concordia diagram for sample SC-2, coarse-grained facies of San Isabel batholith. Model 1 solution based on 4 zircon fractions, probability of fit 40 percent, MSWD = 0.927. Intercepts at 1,362 ± 7 and 36 ± 39 Ma.

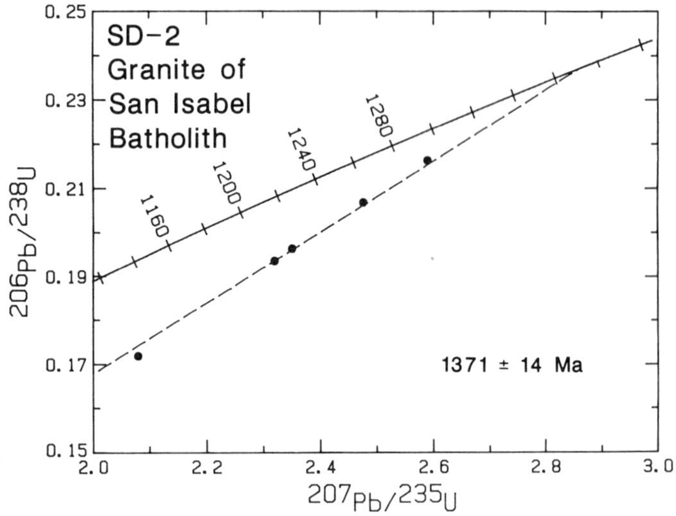

Figure 21. Concordia diagram for sample SD-2, medium-grained facies of San Isabel batholith. Model 2 solution based on 5 zircon fractions and an artificial point to force the chord through 99 Ma, probability of fit 0 percent, MSWD = 57.4. Intercepts at 1,371 ± 14 and 93 ± 38 Ma.

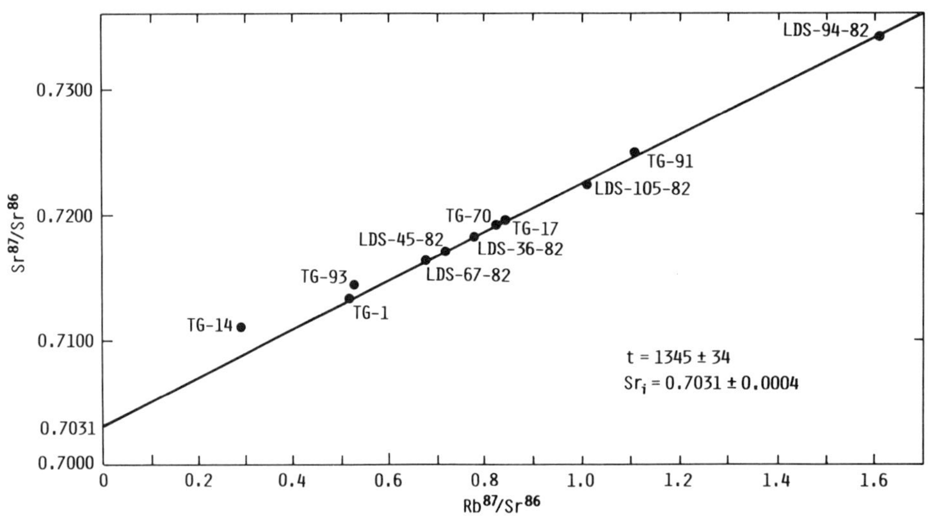

Figure 22. Rb-Sr isochron diagram for samples from San Isabel batholith.

rocks. They yield an age of 1,653 ± 50 Ma (Fig. 9). In a nearby outcrop, a rock of similar composition but with more distinctly gneissic structure (MT-9A) encloses a large inclusion of amphibolite and is, in turn, intruded by a foliated megacrystic granite (MT-9B). Zircons from MT-9A are rounded and relatively clear, resembling those from MT-2, the charnockite from nearby Boneyard Gulch. Analytical data for three fractions of these zircons cluster near concordia (Fig. 10), and regression yields a lower intercept of −616 ± 690 Ma. As discussed earlier, we constrained the regression line to pass through a lower intercept of 99 Ma; the resulting regression yields an age of 1,622 ± 5 Ma. Zircons from MT-9B are more elongate and have well developed, sharply bounded, internal growth zonation, although the external terminations of the crystals are somewhat rounded. Three fractions of these zircons yielded a lower intercept of −240 ± 370 Ma. Constraining the data again to pass through a chord with a lower intercept of 99 Ma yields an age of 1615 ± 3 Ma (Fig. 11).

It is very difficult to interpret the zircon ages of these rocks in the absence of detailed studies of their petrology. It seems likely that the ages indicated by the data from MT-8 and MT-9A indicate a time of metamorphism. The age indicated by zircons from MT-9B, however, probably represents the time of crystallization of the megacrystic granite.

Middle Proterozoic plutons

A number of Middle Proterozoic plutons have intruded older rocks in the southern Front Range and the Wet Mountains. As will be seen below, these were formed during two episodes of igneous activity occurring at 1,430 to 1,470 Ma and 1,350 to 1,370 Ma. These episodes are well known throughout the midcontinent region to the east (Bickford and others, 1986; Van Schmus and others, 1987), elsewhere in Colorado (e.g., Doe and Pearson, 1969; Silver and Barker, 1968; Bickford and others, 1969; Bickford and Cudzilo, 1975), and throughout the southwestern United States (e.g., Steiger and Wasserburg, 1966; Silver, 1978).

Sample BC-1 is from the Granite of Williams Creek (Fig. 1). This body occurs in the southern Wet Mountains as sheets intimately interleaved with paragneiss. Our sample yielded a complex suite of zircons, many of which clearly show cores with overgrowths. We separated the population on the basis of magnetic susceptibility and, additionally, hand picked two clear, uncored fractions. All of these zircon fractions together yield an age of 1,486 ± 36 Ma (Fig. 12). Regression of the two "clear" fractions yields an apparent age of 1,517 ± 7 Ma, whereas the remaining three fractions, which contain cores and overgrowths, yield an apparent age of 1,490 ± 550 Ma. There is considerable scatter in the analytical data. We interpret these results as indicating that the granite of Bear Creek was formed about 1,480 Ma, probably by anatectic melting of older rocks, and has a significant component of inherited zircons. It will require further petrological and geochemical study to confirm this interpretation and to determine the age of the precursor rocks.

Samples CO-7 and CO-9 are from the granite of West McCoy Gulch (Fig. 1; Sassarini. 1984). CO-7 yielded an age of 1,474 ± 7 Ma (Fig. 13), whereas CO-9 yielded an age of 1,460 ± 21 Ma (Fig. 14). The results from both samples are identical within analytical uncertainties and indicate that the pluton was emplaced about 1,470 Ma.

Sample NOB-1 is from a fine-grained granitic sill emplaced within paragneiss in the Hardscrabble Creek area (Fig. 1). The sample was collected by Jeffrey Noblett of Colorado College, who has been studying the petrology of the paragneisses. Initially, we thought that the protolith of this granitic body was a felsic volcanic rock, originally intercalated with the sedimentary protoliths of the paragneisses. However, despite the large uncertainty, its zircon age of 1,441 ± 82 Ma (Fig. 15) precludes this interpretation. Because it is elsewhere intruded by plutons whose ages are ca. 1,700 Ma (e.g., Crampton Mountain–Twin Mountain batholith), the minimum age of the paragneiss complex is somewhat greater than 1,700 Ma. Thus the sampled granite body must be a younger intrusive and not an intercalated volcanic rock.

Samples RG-6 and RV-1C are both from the Oak Creek Pluton (Fig. 1; McCabe and Stone, 1984). The pluton consists mostly of an early, coarse-grained, foliated, two-mica granodiorite to monzogranite that is intruded by small pods and dikes of medium-grained leucogranite. Both our samples are of the major phase. We expected that this pluton would yield a zircon age in the 1,650 to 1,700 Ma range because of its strong foliation and overall petrographic resemblance to the older ("Boulder Creek type") intrusives. When the first sample, RG-6, yielded an age of 1,439 ± 8 Ma (Fig. 16), we were so surprised that we collected a second sample from another major outcrop. This sample, RV-1C, yielded an age of 1,442 ± 7 Ma (Fig. 17), confirming the result of the first analysis. It seems clear that distinguishing older ("Boulder Creek") from younger ("Silver Plume") plutons in the Rocky Mountains on the basis of foliation and petrography can lead to serious errors!

Samples VS-8 and VS-9 were collected from the granite of Cripple Creek–Elevenmile Canyon batholith by W. R. Van Schmus. VS-8 yields an age of 1,423 ± 28 Ma (Fig. 18), and VS-9 yields an age of 1,459 ± 14 Ma (Fig. 19). These results are the same within error and suggest emplacement of the batholith between 1,440 and 1,450 Ma.

Samples SC-2 and SD-2 are from the San Isabel batholith, a major plutonic mass of monzogranite to syenogranite (Fig. 1; Boyer, 1962; Murray, 1970; Griffin and Shaw, 1984). SC-2 (coarse-grained facies) yielded an age of 1,362 ± 7 Ma (Fig. 20), whereas SD-2 (medium-grained facies) yielded an age of 1,371 ± 14 Ma (Fig. 21). We also determined a Rb-Sr isochron (Fig. 22) from analysis of 11 whole-rock samples collected throughout the main body of the San Isabel batholith by T. Griffin and L. Shaw (Fig. 23; Griffin and Shaw, 1984). Analytical data are given in Table 2. Nine of these samples determine an age of 1,345 ± 34 Ma (2 σ) with an initial $^{87}Sr/^{86}Sr$ of 0.7031 ± 0.0004 (2 σ; Fig. 22). The two samples that do not plot on the isochron (TG-14 and TG-93) both contain numerous xenoliths. The fact that both of these samples plot above the isochron determined from the other samples suggests that they were contaminated with radiogenic Sr from included country rock. The agreement of the Rb-Sr isochron age with the zircon age indicates that the Rb-Sr system in the main batholith is not significantly disturbed. The low initial $^{87}Sr/^{86}Sr$ is thus of petrogenetic significance.

Sample MT-3 is from a small stock-like intrusion about 20 km northwest of the main San Isabel batholith. Although the stock was mapped as "San Isabel Granite" (Scott and others, 1978), and the age indicated by analysis of zircons—1359 ± 11 Ma (Fig. 24)—confirms that it is coeval, we prefer the name granite of Bear Basin because of the distinctive petrography of this intrusion and its separation from the main body of the batholith.

As will be developed in more detail in the following discussion section, volcanic and plutonic rocks with ages in the range 1,340 to 1,400 Ma are widespread in the buried basement of the midcontinent region to the east. However, other than a Rb-Sr age of 1,350 ± 50 Ma reported for rhyolite dikes and the Trimble granite of the Needle Mountains in southwestern Colorado (Bickford and others, 1969) we are not aware that this age has been reported previously for rocks in the Rocky Mountains.

The last age determination to be reported here is from sample 80-3, collected by W. R. Van Schmus and W. Premo from syenite of the McClure Mountain alkaline complex. These rocks

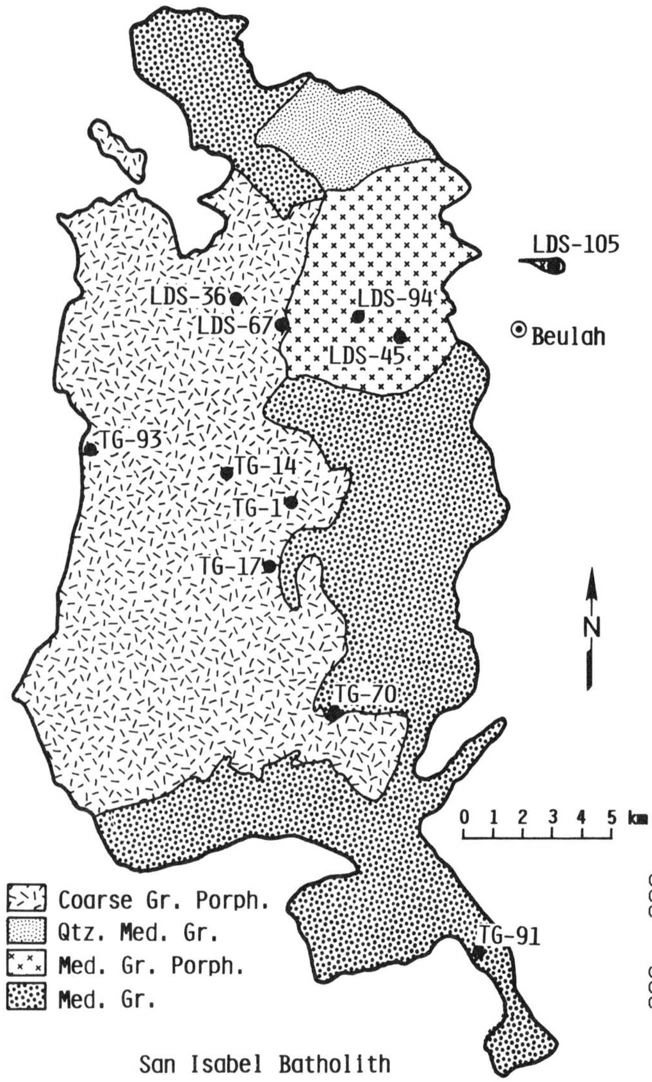

Figure 23. Geologic map of San Isabel batholith showing locations of Rb-Sr samples. After Griffin and Shaw (1984).

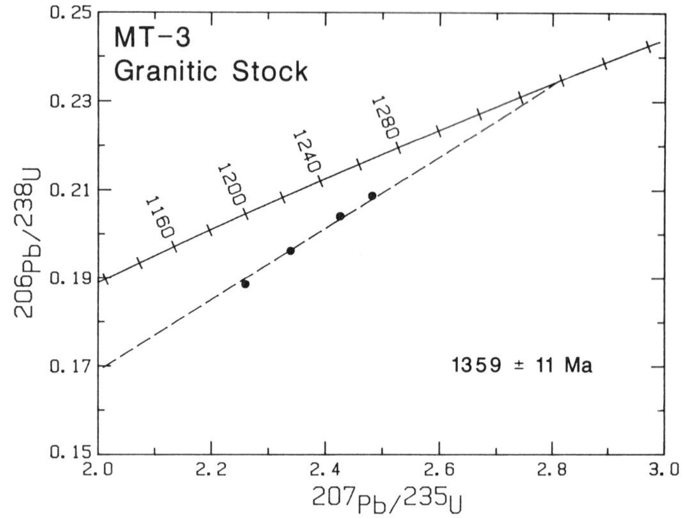

Figure 24. Concordia diagram for sample MT-3, Granite of Bear Basin. Model 2 solution based on 4 zircon fractions and an artificial point to force the chord through 99 Ma, probability of fit 0 percent, MSWD = 27.8. Intercepts at 1,359 ± 11 and 98 ± 29 Ma.

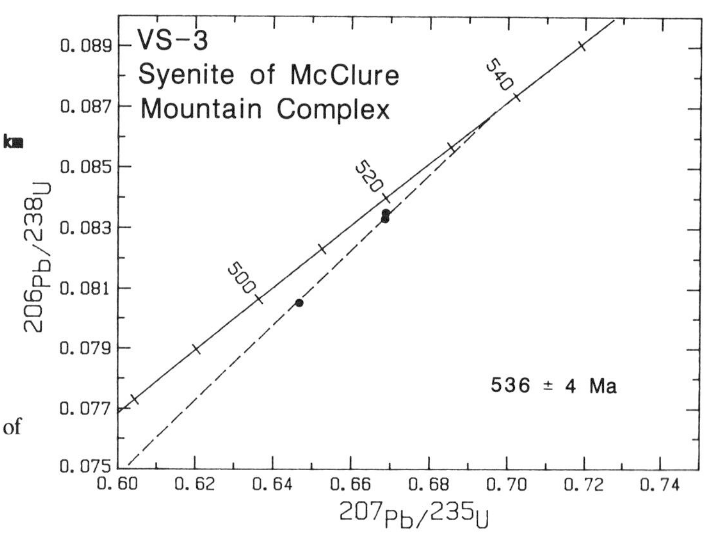

Figure 25. Concordia diagram for sample 80-3, syenite of McClure Mountain alkaline complex. Model 2 solution based on 3 zircon fractions and an artificial point to force the chord through the origin, probability of fit 11 percent, MSWD = 2.19. Intercepts at 536 ± 4 and 0 ± 7 Ma. See text for interpretation.

and the similar occurrences at Gem Park and Democrat Creek in the northern Wet Mountains have been studied by Shawe and Parker (1967), Parker and Sharp (1970), and Kilbane and Cullers (1979). The McClure Mountain Complex and the similar Iron Hill Complex near Powderhorn in the Gunnison area of Colorado (Olson and others, 1977) were first dated by the Rb-Sr method by Fenton and Faure (1970), who reported Cambrian ages. Later determinations by K-Ar, Rb-Sr, and fission-track methods were published by Olson and others (1977), who reported ages ranging from 479 to 536 Ma for the McClure Mountain Complex. More recently, Armbrustmacher and Hedge (1982) reported Rb-Sr isochron ages of 535 ± 5 Ma for the McClure Mountain Complex and 511 ± 8 Ma for the complex at Democrat Creek. Our sample 80-3 yielded an age of 536 ± 4 Ma (Fig. 25), in complete agreement with the Rb-Sr age of Armbrustmacher and Hedge (1982).

DISCUSSION

Early Proterozoic

The Early Proterozoic rocks of the Wet Mountains–southern Front Range region contrast strongly with the volcanogenic rocks of the Gunnison-Salida region to the west (Bickford and others, this volume) in that the country rocks are mostly of metasedimentary origin and are invaded by large masses of plutonic granite. The age range of the plutons dated, 1,700 to 1,665 Ma, is represented to the west by such plutons as the granite of Cochetopa Canyon (1,676 Ma; Olson and Steven, 1976; Bickford and others, this volume), the granite of Wood Gulch (1,701 Ma; Olson, 1976), and the pluton north of the volcanogenic section in the Salida area (Boardman, 1976; Bickford and Boardman, 1984; Bickford and others, this volume), but these plutons intrude older (1,780 to 1,715 Ma) volcanogenic rocks including basalt, rhyolite to dacite, and thick volcaniclastic sedimentary rocks. In contrast, the host to the Early Proterozoic plutons in the Wet Mountains–southern Front Range region is a sequence of metasedimentary rocks. Little detailed geochemical study of these rocks has been done, but they probably represent a thick accumulation of pelitic to psammitic sediment, possibly intruded by dikes of basaltic composition.

Rocks in the Wet Mountains–southern Front Range region are mostly in upper amphibolite facies, whereas those farther west are largely in lower amphibolite to upper greenschist facies. The abundant presence of migmatite in the Wet Mountains indicates in situ partial melting and suggests temperatures of about 650°C under water-saturated conditions. Field relations with the Early Proterozoic (i.e., 1,650 to 1,700 Ma) plutons suggest that in most areas high-grade metamorphism and anatexis occurred prior to emplacement of these plutons, i.e., before 1,700 Ma. Deformation and metamorphism between 1,740 and 1,700 Ma is documented in the Gunnison-Salida area to the west (Bickford and others, this volume; Reed and others, 1987).

The occurrence of metamorphic rocks, including some in granulite facies, yielding probable metamorphic zircon ages ranging from 1,694 ± 10 Ma (charnockite, sample MT-2) to about 1,620 Ma (tonalitic gneisses, samples MT-8 and MT-9A) in the Mt. Tyndall Quadrangle is still very poorly understood. Although the area was mapped in great detail (Brock and Singewald, 1968), there has been no modern study of the P-T conditions of metamorphism. It is not known whether the granulites are part of the generally high-grade country rocks into which the several generations of plutons have been emplaced or are large-scale inclusions within those plutonic rocks. About all that can be said now is that metamorphism, and perhaps major deformation, occurred at least locally between 1,600 and 1,700 Ma.

Middle Proterozoic and Cambrian

Middle Proterozoic plutonic rocks in the Wet Mountains–southern Front Range region fall into two age groups, 1,430 to 1,470 Ma and 1,350 to 1,370 Ma. The age difference between these two groups is much greater than the analytical uncertainties quoted. Thus, two distinct periods of igneous activity, both of which are well known elsewhere in the continent, are represented.

The older group of plutons, formed 1,430 to 1,470 Ma, is evidently part of the extensive "anorogenic" igneous activity (Anderson, 1983) that is well known throughout the western and southwestern United States (e.g., Silver and others, 1977), the midcontinent region (e.g., Bickford and others, 1981a, 1986; Van Schmus and Bickford, 1981; Van Schmus and others, 1975), and Labrador (Emslie, 1978). This widespread igneous activity has been called the "transcontinental anorogenic plutonic perforation of North America" by Silver and others (1977). Although plutons make up most of the dated occurrences of rocks of this age, an extensive terrane of coeval silicic volcanic rocks is known from the St. Francois Mountains of southeastern Missouri (Bickford and Mose, 1975; Bickford and others, 1981b; Sides and others, 1981; Cullers and others, 1981) and the buried basement of the eastern midcontinent. Bickford and others (1986) and Van Schmus and others (1987) have called these rocks the Eastern Granite-Rhyolite Province.

Rocks coeval with the younger group of plutons, the San Isabel batholith and its satellitic stocks which formed 1,340 to 1,370 m.y. ago, are also well known to the east. Silicic volcanic rocks and epizonal plutons of this age make up most of the basement of southwestern Missouri, southern Kansas, Oklahoma, and the Texas panhandle (Bickford and others, 1981a; Thomas and others, 1984) and make up the Western Granite-Rhyolite Province of Bickford and others (1986) and Van Schmus and others (1987).

Thomas and others (1984), Bickford and others (1986), and Van Schmus and others (1987) have argued that the granite-rhyolite terranes are veneers, probably only a few kilometers thick, that were derived by partial melting of rocks formed 1,600 to 1,800 m.y. ago and rest upon these rocks. Keys to this argument are: (1) the Sm-Nd data of Nelson and DePaolo (1985), which showed that the model ages of rocks representative of the 1,340 to 1,480 Ma events indicate separation from the mantle between 1,600 and 1,900 Ma, despite their lower crystallization ages; and (2) the presence of the 1,360-Ma San Isabel batholith in clear intrusive contact with rocks whose ages are greater than 1,700 Ma. The model of Thomas and others (1984) suggests that the rhyolite–shallow granite terranes of the plains may have extended to the west at least as far as the Wet Mountains but have been stripped away by Laramide and Tertiary uplift, exposing the older rocks that lay beneath them. In this model, the San Isabel batholith is a mesozonal body that is perhaps typical of many within the deeper crust to the east beneath the veneer of rhyolite and shallow granite. The presence of 1,350 Ma rhyolite porphyry dikes in the Needle Mountains of southwestern Colorado (Bickford and others, 1969) suggests that the rhyolitic veneer may have extended even farther to the west.

This discussion will not be complete without some mention of the Pike's Peak batholith and the Cambrian alkaline complexes. Both occurrences are well documented but poorly understood. The Pike's Peak batholith, whose Rb-Sr age is about 1,020 Ma (recalculated from data of Hedge, 1970), is coeval with high-grade metamorphic and igneous rocks of the Grenville Province on the eastern margin of the continent, with the igneous rocks of the Llano Uplift region of Texas (Zartman, 1964, 1965; Garrison and others, 1979), and with the dominantly basaltic rocks of the

TABLE 1. ANALYTICAL DATA FOR ZIRCON FRACTIONS

Fraction	Concentrations		Pb Isotopic Composition			Radiogenic Ratios			Ages		
	U (ppm)	Pb (ppm)	Pb^{204}/Pb^{206}	Pb^{207}/Pb^{206}	Pb^{208}/Pb^{206}	Pb^{206}/U^{238}	Pb^{207}/U^{238}	Pb^{207}/Pb^{207}	Pb^{206}/U^{238}	Pb^{207}/U^{235}	Pb^{207}/Pb^{206}
Early Proterozoic Plutons											
RG 25 E	350	107	0.00013	0.10619	0.08520	0.29829	4.2939	0.10440	1682.8	1692.2	1703.9
RG 25 F	328	101	0.00022	0.10744	0.09013	0.29792	4.2913	0.10447	1681.0	1691.7	1705.1
RG 25 G	284	87	0.00023	0.10758	0.09777	0.29777	4.2871	0.10442	1680.3	1690.9	1703.9
RG 25 D	372	113	0.00015	0.10647	0.08626	0.29539	4.2524	0.10441	1668.4	1684.2	1703.9
RG 25 C	346	106	0.00027	0.10805	0.09283	0.29492	4.2457	0.10441	1666.1	1682.9	1703.9
RG 4 1	432	128	0.00025	0.10756	0.07209	0.29040	4.1677	0.10409	1643.5	1667.7	1698.2
RG 4 H	487	145	0.00061	0.11216	0.09232	0.28536	4.0846	0.10381	1618.3	1651.2	1693.4
RG 4 F	719	198	0.00029	0.10711	0.06696	0.27220	3.8684	0.10307	1552.0	1607.1	1680.1
RG 4 D	818	210	0.00033	0.10682	0.06879	0.25264	3.5621	0.10266	1452.1	1541.1	1665.6
VS 75A1	432	126	0.00044	0.11012	0.08220	0.28173	4.0438	0.10410	1600.1	1643.1	1698.4
VS 75B	538	147	0.00068	0.11299	0.09255	0.26182	3.7434	0.10370	1499.1	1580.7	1691.4
VS 750	687	162	0.00121	0.11989	0.12030	0.21692	3.0885	0.10326	1265.6	1429.8	1683.6
VS 75F	775	167	0.00180	0.12686	0.14842	0.19060	2.6837	0.10212	1124.6	1324.0	1662.9
MT 2 NM(-2)	141	43	0.00026	0.10734	0.07913	0.29877	4.2758	0.10379	1685.2	1688.7	1693.0
MT 2 M(-2)	144	43	0.00012	0.10551	0.07488	0.29787	4.2643	0.10383	1680.8	1686.5	1693.8
MT 2 M(-1)	145	44	0.00016	0.10598	0.07618	0.29680	4.2474	0.10379	1675.4	1683.2	1693.0
MT 2 M(0)	150	45	0.00013	0.10522	0.07378	0.29549	4.2160	0.10348	1668.9	1677.1	1687.5
CO 15 G	833	215	0.00022	0.10482	0.08982	0.25079	3.5180	0.10174	1442.6	1531.3	1656.1
CO 15 F	990	243	0.00035	0.10598	0.09774	0.23620	3.2949	0.10117	1366.9	1479.9	1645.7
CO 15 E	1231	282	0.00038	0.10602	0.09819	0.21961	3.0524	0.10080	1279.8	1420.8	1639.1
CO 15 D	1365	284	0.00043	0.10640	0.10191	0.19884	2.7556	0.10051	1169.1	1343.6	1633.6
MM B NM(2)	4907	1119	0.00021	0.10301	0.05310	0.22858	3.1538	0.10007	1327.1	1445.9	1625.4
MM B M(2)	4230	716	0.00028	0.10073	0.04350	0.17136	2.2870	0.09680	1019.6	1208.3	1563.3
MM B M(4)	3962	525	0.00035	0.09818	0.04440	0.13423	1.7271	0.09332	811.9	1018.7	1494.5
MM B M(8)	4011	376	0.00051	0.09368	0.05750	0.09387	1.1205	0.08657	578.4	763.2	1351.2
VS 6 A'	320	92	0.00043	0.10785	0.11097	0.27420	3.8545	0.10195	1562.1	1604.2	1660.2
VS 6 A	329	93	0.00027	0.10589	0.11022	0.27076	3.8131	0.10214	1544.7	1595.5	1663.4
VS 6 B	322	90	0.00038	0.10725	0.11432	0.26526	3.7304	0.10199	1516.7	1577.9	1660.6
VS 6 C	336	92	0.00055	0.10922	0.12496	0.25655	3.5977	0.10171	1472.1	1549.0	1655.6
VS 6 D	359	91	0.00059	0.10978	0.12237	0.23803	3.3356	0.10163	1376.4	1489.4	1654.1
MR 8 NM(-2)	369	107	0.00047	0.10821	0.13660	0.26982	3.7878	0.10182	1539.9	1590.2	1657.4
MT 8 M(-2)	555	150	0.00084	0.11169	0.14633	0.24660	3.4068	0.10020	1420.9	1506.0	1627.7
MT 8 M(-1)	641	147	0.00076	0.11111	0.14849	0.20854	2.8929	0.10061	122.0	1380.1	1635.5
MT 8 M(0)	704	127	0.00099	0.11320	0.15423	0.16369	2.2420	0.09934	977.2	1194.3	1611.7
MT 9A NM(-3)	185	52	0.00016	0.10215	0.07604	0.27826	3.8365	0.01000	1582.6	1600.4	1623.8
MT 9A M(-3)	193	55	0.00016	0.10207	0.07883	0.27894	3.8395	0.09983	1586.0	1601.1	1620.9
MT 9A M(-1)	197	57	0.00016	0.10178	0.08304	0.28414	3.9034	0.09963	1612.2	1614.2	1617.3
MT 9B NM(-3)	433	115	0.00014	0.10109	0.05605	0.26736	3.6547	0.09914	1527.4	1561.5	1608.2
MT 9B M(-3)	448	118	0.00014	0.10126	0.05839	0.26283	3.5988	0.09931	1504.3	1549.3	1611.0
MT 9B M(-1)	468	121	0.00012	0.10102	0.06007	0.25988	3.5594	0.09934	1489.2	1540.5	1611.9
Middle Proterozoic Plutons											
BC 1 (bulk)											
NM(1)	837	186	0.00234	0.12473	0.15631	0.19634	2.4956	0.09219	1155.6	1270.8	1471.1
M(1)	526	132	0.00490	0.15932	0.25825	0.19125	2.4025	0.09111	1128.1	1243.3	1448.8
Bulk	944	199	0.00540	0.16574	0.29609	0.15289	1.9077	0.09050	917.1	1083.8	1436.1
BC 1 (clear)											
NM(6)	761	175	0.00386	0.14642	0.23192	0.18200	2.3275	0.09275	1077.9	1220.7	1482.8
M(6)	866	164	0.00797	0.20087	0.42917	0.11738	1.4534	0.08980	715.5	911.3	1421.5

TABLE 1. ANALYTICAL DATA FOR ZIRCON FRACTIONS (continued)

Fraction	Concentrations		Pb Isotopic Composition			Radiogenic Ratios			Ages		
	U (ppm)	Pb (ppm)	Pb^{204}/Pb^{206}	Pb^{207}/Pb^{206}	Pb^{208}/Pb^{206}	Pb^{206}/U^{238}	Pb^{207}/U^{238}	Pb^{207}/Pb^{207}	Pb^{206}/U^{238}	Pb^{207}/U^{235}	Pb^{207}/Pb^{206}
Middle Proterozoic Plutons (continued)											
CO 7 MN(1)	1176	248	0.00127	0.10864	0.14258	0.19177	2.4038	0.09091	1131.0	1243.7	1444.9
CO 7 M(1)	1536	277	0.00145	0.11004	0.14879	0.16251	2.0135	0.08986	970.7	1120.1	1422.7
CO 7 M(3)	2068	328	0.00197	0.11630	0.17240	0.13838	1.6928	0.08872	835.5	1005.8	1398.0
CO 9 G	1562	314	0.00056	0.09782	0.12717	0.18904	2.3462	0.09001	1116.2	1226.4	1425.8
CO 9 F	1962	369	0.00081	0.10054	0.13391	0.17502	2.1534	0.08923	1039.7	1166.1	1409.4
CO 9 E(3D)	1749	325	0.00063	0.09803	0.13273	0.17370	2.1358	0.08918	1032.5	1160.5	1407.8
CO 9 E(3C)	1741	323	0.00063	0.09803	0.13273	0.17353	2.1337	0.08918	1031.5	1159.8	1407.8
CO 9 D	2051	360	0.00077	0.09936	0.13951	0.16246	1.9833	0.08854	970.5	1109.8	1394.1
CO 9 C	2234	361	0.00074	0.09872	0.13343	0.15067	1.8347	0.08831	904.7	1058.0	1389.5
NOB 1 NM(-2)	20	7	0.00575	0.17151	0.40157	0.23276	2.8996	0.09035	1348.9	1381.8	1432.8
NOB 1 M(-1)	263	65	0.00138	0.10761	0.23063	0.21039	2.5624	0.08833	1230.9	1290.0	1389.8
NOB 1 M(0)	447	106	0.00216	0.11813	0.24240	0.19499	2.3631	0.08789	1148.4	1231.5	1380.1
NOB 1 M(1)	500	112	0.00220	0.11874	0.23685	0.18578	2.2540	0.08800	1098.1	1198.1	1382.4
RG 6 1	238	58	0.00016	0.09273	0.09288	0.23110	3.0095	0.09053	1329.4	1410.0	1436.7
RG 6 G	269	69	0.00011	0.10527	0.13661	0.23814	2.9712	0.09049	1377.0	1400.3	1435.9
RG 6 H	271	66	0.00031	0.09484	0.09717	0.23690	2.9543	0.09045	1370.6	1396.0	1435.2
RG 6 F	300	72	0.00033	0.09498	0.10337	0.23239	2.8971	0.09042	1347.0	1381.2	1434.4
RV 1C NM(-3)	122	32	0.00056	0.09834	0.14678	0.24410	3.0481	0.09056	1408.0	1419.8	1437.4
RV 1C M(0)	134	36	0.00149	0.11155	0.18229	0.23608	2.9558	0.09080	1366.3	1396.3	1442.3
VS 8 A	270	62	0.00093	0.10300	0.17725	0.20502	2.5436	0.08998	1202.2	1284.6	1425.3
VS 8 C	280	61	0.00067	0.09980	0.15817	0.20082	2.5027	0.09039	1179.7	1272.8	1433.9
VS 8 E	343	70	0.00141	0.10991	0.17606	0.18205	2.2668	0.09031	1078.2	1202.0	1432.0
VS 9 A	212	51	0.00025	0.09429	0.14839	0.22431	2.8088	0.09082	1304.6	1357.9	1442.8
VS 9 C	260	58	0.00027	0.09472	0.15159	0.20985	2.6303	0.09090	1228.0	1309.1	1444.5
VS 9 E	284	59	0.00050	0.09833	0.15789	0.19225	2.4210	0.09133	1133.6	1248.8	1453.7
SC 2 H	245	52	0.00042	0.09284	0.13459	0.20201	2.4219	0.08695	1186.1	1249.1	1359.2
SC 2 G	248	51	0.00053	0.09428	0.14621	0.19286	2.3099	0.08687	1136.8	1215.3	1357.7
SC 2 F	285	55	0.00073	0.09690	0.15364	0.17844	2.1327	0.08668	1058.4	1159.5	1353.4
SC 2 E	308	60	0.00159	0.10907	0.18689	0.17193	2.0576	0.08680	1022.7	1134.8	1356.1
SD 2 H	159	37	0.00091	0.09961	0.15908	0.21629	2.5895	0.08683	1262.3	1297.7	1356.9
SD 2 G	175	40	0.00110	0.10228	0.16421	0.20682	2.4759	0.08683	1211.8	1265.0	1356.5
SD 2 F	202	50	0.00358	0.13697	0.25360	0.19635	2.3503	0.08682	1155.7	1227.7	1356.5
SD 2 E	170	39	0.00239	0.12035	0.22365	0.19356	2.3197	0.08692	1140.6	1218.3	1358.8
SD 2 D	223	48	0.00337	0.13496	0.24963	0.17189	2.0795	0.08774	1022.5	1142.1	1376.9
MT 3 NM(-1)	164	50	0.00605	1.17096	0.36764	0.20872	2.4817	0.08624	1222.0	1266.7	1343.4
MT 3 M(-1)	208	48	0.00093	0.09922	0.19533	0.20404	2.4257	0.08622	1197.0	1250.2	1343.4
MT 3 M(0)	278	65	0.00173	0.11072	0.22961	0.19619	2.3390	0.08647	1154.8	1224.2	1348.8
MT 3 M(1)	333	71	0.00091	0.09967	0.20026	0.18854	2.2598	0.08693	1113.5	1199.8	1359.0
Syenite of McClure Mountain Alkaline Complex											
VS 3 A'	87	8	0.00081	0.06984	0.18464	0.08351	0.6688	0.05809	517.0	520.0	532.8
VS 3 B	95	8	0.00050	0.06556	0.17457	0.08331	0.6686	0.05821	515.8	519.9	537.5
VS 3 D	102	9	0.00165	0.08231	0.21607	0.08052	0.6467	0.05825	499.2	506.5	539.5

Notes: The designation A, B, C, ... for some samples indicates decreasingMagnetic susceptibility. For other samples NM or M followed by a number refer to zircon fractions that are magnetic or non-magnetic on the Frantz separator at the side-tilt indicated by the number. All ratios are atomic ratios.

TABLE 2. Rb AND Sr ANALYTICAL DATA FOR SAN ISABEL BATHOLITH

Sample	Rb*	Sr*	Rb†	Sr†	$^{87}Rb/^{86}Sr$§	$^{87}Sr/^{86}Sr$§
Coarse-grained Porphyritic Phase						
TG-1	106	589	97	542	0.5184	0.7133
TG-14	77	637	63	623	0.2918	0.7110
TG-17	138	470	125	429	0.8409	0.7194
TG-93	94	469	83	458	0.5255	0.7145
LDS-36	139	533	137	511	0.7780	0.7181
LDS-67	133	567	122	520	0.6774	0.7162
Medium-grained Porphyritic Phase						
LDS-45	127	545	124	503	0.7162	0.7168
LDS-94	214	334	166	299	1.6082	0.7342
LDS-105	139	436	144	414	1.0100	0.7222
Medium-grained Phase						
TG-70	145	432	127	448	0.8206	0.7191
TG-91	140	377	129	337	1.1088	0.7248

*Concentrations (ppm) determined by x-ray fluorescence.
†Concentrations (ppm) determined by isotope dilution analysis.
§Atomic ratio.

Midcontinent rift system (Chaudhuri and Faure, 1967; Silver and Green, 1972; Green, 1982; Van Schmus and Hinze, 1985), but its isolated occurrence far to the west of these coeval terranes is not understood. The Cambrian alkaline complexes have been correlated with igneous activity along the so-called 38th parallel lineament of Heyl (1972); it has also been pointed out that the two occurrences in Colorado—the rocks in the northern Wet Mountains mentioned here and those in the Powderhorn area farther west—are on strike with the extension of the Middle Cambrian Wichita Mountains igneous complex of southern Oklahoma (e.g., Ham and others, 1964; Gilbert, 1983). Rocks in the Wichita Mountains have yielded essentially identical zircon U-Pb ages (about 550 Ma; Bowring and Hoppe, 1982) as that reported here from the McClure Mountain Complex. Hoffman and others (1974) and numerous others have interpreted the Wichita Mountains and the deep Anadarko Basin as an aulacogen. In this model, the igneous suite formed during initial rifting. That this rift extended as far northwest as Colorado is intriguing, but speculative. The answers will probably come from extensive geophysical study.

ACKNOWLEDGMENTS

This research was supported by NSF grant EAR-8218463 (MEB) and EAR-8208564 (RLC). The manuscript benefited greatly from reviews by J. Aleinikoff, T. Krogh, and J. Grambling.

REFERENCES CITED

Anderson, J. L., 1983, Proterozoic anorogenic granite plutonism of North America, *in* Medaris, L. G., Jr., Mickelson, D. M., Byers, C. W., and Shanks, W. G., eds., Proterozoic geology; Selected papers from an international Proterozoic symposium: Geological Society of America Memoir 161, p. 133–154.

Armbrustmacher, T. J., and Hedge, C. E., 1982, Genetic implications of minor-element and Sr-isotope geochemistry of alkaline rock complexes in the Wet Mountains area, Fremont and Custer Counties, Colorado: Contributions to Mineralogy and Petrology, v. 79, p. 424–435.

Bickford, M. E., and Boardman, S. J., 1984, A Proterozoic volcano-plutonic terrane, Gunnison and Salida areas, Colorado: Journal of Geology, v. 92, p. 657–666.

Bickford, M. E., and Cudzilo, T. F., 1975, U-Pb ages of zircons from Vernal Mesa-type quartz monzonite, west-central Colorado: Geological Society of America Bulletin, v. 86, p. 1432–1434.

Bickford, M. E., and Mose, D. G., 1975, Geochronology of Precambrian rocks in the St. Francois Mountains, southeastern Missouri: Geological Society of America Special Paper 165, 48 p.

Bickford, M. E., Wetherill, G. W., Barker, F., and Lee-Hu, Chin-Nan, 1969, Precambrian Rb-Sr geochronology in the Needle Mountains, southwestern Colorado: Journal of Geophysical Research, v. 74, p. 1660–1676.

Bickford, M. E., Harrower, K. L., Hoppe, W. J., Nelson, B. K., Nusbaum, R. L., and Thomas, J. J., 1981a, Rb-Sr and U-Pb geochronology and distribution of rock types in the Precambrian basement of Missouri and Kansas: Geological Society of America Bulletin, Part 1, v. 92, p. 323–341.

Bickford, M. E., Sides, J. R., and Cullers, R. L., 1981b, Chemical evolution of magmas in the Proterozoic terrane of the St. Francois Mountains, southeastern Missouri; 1, Field, petrographic, and major element data: Journal of Geophysical Research, v. 86, p. 10365–10386.

Bickford, M. E., Van Schmus, W. R., and Zietz, I., 1986, Proterozoic history of the midcontinent region of North America: Geology, v. 14, p. 492–496.

Boardman, S. J., 1976, Geology of the Precambrian metamorphic rocks of the Salida area, Chaffee County, Colorado: The Mountain Geologist, v. 13, p. 89–100.

Bowring, S. A., and Hoppe, W. J., 1982, U/Pb zircon ages from Mount Sheridan gabbro, Wichita Mountains: Oklahoma Geological Society Guidebook 21, p. 54–59.

Bowring, S. A., Reed, J. C., Jr., and Condie, K. C., 1984, U-Pb geochronology of Proterozoic volcanic and plutonic rocks, Sangre de Cristo Mountains, N.M.: Geological Society of America Abstracts with Programs, v. 16, p. 216.

Boyer, R. E., 1962, Petrology and structure of the southern Wet Mountains, Colorado: Geological Society of America Bulletin, v. 73, p. 1047–1070.

Brock, M. R., and Singewald, Q. D., 1968, Geologic map of the Mount Tyndall Quadrangle, Custer County, Colorado: U.S. Geological Survey Geologic Quadrangle Map GQ–596, scale 1:24,000.

Cameron, A. E., Smith, D. H., and Walker, R. L., 1969, Mass spectrometry of nannogram-size samples of lead: Analytical Chemistry, v. 41, p. 525–526.

Christman, R. A., Brock, M. R., Pearson, R. C., and Singewald, Q. D., 1959, Geology and thorium deposits of the Wet Mountains, Colorado; A progress report: U.S. Geological Survey Bulletin 1072–H, p. 491–533.

Cullers, R. L., Koch, R. J., and Bickford, M. E., 1981, Chemical evolution of magmas in the Proterozoic terrane of the St. Francois Mountains, southeastern Missouri; 2, Trace element data: Journal of Geophysical Research, v. 86, p. 10388–10401.

Doe, B. R., and Pearson, R. C., 1969, U-Th-Pb geochronology of zircons from the St. Kevin Granite northern Sawatch Range, Colorado: Geological Society of America Bulletin, v. 80, p. 2495–2502.

Emslie, R. F., 1978, Anorthosite massifs, rapakivi granites, and Late Proterozoic rifting of North America: Precambrian Research, v. 7, p. 61–98.

Fenton, M. D., and Faure, G., 1970, Rb-Sr whole-rock age determinations of the Iron Hill and McClure Mountain carbonatite-alkalic complexes, Colorado: Mountain Geologist, v. 7, p. 268–275.

Garrison, J. R., Jr., Long, L. E., and Richman, D. L., 1979, Rb-Sr and K-Ar geochronologic and isotopic studies, Llano Uplift, central Texas: Contributions to Mineralogy and Petrology, v. 69, p. 361–374.

Gilbert, M. C., 1983, Timing and chemistry of igneous events associated with the southern Oklahoma aulacogen: Tectonophysics, v. 94, p. 439–455.

Green, J. C., 1982, Geology of the Keweenawan extrusive rocks *in* Wold, R. J., and Hinze, W. J., eds., Geology and tectonics of the Lake Superior basin: Geological Society of America Memoir 152, p. 47–56.

Griffin, T., and Shaw, L., 1984, Petrology of the San Isabel batholith, southern Wet Mountains, Colorado: Geological Society of America Abstracts with Programs, v. 16, p. 223.

Ham, W. E., Denison, R. E., and Merritt, C. A., 1964, Basement rocks and structural evolution of southern Oklahoma: Oklahoma Geological Survey Bulletin 95, 302 p.

Hedge, C. E., 1970, Whole-rock Rb-Sr age of the Pikes Peak batholith, Colorado, *in* Geological Survey Research, 1970: U.S. Geological Survey Professional Paper 700–B, p. B86–B89.

Heyl, A. V., Jr., 1972, The 38th Parallel Lineament and its relationship to ore deposits: Economic Geology, v. 67, p. 879–894.

Hoffman, P. F., Dewey, J. F., and Burke, K., 1974, Aulacogens and their genetic relationship to geosynclines with a Proterozoic example from the Great Slave Lake, Canada, *in* Dott, R. H., and Shaver, R. H., eds., Modern and ancient geosynclinal sedimentation: Society of Economic Paleontologists and Mineralogists Special Publication 19, p. 38–55.

Kilbane, N., and Cullers, R. L., 1979, Petrogenesis of the McClure Mountain mafic-ultramafic and alkaline complex, Fremont County, Colorado: Geological Society of America Abstracts with Programs, v. 11, p. 150.

Knepper, D. H., 1972, Stratigraphy and macroscopic structure of the metasedimentary rocks of the Blue Ridge area, Fremont County, Colorado [M.S. thesis]: Lawrence, University of Kansas, 53 p.

Krogh, T. E., 1973, A low contamination method for the hydrothermal decomposition of zircon and extraction of U-Pb for isotopic age determinations: Geochimica et Cosmochimica Acta, v. 37, p. 485–494.

Logan, J. M., 1966, Structure and petrology of the eastern margin of the Wet Mountains, Colorado [Ph.D. thesis]: Norman, University of Oklahoma, 237 p.

Ludwig, K. R., 1983, Plotting and regression programs for isotope geochemists, for use with HP-86/87 microcomputers: U.S. Geological Survey Open File Report 83-849, 94 p.

Ludwig, K. R., 1980, Calculation of uncertainties of U-Pb isotopic age data: Earth and Planetary Science Letters, v. 46, p. 212–220.

McCabe, M., and Stone, J., 1984, The petrology of foliated plutons at Garell Peak and Oak Creek, northern Wet Mountains, Colorado: Geological Society of America Abstracts with Programs, v. 16, p. 247.

Murray, M. M., 1970, Petrology, structure, and origin of the San Isabel batholith, Wet Mountains, Colorado [Ph.D. thesis]: Houston, Texas, Rice University, 54 p.

Nelson, B. K., and DePaolo, D. J., 1985, Rapid production of continental crust 1.7–1.9 b.y. ago; Nd isotopic evidence from the basement of the North American midcontinent: Geological Society of America Bulletin, v. 96, p. 746–754.

Olson, J. C., 1976, Geologic map of the Houston Gulch Quadrangle, Gunnison and Saguache Counties, Colorado: U.S. Geological Survey Map GQ–1287, scale 1:24,000.

Olson, J. C., and Steven, T. A., 1976, Geologic map of the Sawtooth Mountain Quadrangle, Saguache County, Colorado: U.S. Geological Survey Map MF–733, scale 1:24,000.

Olson, J. C., Marvin, R. C., Parker, R. L., and Mehnert, H. H., 1977, Age and tectonic setting of lower Paleozoic alkalic and mafic rocks, carbonatites, and thorium veins in south-central Colorado: U.S. Geological Survey Journal of Research, v. 5, p. 673–687.

Parker, R. L., and Sharpe, W. N., 1970, Mafic-ultramafic igneous rocks and associated carbonatites of the Gem Park complex, Custer and Fremont Counties, Colorado: U.S. Geological Survey Professional Paper 649, 24 p.

Reuss, R. L., 1974, Precambrian quartzite-schist sequence in Wilson Park, Fremont County, Colorado: Mountain Geologist, v. 11, p. 45–58.

Sassarini, N. A., 1984, Petrology of Middle Proterozoic granites of the West McCoy Gulch, Texas Creek and Cotopaxi area, Fremont County, Colorado: Geological Society of America Abstracts with Programs, v. 16, p. 253.

Scott, G. R., Taylor, R. B., Epis, R. C., and Wobus, R. A., 1978, Geologic map of the Pueblo 1° by 2° Quadrangle, south-central Colorado: U.S. Geological Survey Miscellaneous Investigations Series Map I–1022, scale 1:250,000.

Shawe, D. R., and Parker, R. L., 1967, Mafic-ultramafic layered intrusion at Iron Mountain, Fremont County, Colorado: U.S. Geological Survey Bulletin, 1251–A, 29 p.

Shuster, R. D., and Bickford, M. E., 1985, Chemical and isotopic evidence for the petrogenesis of the northeastern Idaho batholith: Journal of Geology, v. 93, p. 727–742.

Sides, J. R., Bickford, M. E., Shuster, R. D., and Nusbaum, R. L., 1981, Calderas in the Precambrian terrane of the St. Francois Mountains, southeastern Missouri: Journal of Geophysical Research, v. 86, p. 10349–10364.

Silver, L. T., 1978, Precambrian formations and Precambrian history in Cochise County, southeastern Arizona, *in* Callendar, J. F., Wilt, J. C., and Clemons, R. E., eds., Land of Cochise: New Mexico Geological Society 29th Field Conference Guidebook, p. 157–163.

Silver, L. T., and Barker, F., 1968, Geochronology of Precambrian rocks of the Needle Mountains, southwestern Colorado; Part 1, U-Pb results: Geological Society of America Abstracts for 1967, Special Paper 115, p. 204–205.

Silver, L. T., and Green, J. C., 1972, Time constants for Keweenawan igneous activity: Geological Society of America Abstracts with Programs, v. 4, p. 665–666.

Silver, L. T., Bickford, M. E., Van Schmus, W. R., Anderson, J. L., Anderson, T. H., and Medaris, L. G., 1977, The 1.4–1.5 b.y. transcontinental anorogenic perforation of North America: Geological Society of America Abstracts with Programs, v. 9, p. 1176.

Steiger, R. H., and Jaeger, E., 1977, Subcommission on geochronology: Convention on use of decay constants in geo- and cosmo-chronology: Earth and Planetary Science Letters, v. 36, p. 359–362.

Steiger, R. H., and Wasserburg, G. J., 1966, Systematics in the Pb^{208}-Th^{232}, Pb^{207}-U^{235}, Pb^{206}-U^{238} systems: Journal of Geophysical Research, v. 71, p. 6065–6068.

Thomas, J. J., Shuster, R. D., and Bickford, M. E., 1984, A terrane of 1,350–1,400 m.y. old silicic volcanic and plutonic rocks in the buried Proterozoic of the midcontinent and in the Wet Mountains, Colorado: Geological Society of America Bulletin, v. 95, p. 1150–1157.

Van Schmus, W. R., and Bickford, M. E., 1981, Proterozoic chronology and evolution of the midcontinent region, North America, *in* Kroner, A., ed., Amsterdam, Elsevier Pub. Co., Precambrian plate tectonics: p. 261–296.

Van Schmus, W. R., and Hinze, W. J., 1985, The midcontinent rift system: Annual Reviews of Earth and Planetary Science, v. 13, p. 345–383.

Van Schmus, W. R., Medaris, L. G., and Banks, P. O., 1975, Geology and age of the Wolf River batholith, Wisconsin: Geological Society of America Bulletin, v. 86, p. 907–914.

Van Schmus, W. R., Bickford, M. E., and Zietz, I., 1987, Early and Middle Proterozoic provinces in the central United States, *in* Kroner, A., ed., Proterozoic lithospheric evolution: American Geophysical Union Geodynamics Series, v. 17, p. 43–68.

Vera, R. H., and Van Schmus, W. R., 1974, Geochronology of some Precambrian rocks of the southern Front Range, Colorado: Geological Society of America Bulletin, v. 85, p. 77–82.

Zartman, R. E., 1964, A geochronologic study of the Lone Grove pluton from the Llano Uplift, Texas: Journal of Petrology, v. 5, p. 359–408.

——, 1965, Rubidium-strontium age of some metamorphic rocks from the Llano Uplift, Texas: Journal of Petrology, v. 6, p. 28–36.

MANUSCRIPT ACCEPTED BY THE SOCIETY SEPTEMBER 26, 1988

Printed in U.S.A.

Geological Society of America
Special Paper 235
1989

Proterozoic geology of the Needle Mountains; A summary

Barbara J. Tewksbury
Department of Geology, Hamilton College, Clinton, New York 13323

ABSTRACT

The Early and Middle Proterozoic rocks in the Needle Mountains include three distinct rock sequences: (1) multiply deformed bimodal metavolcanic rocks, related sedimentary rocks, and plutonic units, all metamorphosed to medium grade; (2) multiply deformed clastic sedimentary rocks metamorphosed to low grade; and (3) weakly foliated to unfoliated plutonic rocks. Although many important questions remain unresolved, work by a number of individuals suggests the following history. Mafic and silicic volcanics, sediments, and intrusives of the Irving Formation and Twilight Gneiss accumulated at least 1,760 Ma and were multiply deformed and metamorphosed shortly after accumulation. Small post-tectonic plutons of the Tenmile and Bakers Bridge Granites invaded the metavolcanic sequence between 1,680 and 1,700 Ma. The timing of deposition of the conglomerates, clean quartzites, and pelites of the Vallecito Conglomerate and the Uncompahgre Formation with respect to events in the metavolcanic sequence is poorly constrained, largely because shear zones lie along crucial contacts. Recent work suggests that much if not all of this clastic sedimentary sequence may be parautochthonous and was probably deposited sometime after 1,680 to 1,700 Ma.

Despite uncertainties in timing of deposition, it is clear that a second event of deformation and metamorphism affected the region, producing polyphase features in both the metavolcanic sequence and the clastic sedimentary sequence, as well as a weak foliation in the Tenmile Granite. The Eolus Granite and related intrusions, ranging in age from 1,320 to 1,450 Ma, crosscut all structural features in the region and provide a minimum age limit for the latest deformation in the region. This younger event is recorded nowhere else in Colorado but may correlate with events in northern New Mexico.

INTRODUCTION

Proterozoic rocks ranging in age from 1,780 to 1,320 Ma are exposed over an area of about 1,000 km^2 in the Needle Mountains region of southwestern Colorado (Fig. 1). These rocks contain a sequence of metavolcanic and associated plutonic rocks, as well as exposures of ca. 1,400 Ma anorogenic granites. They also contain a spectacular sequence of clastic metasedimentary rocks including metaconglomerate, quartzite, and metapelite.

Understanding Proterozoic rocks in the Needle Mountains is important for two reasons. First, these units are the only ones in Colorado that show unequivocal evidence of deformation between 1,680 and 1,450 Ma. Second, controversy exists over whether the metavolcanic sequence forms the basement in the Needle Mountains or whether portions of the metasedimentary sequence are, in fact, the oldest units in the area. Establishing this sequence is important in providing constraints on regional models for the history of the Proterozoic margin of North America.

The purpose of this chapter is to summarize the current level of understanding of Proterozoic events in the Needle Mountains region, and to place these events in a general regional context. This summary necessarily leans heavily on work done by a number of individuals: Cross and others (1905); Barker (1969); Burns and others (1980); Gibson and Simpson (1986); Harris and others (1987); Gonzales and Ruiz (1982); Tewksbury (1985a, 1985b, 1986a, and 1986b); and ongoing studies by C. Harris, J. Ellingson, R. Gibson, D. Gonzales, and B. Tewksbury.

Tewksbury, B. J., 1989, Proterozoic geology of the Needle Mountains; A summary, *in* Grambling, J. A., and Tewksbury, B. J., Proterozoic geology of the southern Rocky Mountains: Boulder, Colorado, Geological Society of America Special Paper 235.

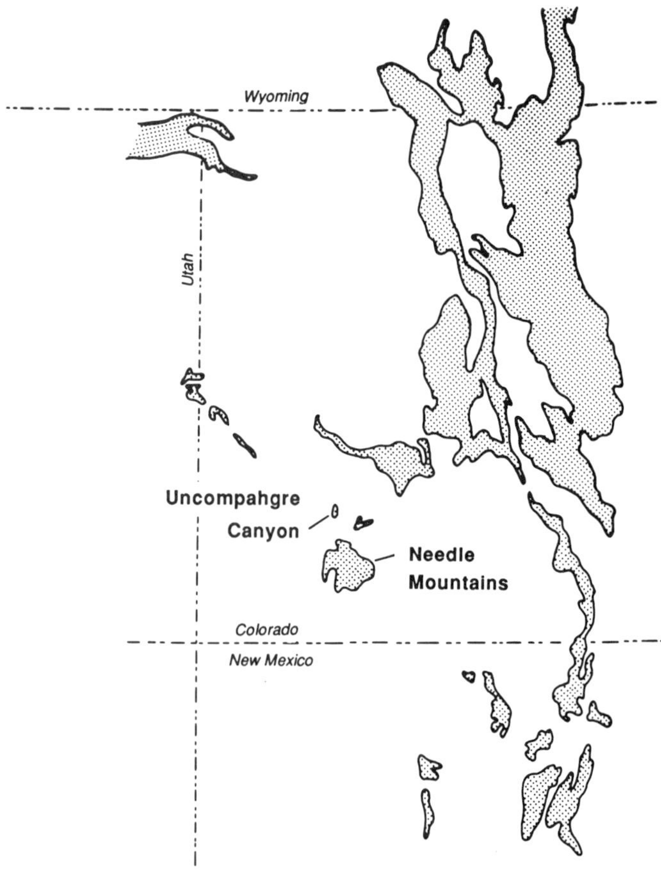

Figure 1. Exposures of Precambrian bedrock in Colorado and vicinity.

LITHOLOGIC UNITS

Proterozoic rocks of the Needle Mountains can be grouped into three sequences on the basis of protoliths: a metavolcanic sequence, a metasedimentary sequence, and a plutonic sequence (Fig. 2). Both the metavolcanic and metasedimentary sequences, and some members of the plutonic sequence, have undergone deformation and metamorphism. The Irving Formation and Twilight Gneiss compose the metavolcanic sequence and include not only metavolcanic rocks, but also related metamorphosed volcaniclastic sediments and plutons. The metasedimentary sequence consists of the Uncompahgre Formation and Vallecito Conglomerate, with protoliths including conglomerate, quartz sandstone, and mudstone. The plutonic sequence consists of small plutons in the Tenmile Granite, Bakers Bridge Granite, Electra Lake Gabbro, and Trimble Granite, and the large composite batholith of the Eolus Granite.

Metavolcanic sequence

The Irving Formation includes amphibolite, biotite- and epidote-bearing schist and gneiss, greenstone, quartzofeldspathic gneiss, feldspathic quartzite, metagraywacke, banded iron formation, and rare calc-silicate rock. Much of the sequence appears to have been volcanic, with metabasalts containing rare well-preserved pillow structures, and silicic metavolcanics preserving outlines of relict phenocrysts. Related plutonic and volcaniclastic rocks provided protoliths for the remaining lithologies. Volcanic portions of the Irving Formation are dominantly bimodal, but Barker (1969) reports the presence of some andesites. Chemical data indicate that at least some of the basalts are normative olivine tholeiites (Barker, 1969). Detailed metamorphic studies have not been done on the Irving, but mineral assemblages are consistent with amphibolite-facies metamorphism partly retrograded to greenschist assemblages (Barker, 1969).

The Twilight Gneiss is a trondhjemitic to quartz monzonitic gneiss with variable percentages of amphibolite. It is unclear whether the Twilight Gneiss was originally a sequence of interlayered silicic and basaltic volcanics or a basaltic volcanic pile invaded by trondhjemitic magmas (Barker, 1969).

U/Pb zircon and Rb/Sr geochronology indicates that the Twilight Gneiss accumulated and was deformed and metamorphosed in the interval 1,760 to 1,780 Ma* (Silver and Barker, 1968; Barker and others, 1969). The Irving Formation in the West Needle Mountains has not been dated radiometrically but has been considered coeval with the Twilight Gneiss (Barker, 1969). Whether this is reasonable depends on the nature of the contact between the Twilight and the Irving, and field evidence suggests that the contact is probably not as simple as presented in Figure 2. Both Tewksbury (1985a) and Harris and others (1987) have noted that the Irving contains a well-developed transposed foliation with intrafolial folds. The foliation has a persistent east-west strike and very steep dip in portions of the West Needle Mountains near the contact, and these attitudes are nearly perpendicular to the mapped Irving–Twilight contact. The presence of a transposed foliation perpendicular to a presumably older, straight contact suggests that the relationship between the Irving and the Twilight in the West Needle Mountains may in part be structural.

Metasedimentary sequence

The Uncompahgre Formation is exposed both in the Needle Mountains and in Uncompahgre Canyon, approximately 20 km to the north. Both outcrop areas contain mappable units of quartzite and metapelite (Fig. 2). The former consist largely of quartzite, with rare pebbly layers, thin pelitic layers, and coarse conglomeratic lenses interspersed. Conglomeratic layers are confined to the lower portions of the section in the Animas Canyon/West Needle Mountains area. Clasts include quartz pebbles, hematitic chert, and jasper. Although these conglomerates lie directly on the Irving Formation, they lack clasts of lithologies typical of the Irving Formation and do not contain mafic matrix material.

*Radiometric dates have been recalculated to reflect decay constants as published in Steiger and Jaeger (1977).

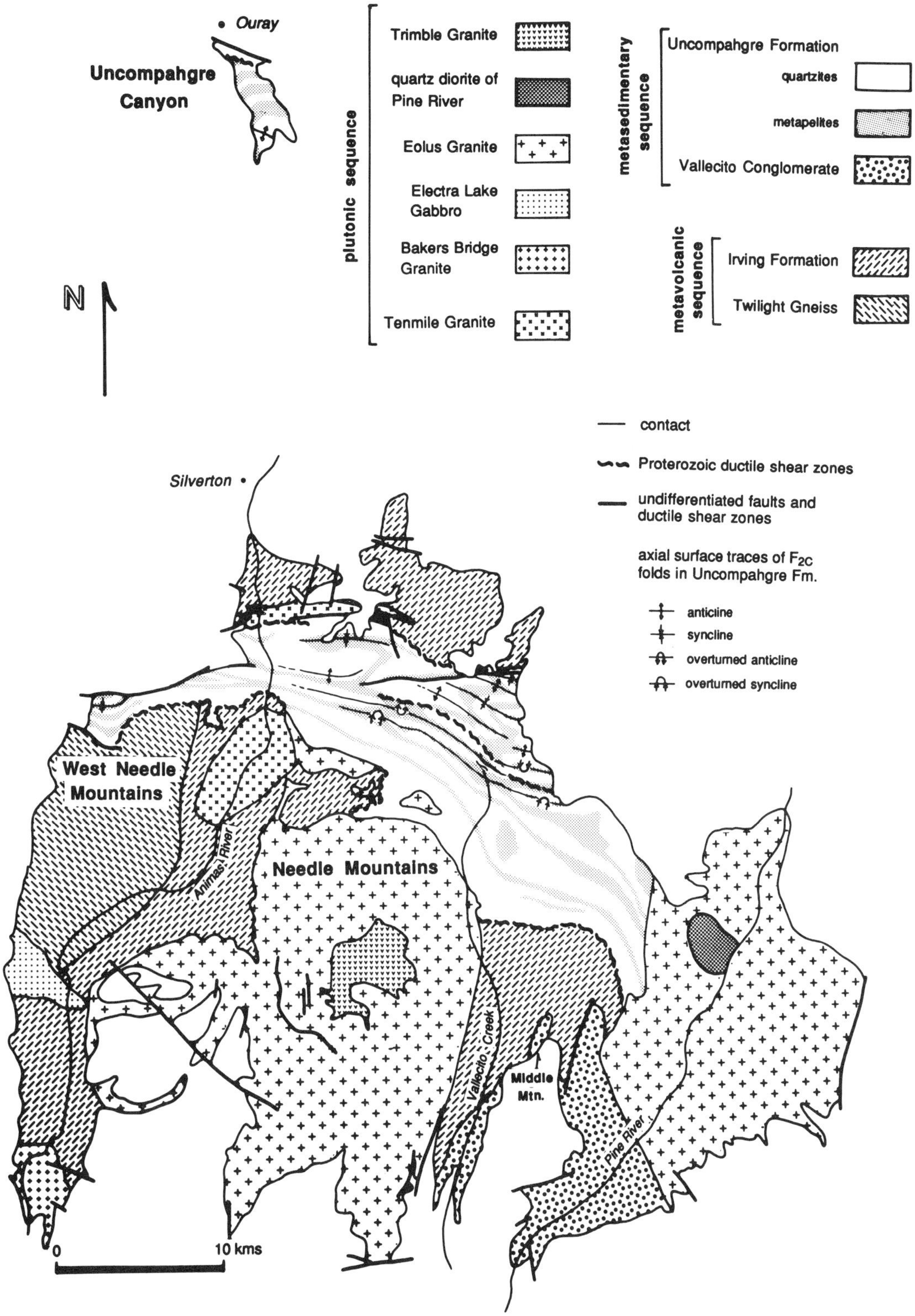

Figure 2. General geology of the Needle Mountains region. Geology of the Uncompahgre Formation in the Needle Mountains from Tewksbury (1985a and 1986a); geology of the Uncompahgre Formation, Irving Formation, and Twilight Gneiss in the West Needle Mountains from Harris and others (1987); geology of the Uncompahgre Formation in Uncompahgre Canyon from Tewksbury (1985b). The remainder is modified from Barker (1969) and unpublished work by Tewksbury.

The metapelite map units contain pelitic material interlayered with psammitic material on scales from meters to millimeters. Pelitic layers are predominantly slate and phyllite; psammitic layers range from fine-grained quartzites to micaceous quartzite and metasiltstone. One sequence of metapelite in Animas Canyon is unusually aluminous, containing many layers that consist almost entirely of andalusite. High aluminum concentrations in these layers may be due to the presence of kaolinitic clay drapes in the protolith, reflecting intense chemical weathering of the source region, as suggested by Soegaard and Eriksson (1986) for aluminous rocks in northern New Mexico. Rare calc-silicate assemblages occur as lenses in some of the metapelite units and contain actinolite-grossularite-calcite-clinozoisite-quartz-sphene.

Texture in Uncompahgre metapelites changes from very fine-grained and sparsely porphyroblastic in Uncompahgre Canyon and the West Needle Mountains to medium-grained and coarsely porphyroblastic in the southeastern Needle Mountains. Although detailed metamorphic studies have not been done, syntectonic mineral assemblages are consistent with greenschist-facies metamorphism. The most common assemblage is quartz-sericite-carbonaceous material ± chlorite ± andalusite. Appearance of post-tectonic porphyroblasts accompanies the southeastward change from a foliated to a granoblastic fabric. Southeastward from the central part of the Needle Mountains, chloritoid, biotite, garnet, and staurolite successively appear, and Barker (1969) reports plagioclase, microcline, and sillimanite in the most southeasterly portions of the Uncompahgre Formation. It is unclear whether a progressive increase in metamorphic grade toward the southeast was a feature of both syntectonic and post-tectonic metamorphism or only of post-tectonic recrystallization.

The Vallecito Conglomerate is exposed in three physically separated areas: along Vallecito Creek (including the type section), in the Pine River area, and at Middle Mountain (Fig. 2). The Vallecito Conglomerate consists in large part of moderately well-sorted pebble conglomerate, with lesser amounts of poorly sorted boulder conglomerate, quartzite, and foliated conglomerate (Ethridge and others, 1984). Foliated conglomerate is restricted to exposures on Middle Mountain.

Unlike conglomerates in the Uncompahgre Formation, some of the conglomerates in the Vallecito contain clasts of lithologies characteristic of the adjacent Irving Formation, e.g., amphibolite, feldspathic gneiss, banded iron formation, and muscovite schist. Clasts in the foliated conglomerates of Middle Mountain include nearly all of these rock types, set in a matrix containing abundant mafic material. Other portions of the Vallecito are dominated by vein quartz, quartzite, jasper, and hematitic chert clasts similar to those in the Uncompahgre. Andalusite has been reported from the Vallecito (Burns and others, 1980). No work has been done to determine metamorphic grade, and no radiometric dates are available on the Vallecito Conglomerate.

Plutonic sequence

The oldest of the plutonic rocks is the Tenmile Granite. In addition to the pluton shown in Figure 2, related smaller bodies and dikes intrude the Irving Formation in the West Needle Mountains both north and south of the Uncompahgre Formation, but nowhere does the Tenmile Granite intrude the Uncompahgre Formation. The Tenmile Granite is a foliated quartz monzonite that gives variable ages. U/Pb zircon dates of about 1,700 Ma are reported by Silver and Barker (1968); Rb/Sr whole-rock ages range from 1,680 to 1,640 Ma in different plutons (Bickford and others, 1969). The susceptibility of Rb/Sr systematics to resetting suggests that the U/Pb zircon ages may be the more reliable.

The Bakers Bridge Granite is a stock of massive granite and alaskite intruded into the southwestern part of the Irving Formation along the Animas River. U/Pb zircon dates of approximately 1,700 Ma (Silver and Barker, 1968), and Rb/Sr whole-rock ages of approximately 1,670 Ma (Bickford and others, 1969) suggest that this body is broadly coeval with the Tenmile Granite.

The Eolus Granite is a complex batholithic body ranging in composition from monzodiorite to granite. Details of the Eolus Granite are presented in Collier (this volume). The Eolus Granite intrudes deformed units of the Irving Formation, Uncompahgre Formation, and Vallecito Conglomerate. U/Pb zircon dates and Rb/Sr whole-rock dates fall between 1,430 and 1,450 Ma (Silver and Barker, 1968; Bickford and others, 1969). The Electra Lake Gabbro, a complex of gabbros and related rocks intruding the Irving Formation and Twilight Gneiss, is the same age as the Eolus Granite (Silver and Barker, 1968; Bickford and others, 1969) and may be genetically related. The quartz diorite of Pine River is probably an early phase of the batholith (Barker, 1969).

DEFORMATION

Structural data are available for only some of the deformed sequences. The Uncompahgre Formation, Twilight Gneiss, and Irving Formation have been extensively studied in the West Needle Mountains (Gibson and Simpson, 1986; Harris and others, 1986 and 1987). Structural analyses have also been published for the Uncompahgre Formation in the Needle Mountains and in the type section at Uncompahgre Canyon (Tewksbury, 1985a and 1986a). The remaining units in the Needle Mountains have seen little structural study, but limited data are available for the Vallecito Conglomerate (Burns and others, 1980) and portions of the Vallecito Conglomerate and Irving Formation at Middle Mountain (Tewksbury, unpublished data).

The metavolcanic sequence experienced two polyphase deformational events during the Proterozoic, while the metasedimentary sequence experienced only one. The terminology used by Harris and others (1987) to designate phases of these events has been adopted in this paper. Figure 3 provides a chart for correlating this terminology with that in previous papers.

Metavolcanic sequence in the West Needle Mountains

Work by Gibson and Simpson (1986) and Harris and others (1987) suggests four phases of deformation in the Irving Formation and Twilight Gneiss in the West Needle Mountains (Fig. 3).

D_{1B} and D_{2B} pre-date deformation in both the adjacent Uncompahgre Formation and the Tenmile Granite. D_{3B} and D_{4B} are related to deformation seen in both the Uncompahgre and the Tenmile Granite. D_{1B} produced isoclinal F_{1B} folds with related axial planar foliation. D_{2B} coaxially refolded F_{1B} folds and transposed both S_0 and S_{1B} into parallelism with S_{2B}. F_{1B} and F_{2B} folds are upright and have subhorizontal hinges, and all three foliations generally dip steeply and strike west-northwest to north-northeast. D_{3B} produced open east-trending folds with gently plunging to vertical hinges. The related S_{3B} foliation is also present in the Tenmile Granite. F_{4B} folds plunge steeply and appear to be related to conjugate strike-slip shear zones (Gibson and Simpson, 1986; Harris and others, 1987).

Uncompahgre Formation

Rocks of the Uncompahgre Formation are multiply folded, and recent workers have all suggested that folding was intimately related to shearing, both within the sequence and between Uncompahgre rocks and the adjacent metavolcanic sequence (Tewksbury, 1985a; Harris and others, 1986 and 1987; Gibson and Simpson, 1986).

Two major coaxial deformational events and one minor cross-folding event affected the Needle Mountains portion of the Uncompahgre Formation (Fig. 3). D_{1C} produced isoclinal F_{1C} folds and a pervasively developed S_{1C} slaty cleavage typically parallel or at a low angle to S_0. Abundant evidence for D_{1C} exists at the mesoscopic and microscopic scales, but no macroscopic folding has been recognized. Although transposition of S_0 during D_{1C} was extensive, it appears to have been localized within the metapelite units. The entire stratigraphic sequence shows no evidence of transposition as a whole (Tewksbury, 1986a).

D_{2C} produced a set of macroscopic folds of both S_0 and S_{1C} (Fig. 2). The attitude of these folds is 10,S75E in the southern portion of the Uncompahgre Formation in the Needle Mountains and 35,S75E in the northern portion (Fig. 4). The macroscopic folds range from open to isoclinal; most are asymmetric, and those that are overturned verge south-southwest. Mesoscopic and microscopic structures include well-developed conjugate F_{2C} folds with related axial plane crenulation cleavage (S_{2C}). F_{2C} mesoscopic folds are typically open and coaxial with F_{1C} folds in the southern portion of the Uncompahgre Formation in the Needle Mountains but not in the northern portion. Conjugate D_{2C} mesoscopic structures appear to have developed late in the D_{2C} phase, after much of the F_{2C} macroscopic folding was complete and after thrusting cut the D_{2C} macroscopic folds and rotated the northern portion of the Uncompahgre in the Needle Mountains relative to the southern portion (Tewksbury, 1986a).

D_{3C} structures are only weakly developed and have attitudes nearly perpendicular to both D_{1C} and D_{2C}. Where measurable, F_{3C} crenulation cleavages are vertical and strike northeast-southwest, directly across the main structural grain of the belt.

Work by Harris and others (1986 and 1987) reveals similar patterns in the West Needle Mountains. They report D_{1C} deformation zones containing bedding-parallel S_{1C} and tight to isoclinal F_{1C} mesoscopic folds, and S_{1C} slaty cleavage related to shearing within the pelitic layers. D_{2C} produced open to tight, east-trending F_{2C} folds and a related S_{2C} spaced cleavage. In the southwestern part of the West Needle Mountains, late, steeply plunging F_{3C} cross folds refold all structures.

Although metamorphic grade and intensity of deformation are higher in the Uncompahgre Formation in the Needle Mountains than in the type section in Uncompahgre Canyon, the two are strikingly similar in deformational style and history. In the Uncompahgre Formation type section, D_{1C} appears as a pervasive S_{1C} slaty cleavage developed parallel or at a low angle to S_0.

deformation	brackets on timing	phase (designation of Harris et al., 1987)	phase (designation of Tewksbury, 1985a, 1986a)	structures
D_{BC} - *affecting Irving Fm., Twilight Gneiss, & Uncompahgre Fm.*	1680-1450 Ma	D_{3C}/D_{4B}	D_3	**Uncompahgre Fm.** - sporadic development of macroscopic & mesoscopic open F_{3C} folds with axial planar S_{3C} spaced cleavage; absent in Uncompahgre Canyon **Irving Fm./Twilight Gneiss** - mesoscopic F_{4B} folds
		D_{2C}/D_{3B}	D_2	**Uncompahgre Fm.** - macroscopic and mesoscopic tight to open F_{2C} folds; axial planar S_{2C} spaced cleavage; conjugate structures in the Needle Mountains portion **Irving Fm./Twilight Gneiss** - macroscopic tight to open F_{3B} folds with axial planar S_{3B} foliation
		D_{1C}	D_1	**Uncompahgre Fm.** - mesoscopic tight to isoclinal F_{1C} folds, S_{1C} axial plane slaty cleavage
D_B - *affecting Irving Fm. & Twilight Gneiss*	1740-1700 Ma	D_{2B}		isoclinal folds coaxial with F_{1B}, weak axial planar S_{2B}
		D_{1B}		isoclinal folds, axial planar S_{1B}

Figure 3. Characteristics of deformational events in the Needle Mountains, using the terminology of Harris and others (1987). D_B refers to deformation in the metavolcanic sequence (the "basement" of Harris and others, 1987); D_{BC} refers to deformation in both the metavolcanic and metasedimentary sequences (the "basement" and "cover").

F_{1C} mesoscopic folds occur as intrafolial folds near contacts between metapelite and quartzite. D_{2C} folded both S_0 and S_{1C} around open macroscopic structures (Fig. 2). Related S_{2C} crenulation cleavage does not show the conjugate geometry common in the Needle Mountains.

D_{2C} structures are pervasive in the Needle Mountains and less well developed in the type section, but geometries are similar. F_{2C} fold orientations in the type section trend 40,N75W, and axial surfaces dip vertically or steeply northeast (Fig. 4). As in the Needle Mountains, D_{2C} folds in the type section are cut by north-dipping thrusts. Unlike the Needle Mountains portion, the type section shows no evidence of a D_{3C} event (Tewksbury, 1985a).

Vallecito Conglomerate

Foliated metaconglomerates at Middle Mountain display a persistent S_1 foliation at high angles to S_0 and a prominent stretching lineation trending south-southwest and plunging 40 to 50° (Fig. 4). The same foliations and lineations may be traced into the adjacent Irving Formation below the unconformity. Because no detailed structural studies are available for the Irving Formation, it is not known whether this foliation is the oldest tectonite fabric in the Irving as well as in the Vallecito.

The πS_0 plots of data extracted from Burns and others (1980) for the remainder of the Vallecito Conglomerate indicate moderately tight folds lacking good cylindricity even when data are plotted separately for the Vallecito Creek section and the Pine River section (Fig. 4). Although data are scattered, the dominant north-northeast–south-southwest trends are very similar to those at Middle Mountain.

How the fabrics in the Vallecito Conglomerate and this portion of the Irving Formation correlate with those in the Uncompahgre and the Irving Formations in the areas studied to the northwest is not known. It is clear, however, that the prominent east-west to east-southeast–west-northwest trends in the Uncompahgre and Irving Formation in the central and West Needle Mountains are nearly at right angles to the north-northeast–south-southwest trends in both the Irving Formation at Middle Mountain and the Vallecito Conglomerate as a whole.

STRATIGRAPHIC RELATIONSHIPS

Stratigraphic relationships in the region are controversial, largely because of the paucity of radiometric ages and the structural, not stratigraphic, nature of many of the critical contacts. The crucial arguments concern relationships among the Vallecito Conglomerate, Uncompahgre Formation, and Irving Formation. If either the Uncompahgre Formation or the Vallecito Conglomerate is older than the Irving Formation, crustal evolution models (as suggested, for example, by Soegaard and Eriksson, 1986) involving development of transitional volcano-plutonic crust progressively stabilized and blanketed with a sedimentary succession must be seriously re-evaluated. The following section first considers the relative ages of: (1) the Vallecito Conglomerate and the Irving Formation, (2) the Uncompahgre and Irving Formations, and (3) the Uncompahgre Formation and Vallecito Conglomerate.

Vallecito Conglomerate/Irving Formation

The Vallecito Conglomerate has been interpreted as both older (Barker, 1969; Ellingson and others, 1982) and younger (Cross and others, 1905; Burns and others, 1980; Burns and Wobus, 1983) than the Irving Formation. Cross bedding in the Vallecito facing toward the Irving near the contact in the valley of Vallecito Creek led Barker (1969) to suggest that the Vallecito is older, but the contact itself is so poorly exposed that the possibility of a tectonic contact cannot be ruled out.

An untectonized contact between the two units may be exposed at Middle Mountain (Fig. 2). There, Burns and others (1980) and Tewksbury (unpublished) have recognized that portions of the contact are depositional. The contact is well exposed and shows no evidence of shear fabrics. Tops in the metaconglomerate clearly face away from the contact, suggesting that the metaconglomerate lies stratigraphically above the Irving. The presence of abundant mafic detritus in both clasts and matrix of the metaconglomerate, including jasper clasts with a jasper source immediately below the contact in the Irving, is consistent with this interpretation.

Although the metaconglomerates at Middle Mountain are somewhat different from the rest of the Vallecito Conglomerate, both in lithology and intensity of fabric development, they probably represent a locally derived, more poorly sorted and less mature (and perhaps the oldest) portion of the Vallecito, rather than a different formation entirely. If the relationship at Middle Mountain is generalized to the remainder of the Vallecito Conglomerate, the Vallecito Conglomerate cannot be the oldest unit in the stratigraphic column in the Needle Mountains.

Uncompahgre Formation/Irving Formation

Although the contact between the Irving Formation/Twilight Gneiss and the Uncompahgre Formation is very well exposed in a number of places, the nature of the contact has been debated. Barker (1969) mapped all contacts between the Uncompahgre Formation and the metavolcanic sequence as depositional contacts. Since that time, work by Tewksbury (1985a) has indicated that the contact is a shear zone ranging up to 12 m in thickness and involving lithologies derived both from the Uncompahgre and Irving Formations. Displacement magnitudes along the contacts have not been determined, and it is unclear whether the Uncompahgre sedimentary sequence is allochthonous or parautochthonous. Tewksbury (1985a) suggested that the Uncompahgre Formation is allochthonous and could be as old or older than the Irving Formation, provided that the Uncompahgre protolith had been far enough away to have escaped both the early deformation and higher grade metamorphism that affected

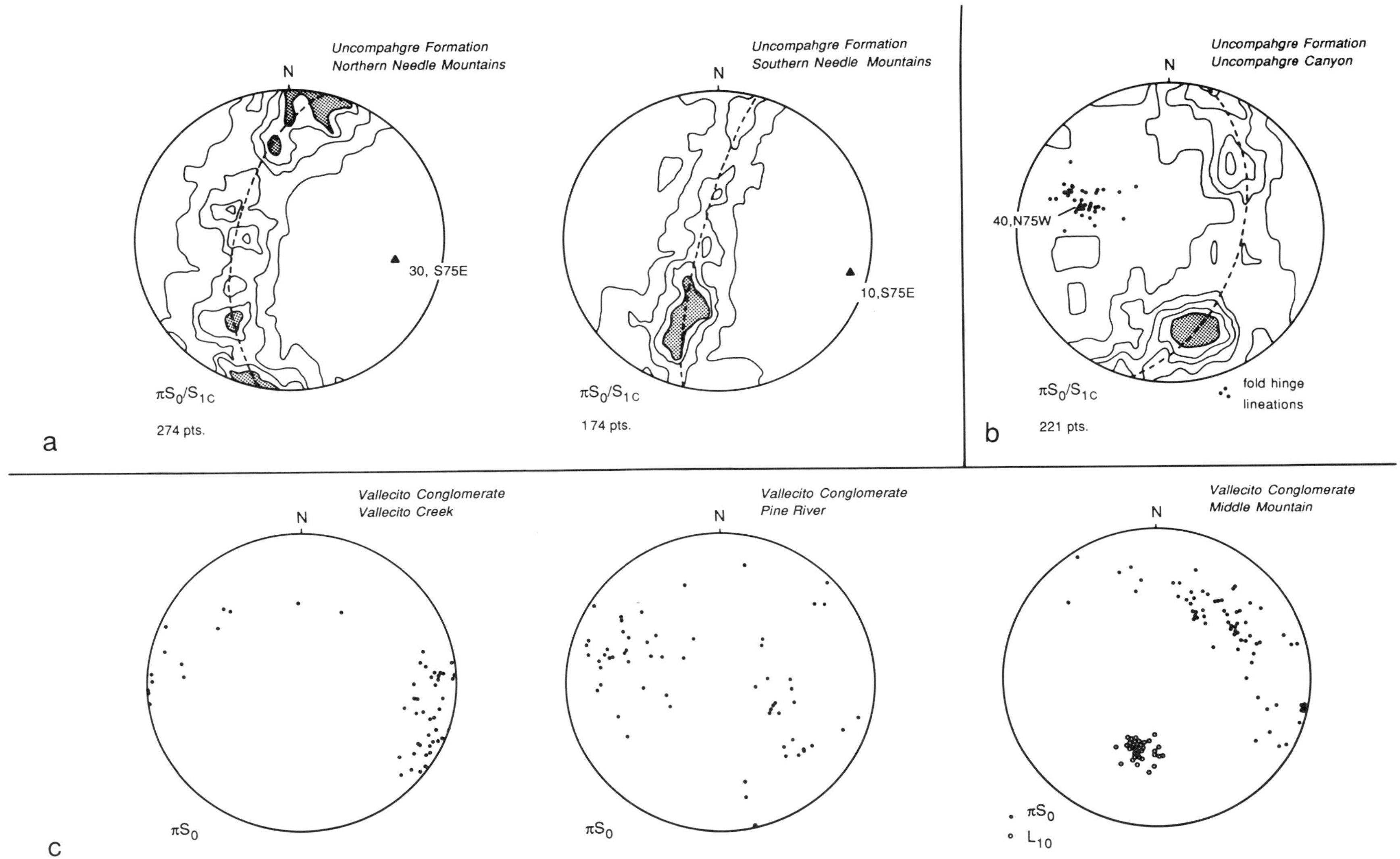

Figure 4. πS_0 diagrams for the Uncompahgre Formation and Vallecito Conglomerate. (a) $\pi S_0/S_{1C}$ for the northern and southern portions of the Uncompahgre Formation in the Needle Mountains. Because S_0 and S_{1C} are parallel in the Uncompahgre Formation, these diagrams are also πS_{1C} plots and reflect the geometry of the F_{2C} macroscopic folds shown in map view in Figure 2. (b) $\pi S_0/S_{1C}$ diagrams for the Uncompahgre Formation in Uncompahgre Canyon. (c) πS_0 diagrams for the Vallecito, Pine River, and Middle Mountain portions of the Vallecito Conglomerate. Data for the Vallecito Creek and Pine River areas were extracted from maps published in Burns and others (1980). Middle Mountain data were collected in the field. Contours in all diagrams are at .6, 3, 4, and 6 percent per 1 percent area.

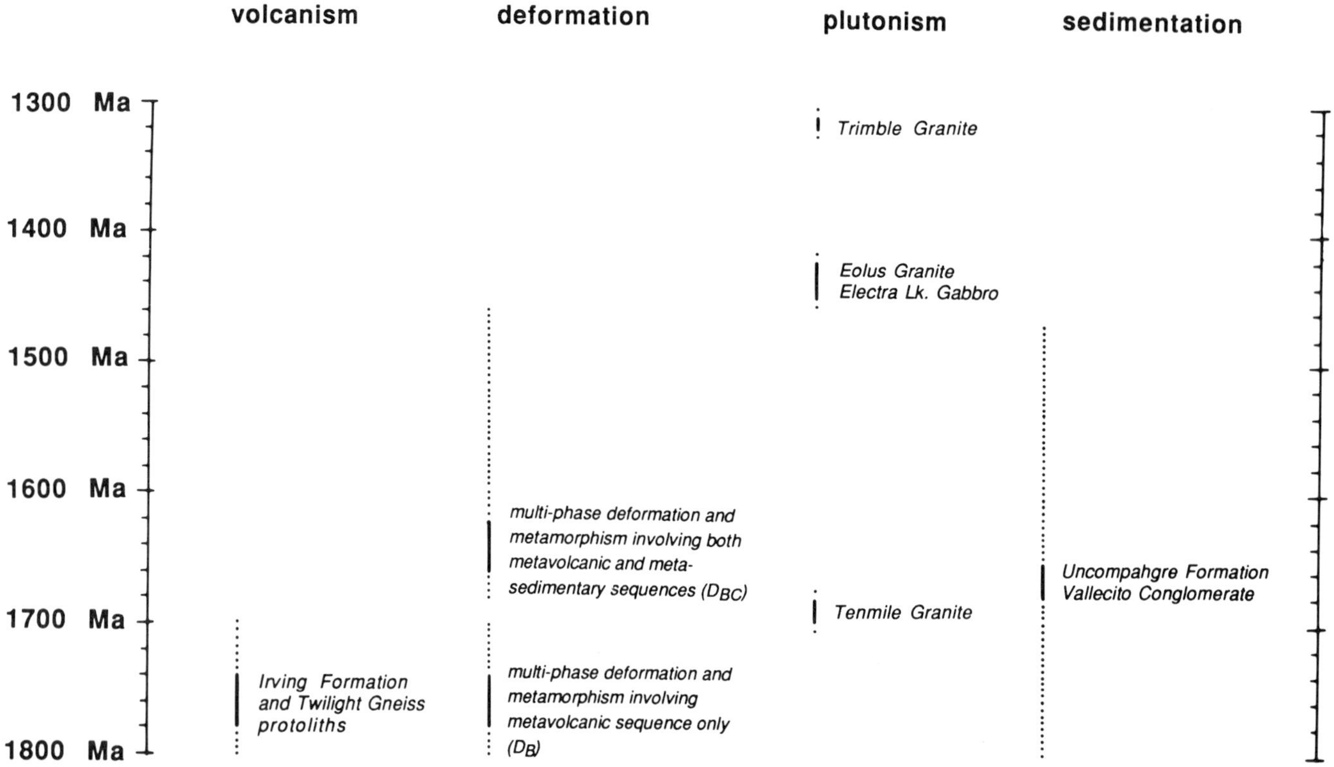

Figure 5. Summary of the timing of events in the Needle Mountains. Solid lines indicate probable time ranges on the basis of the most plausible current models. Dotted lines reflect the total possible time range for each event. "Volcanism" includes related plutonic activity and volcaniclastic sedimentation. "Plutonism" includes only post-tectonic plutons. "Sedimentation" includes only stable margin sedimentation.

the Irving. Tewksbury (1985a) envisioned the Uncompahgre Formation as an allochthonous sheet emplaced along a sole thrust of variable dip, with imbricate internal north-dipping thrusts and related folds, and a northern thrust contact of similar attitude, all produced by protracted southward net tectonic transport during D_{1C}-D_{3C}.

Harris and others (1987) have recently argued that the Uncompahgre Formation is parautochthonous and must be younger than the Irving. They have shown that, although the Uncompahgre Formation moved south over the metavolcanic sequence along the southern contact during later phases of D_{BC} deformation, it moved up relative to the metavolcanic sequence along the northern contact. They propose that these relationships are better explained by complex slip along an unconformity during shortening of a basement/cover sequence than by major south-directed thrusting of an allochthon. They envision limited north-directed, thin-skinned thrusting (D_{1C}), followed by folding of both basement and cover into a complex cusp-shaped fold and shear zone system (D_{2C} and D_{3C}; Harris and others, 1987).

Plutons of the Tenmile Granite are truncated by the shear zones bounding the Uncompahgre Formation. The Tenmile contains a D_{BC} fabric, indicating clearly that deformation in the Uncompahgre Formation postdates intrusion of the Tenmile Granite at 1,680 to 1,700 Ma. Although the time of deposition of the Uncompahgre Formation has not been clearly established, Harris and others' (1987) conclusion that the Uncompahgre Formation is parautochthonous suggests that deposition of the Uncompahgre Formation probably also postdates intrusion of the Tenmile Granite. Confirmation must await either radiometric dates or determination of displacement magnitudes along the shear zones.

Vallecito Conglomerate/Uncompahgre Formation

Although the Vallecito Conglomerate and Uncompahgre Formation share no common contacts, they have been viewed by many workers (e.g., Cross and others, 1905; Harris and others, 1987) as part of the same sedimentary package on the basis of similarities between much of the Vallecito Conglomerate and the conglomeratic horizons in the Uncompahgre Formation. Evidence presented in earlier sections suggests that deposition of both sequences quite likely postdated the metavolcanic sequence, supporting the possibility that they are related. The disparity in attitudes of dominant structural features, however, makes it more difficult to correlate the histories of the two units.

DISCUSSION

Figure 5 summarizes Proterozoic events in the Needle Mountains region. Timing of many of these events is poorly constrained because of the paucity of radiometric dates and prevalence of structural contacts in the region, and possible ranges for events are shown with dotted lines. The following section describes a plausible history based on current models.

Earliest events recorded in the Needle Mountains occurred between 1,780 and 1,760 Ma and involved accumulation, metamorphism, and deformation of mafic to silicic volcanic rocks, intrusives, and volcaniclastic sediments of the Irving Formation and Twilight Gneiss. Subsequent deformation was polyphase and involved repeated isoclinal folding and amphibolite-facies metamorphism. Deformation ceased before intrusion of the Tenmile Granite at 1,680 to 1,700 Ma.

The timing of deposition of the Vallecito Conglomerate and the Uncompahgre Formation is not well constrained, because possible age ranges depend heavily on models invoked both for the nature of contacts and the style of deformation. Arguments advanced in previous sections indicate the strong likelihood that the metavolcanic sequence formed the basement upon which both the Vallecito Conglomerate and the Uncompahgre Formation were deposited sometime after 1,680 to 1,700 Ma, but before intrusion of the Eolus Granite at 1,450 to 1,430 Ma. If this is the case, the metasedimentary sequence in the Needle Mountains is unique in Colorado. Correlatives may exist southward in New Mexico, where rocks of similar character in the Ortega Group were deposited after 1,700 Ma (Soegaard and Eriksson, 1986).

The second event of deformation and metamorphism involved both the metavolcanic sequence and the overlying sediments in a major event of north-south crustal shortening that occurred sometime between 1,680 Ma and intrusion of the post-tectonic Eolus Granite and related plutons between 1,450 and 1,320 Ma. This event has not been recognized elsewhere in Colorado, but may correlate with events in New Mexico and Arizona.

ACKNOWLEDGMENTS

I am indebted to the careful work of the many geologists who have worked in the Needle Mountains over the years, but I would particularly like to acknowledge the fine recent work of Richard Gibson and Charles Harris. I would also like to thank C. Simpson, S. Mosher, J. Ellingson, and J. Grambling for thorough and thoughtful reviews that resulted in many needed improvements in the manuscript.

REFERENCES CITED

Barker, F., 1969, Precambrian geology of the Needle Mountains, southwestern Colorado: U.S. Geological Survey Professional Paper 644–A, 35 p.

Barker, F., Peterman, A., and Hildreth, R. A., 1969, A rubidium-strontium study of the Twilight Gneiss, West Needle Mountains, Colorado: Contributions to Mineralogy and Petrology, v. 23, p. 271–282.

Bickford, M. E., Wetherill, G. W., Barker, F., and Lee-Hu, C-N., 1969, Precambrian Rb/Sr chronology in the Needle Mountains, southwestern Colorado: Journal of Geophysical Research, v. 74, p. 1660–1676.

Burns, L. K., and Wobus, R. A., 1983, Correlations and revisions of Precambrian stratigraphy, Needle Mountains, southwest Colorado, and Tusas Mountains, north-central New Mexico: Geological Society of America Abstracts with Programs, v. 15, p. 424.

Burns, L. K., Ethridge, F. G., and Tyler, N., 1980, Geology and uranium evaluation of the Precambrian quartz-pebble conglomerates of the Needle Mountains, southwestern Colorado: U.S. Department of Energy Report GJBK–118–80, 161 p.

Cross, W., Howe, E., Irving, J. D., and Emmons, W., 1905, Description of the Needle Mountains Quadrangle, Colorado: U.S. Geological Survey Atlas Folio 131, 14 p.

Ellingson, J. A., Gonzales, D. A., and Ruiz, L. N., 1982, The Irving Formation and the Proterozoic sequence in the Needle Mountains, southwestern Colorado: Geological Society of America Abstracts with Programs, v. 14, p. 311.

Ethridge, F. G., Tyler, N., and Burns, L. K., 1984, Sedimentology of a Precambrian quartz-pebble conglomerate, southwest Colorado, *in* Koster, E. H., and Steel, R. J., eds., Sedimentology of gravels and conglomerates: Canadian Society of Petroleum Geologists Memoir 10, p. 165–174.

Gibson, R. G., and Simpson, C., 1986, Kinematic analysis and implications of Proterozoic strike-slip shear zones, West Needle Mountains, Colorado: Geological Society of America Abstracts with Programs, v. 18, p. 613.

Gonzales, D., and Ruiz, L., 1982, Proterozoic geology of the Middle Mountain area, Needle Mountains, Colorado: Geological Society of America Abstracts with Programs, v. 14, p. 312.

Harris, C. W., Gibson, R. G., Simpson, C., and Eriksson, K. A., 1986, Early to Middle Proterozoic polyphase deformation, West Needle Mountains, southwest Colorado; Preliminary results: Geological Society of America Abstracts with Programs, v. 18, p. 360.

—— , 1987, Proterozoic cuspate basement-cover structure, Needle Mountains, Colorado: Geology, v. 15, p. 950–953.

Silver, L. T., and Barker, F., 1968, Geochronology of Precambrian rocks of the Needle Mountains, southwestern Colorado; Part I, U/Pb zircon results: Geological Society of America Special Paper 115, p. 204–205.

Soegaard, K., and Eriksson, K. A., 1986, Transition from arc volcanism to stable-shelf and subsequent convergent-margin sedimentation in northern New Mexico from 1.76 Ga: Journal of Geology, v. 94, p. 47–66.

Steiger, R. H., and Jaeger, E., 1977, Subcommission on geochronology; Convention on the use of decay constants in geo- and cosmochronology: Earth and Planetary Science Letters, v. 36, p. 359–362.

Tewksbury, B. J., 1985a, Revised interpretation of the age of allochthonous rocks of the Uncompahgre Formation, Needle Mountains, Colorado: Geological Society of America Bulletin, v. 96, p. 224–232.

—— , 1985b, Similarities in deformational style as evidence for correlation between the type section and Needle Mountains portion of the Proterozoic Uncompahgre Formation, southwestern Colorado: Geological Society of America Abstracts with Programs, v. 17, p. 267.

—— , 1986a, Conjugate crenulation cleavages in the Uncompahgre Formation, Needle Mountains, Colorado: Journal of Structural Geology, v. 8, no. 2, p. 145–155.

—— , 1986b, Proterozoic geology of the Needle Mountains, southwest Colorado; Comparisons with the Proterozoic geology of central Colorado, *in* Van Schmus, W. R., ed., International Field Conference on Proterozoic Geology and Geochemistry, Central Colorado, Guidebook: International Geological Correlation Project, p. 124–125.

Manuscript Accepted by the Society September 26, 1988

Printed in U.S.A.

Geological Society of America
Special Paper 235
1989

Mid-Proterozoic postorogenic granites, and associated uranium mineralization of the Needle Mountains, southwestern Colorado

James D. Collier
Department of Geology, Fort Lewis College, Durango, Colorado 81301

ABSTRACT

The Eolus batholith of the Needle Mountains, southwestern Colorado, contains two principal map units, the Eolus and Trimble Granites. The Eolus Granite has been dated at 1,460 Ma; the Trimble Granite formed about 1,350 Ma. Thus, the Eolus and Trimble Granites fall within the time range spanned by two well-documented anorogenic periods in North America (see Bickford and others, this volume).

Although the rocks of the Eolus batholith share many chemical traits with anorogenic, or A-type, granites, they differ by having a wider, and on the average lower, range of silica contents, by being calcalkaline rather than alkaline, and by having rare-earth element (REE) patterns more characteristic of synorogenic granites. The Eolus batholith also differs from most anorogenic batholiths by having undergone extensive fractionation, involving both partial-melt fractionation and hornblende-dominated fractional crystallization.

A consequence of this fractionation was progressive enrichment of uranium as the Eolus batholith evolved. In early products of differentiation, uranium is held entirely within relatively nonlabile accessory minerals such as zircon, sphene, apatite, and allanite. In late-stage differentiates, particularly the Trimble Granite, most uranium is found in scattered grains of more readily leachable high-uranium phases such as uraninite and uranothorite. These late-stage differentiates were source rocks for later mineralization.

It is believed that the Eolus batholith formed by melting of an underplated mantle wedge above subducted oceanic plate. Melting may have been initiated by low pressure in an extensional regime, caused by a transition in tectonic style from synorogenic ductile compression to postorogenic brittle extension. Melting was probably localized along preexisting zones of weakness, which would account for synchronous magmatism over a broad transcontinental belt. Differences in distance from a subduction zone, crustal thickness, depth of melting, and other factors may account for the chemical dissimilarities between the postorogenic Eolus batholith and other "anorogenic" batholiths of approximately the same age.

INTRODUCTION

The most widespread outcropping of Proterozoic plutonic rocks in southwest Colorado occurs as two composite intrusions of batholithic proportions in the Needle Mountains (Fig. 1). Larsen and Cross (1956) referred to these bodies collectively as the Pine River batholith. Barker (1969) discontinued use of the term batholith and called the two intrusions the western and eastern bodies of the Eolus Granite. Genetic relationships and field complexities prompted a return to the practice of Larsen and Cross and the collective naming of all intrusive varieties of the two bodies as the Eolus batholith (Collier, 1985).

The Eolus batholith consists of two principal map units, the Eolus and Trimble Granites. A third unit to the west of the batholith, the Electra Lake Gabbro, consists of dioritic and gabbroic rocks that are identical in age to the Eolus Granite. It is

Collier, J. D., 1989, Mid-Proterozoic postorogenic granites and associated uranium mineralization of the Needle Mountains, southwestern Colorado, *in* Grambling, J. A., and Tewksbury, B. J., Proterozoic geology of the southern Rocky Mountains: Boulder, Colorado, Geological Society of America Special Paper 235.

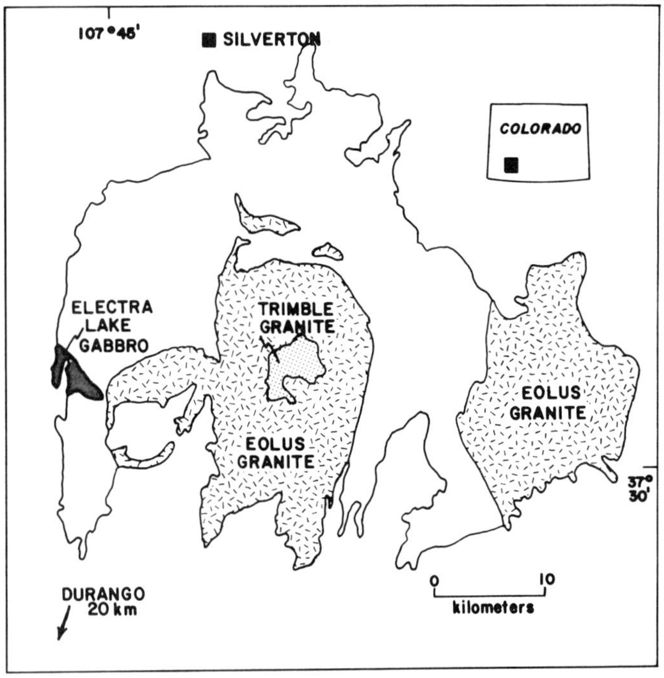

Figure 1. Location map showing the outline of exposed Proterozoic rocks of the Needle Mountains and the granitic rocks of the Eolus batholith.

probably genetically related to the Eolus Granite, but was not included as part of this study.

The Eolus and Trimble Granites intruded relatively young crust formed along a destructive plate margin between about 1,800 and 1,700 Ma (Condie, 1982). The Eolus Granite has been dated at ca. 1,460 Ma (Bickford and others, 1968; Silver and Barker, 1968) and thus falls within the 1,400 to 1,500 Ma anorogenic transcontinental province of North America (Anderson, 1983). The Trimble Granite, dated at approximately 1,350 Ma (Bickford and others, 1968), may be representative of a separate 1,340 to 1,410 Ma magmatic event and, if so, is the most westerly expression of this event.

The Eolus and Trimble Granites share certain mineralogic and chemical characteristics with A-type anorogenic granites, but differ significantly from other 1,400 Ma "anorogenic" granites by possessing features commonly attributed to highly evolved I-type calcalkaline magmas associated with destructive plate margins. One objective of this study was to try to arrive at a speculative tectonic and petrogenetic model that might explain these shared synorogenic and anorogenic traits. A second objective was based on the fact that the petrogenetic evolution of the Eolus batholith resulted in strong uranium enrichment and the production of source and host rocks for major uranium deposits. The fertility of the Eolus batholith as a source for uranium mineralization was investigated in terms of changes in uranium residence from nonlabile to labile phases as well as the petrologic reasons for uranium enrichment.

TEXTURES AND MINERALOGY

The Eolus Granite is a heterogeneous body consisting of numerous small intrusions ranging in composition from monzodiorite to granite. Varieties include biotite-hornblende monzodiorite, biotite-hornblende granodiorite, biotite monzogranite, and biotite granite. Textures range from coarse-grained porphyritic to aplitic. Despite these wide variations, the rocks of the Eolus Granite may be crudely divided into three dominant textural and mineralogic variants. These include: (1) coarse-grained, dominantly porphyritic, hornblende-bearing rocks, (2) coarse-grained, commonly porphyritic, hornblende-free biotite granitoids, and (3) equigranular, fine- to medium-grained biotite aplites.

The hornblende-bearing varieties range in composition from monzodiorite to monzogranite; the most common variety is granodiorite. The rocks are mostly gray and inequigranular, consisting of scattered coarse- to very coarse-grained (1.5 to 5 cm) phenocrysts of microcline set in a medium- to coarse-grained groundmass of quartz, plagioclase (An_{40-45}), biotite, and hornblende. The phenocrysts are usually fresh, microperthitic, and show conspicuous Carlsbad twinning. Plagioclase occurs as irregular-shaped grains and almost always shows some degree of sericitization, although rarely is it strongly altered. Modal plagioclase ranges from 35 to 60 percent; quartz ranges from 15 to 20 percent. Hornblende and biotite together may constitute as much as 35 percent of the rocks, with their relative proportion ranging from hornblende-dominant to biotite-dominant. In some of the more mafic rocks, hornblende occasionally contains cores of clinopyroxene. Biotite is generally quite fresh and only rarely shows evidence of chloritization. Accessory minerals include a few percent magnetite and ilmenite, apatite, and an occasional grain of sphene.

Another dominant group of textural and mineralogical varieties consists of generally inequigranular biotite monzogranites. Unlike the hornblende-bearing varieties, these rocks are commonly pink to brick red in color. They are usually porphyritic and contain variable amounts of coarse- to very coarse-grained phenocrysts of microcline microperthite in a medium- to coarse-grained groundmass of quartz, sodic andesine, and biotite. In addition to lack of hornblende, these rocks differ from the hornblende varieties in their accessory mineral assemblages. Sphene is generally absent, and apatite is less abundant. Zircon and allanite are much more common than in the hornblende-bearing suite.

Although pegmatites are quite rare in the Eolus Granite, aplitic phases are fairly common. Equigranular, fine- to medium-grained, biotite monzogranite or granite aplites are scattered throughout the Eolus Granite and appear to have been intruded at several different times. The aplites are gray to pink in color and generally contain more than 20 percent quartz, less than 15 percent biotite, and variable percentages of plagioclase (An_{25-35}) and microcline.

The Trimble Granite is a small pluton roughly in the center of the western body of the Eolus batholith. It is a fairly homo-

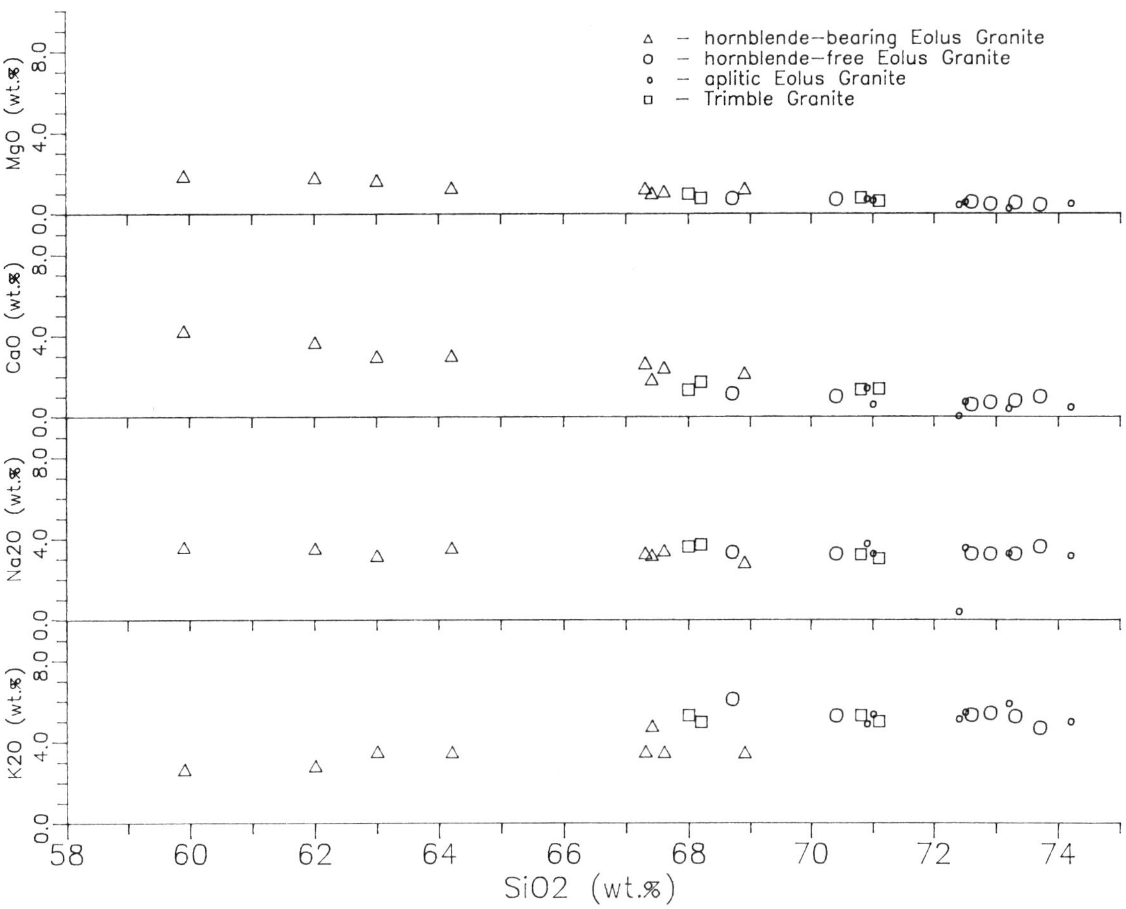

Figure 2. Harker diagrams for granitic rocks of the Eolus batholith.

geneous body of crudely equigranular, fine- to medium-grained biotite monzogranite and granite. Both aplites and pegmatites, although not abundant, are present in the Trimble Granite. The rocks contain 20 to 30 percent quartz, 5 to 15 percent biotite, and variable percentages of microcline and plagioclase (An_{25-35}). Potassium feldspar is always present in greater amounts than plagioclase. Microcline is relatively fresh, whereas plagioclase nearly always shows evidence of sericitization. Even drill core samples taken from depths of more than 500 m showed intense sericitization, suggesting that the alteration was probably deuteric. Biotite is commonly partially or intensely chloritized, and in many samples shows minor replacement by muscovite. The accessory oxides ilmenite and magnetite are generally partially to completely altered to leucoxene and hematite. This ubiquitous deuteric alteration, as well as the presence of pegmatites, suggests that a more substantive volatile phase developed during the crystallization of the Trimble Granite than was the case for the Eolus Granite.

CHEMISTRY

Chemical analyses for 22 samples from the Eolus batholith are included in Table 1. The behavior of major oxides as a function of SiO_2 for samples of Eolus and Trimble Granite is shown in standard Harker diagrams on Figure 2. These diagrams not only show a broad range in silica values, but also—constant-sum effect aside—suggest a genetic relationship between samples.

Comparing SiO_2 versus CaO and $K_2O + Na_2O$ (Fig. 3) gives an extrapolated alkali-lime index of between 56 and 59 percent SiO_2. The range is due to relatively poor correlation of SiO_2 versus $K_2O + Na_2O$ ($r = 0.58$) as opposed to an excellent correlation for SiO_2 versus CaO ($r - 0.99$), as well as the type of regression model used to fit the lines. According to the rock series classification of Peacock (1931), the Eolus batholith may be classified as calcalkaline.

The rocks of the Eolus batholith show a linear progression from metaluminous to peraluminous (Fig. 4), according to the alumina-saturation classification of Shand (1947). This progression is common in anorogenic suites (Anderson, 1983) and can be explained by early fractionation of hornblende. However, comparison with earlier (ca. 1,700 Ma) synorogenic and early postorogenic granites of the Needle Mountains shows insufficient chemical distinction to allow use of alumina-saturation as an indicator of tectonic origin.

The granites of the Eolus batholith have chemical characteristics that are distinct from the S- and I-types first defined by

Chappell and White (1974). Compared to typical I-type granites, the Eolus batholith shows lower abundances of CaO, MgO, TiO_2, and P_2O_5, but relatively high $K_2O + Na_2O$. These characteristics are similar to anorogenic rocks that fall into the A-type classification of Loiselle and Wones (1979), such as those of Nigeria (Kinnaird and others, 1985), Greenland (Gulson and Krogh, 1975), and Sweden (Persson, 1978).

REE data

Although the major-oxide chemistry of the Eolus batholith is similar to that of many A-type granites, differences exist in trace-element concentrations. Chondrite-normalized rare–earth element patterns (Fig. 5) show strong LREE enrichment, moderate to strong negative Eu anomalies, and steep curves with moderate to strong HREE depletion. This contrasts with typical patterns from anorogenic granites which, although LREE enriched, show generally flat curves, high HREE abundances, and consistently large negative Eu anomalies (for example, Plant and others, 1985). The trends for the Eolus batholith are consistent with both models and observed patterns for evolved calcalkaline magmas associated with Cordilleran-type mobile belts (Tarney and Saunders, 1979).

TABLE 1. CHEMICAL ANALYSES OF ROCKS OF THE EOLUS BATHOLITH

Field No.	EG-1	EG-2	EG-3	EG-4	EG-5	EG-6	EG-8	EG-9	EG-10	EG-11	EG-12
					Oxides (wt. %)						
SiO_2	70.4	67.3	67.6	72.5	68.9	72.4	73.2	74.2	68.7	67.4	64.2
Al_2O_3	13.9	13.3	13.7	14.1	12.0	14.7	13.8	12.9	14.9	14.2	14.3
Fe_2O_3	3.32	6.80	5.58	1.63	6.02	2.33	1.20	2.44	3.21	5.51	7.55
MgO	0.70	1.23	1.07	0.54	1.22	0.41	0.24	0.46	0.74	1.00	1.28
CaO	1.00	2.63	2.41	0.73	2.14	0.02	0.40	0.47	1.15	1.84	3.02
Na_2O	3.24	3.27	3.40	3.54	2.80	0.37	3.26	3.14	3.32	3.17	3.55
K_2O	5.27	3.50	3.47	5.43	3.45	5.09	5.86	4.96	6.08	4.76	3.48
TiO_2	0.37	0.83	0.67	0.15	0.74	0.21	0.05	0.22	0.36	0.70	0.94
P_2O_5	0.11	0.20	0.17	<0.05	0.19	<0.05	<0.05	<0.05	0.08	0.16	0.25
MnO	0.04	0.10	0.08	<0.02	0.09	0.04	<0.02	0.03	0.03	0.08	0.11
LOI	0.76	0.83	0.83	0.66	0.89	3.46	0.52	0.72	0.81	0.69	0.62
Total	99.10	99.99	98.98	99.35	98.44	99.08	98.60	99.59	99.38	99.51	99.30
					Trace Elements (ppm)						
U-N	6.1	3.6	2.6	8.7	3.4	5.2	11.0	7.0	3.4	2.8	3.5
Th-N	30	7.6	7.4	30	8.8	42	38	11.0	7.1	9.0	28
Zr-N	450	760	630	220	63	340	133	316	384	32	261
Ni	3	7	5	4	5	4	4	5	3	6	6
Ba	620	1200	1100	590	1000	810	108	229	1116	1602	1188
Cr-N	210	280	230	180	220	250	267	164	261	196	124
Cu	6	12	11	2	14	13	6	2	8	10	13
Li	34	31	34	22	48	59	7	21	27	43	25
Nb	11	17	14	24	16	13	5	15	10	14	19
Pb	15	12	16	28	16	16	31	30	25	15	18
Sc-N	6.30	18.00	13.00	2.80	17.00	3.20	1.46	7.81	13.20	17.90	4.60
Sr	91	180	180	96	140	48	43	60	131	206	202
V	18	45	37	11	40	15	2	5	17	34	49
Y	25	27	25	14	33	9	9	12	33	24	28
Zn	58	90	76	40	85	57	30	41	46	76	97
Rb-N	250	100	95	270	120	280	225	196	161	86	210
Cs-N	11.0	3.8	4.4	5.6	5.3	17.0	5.0	4.4	5.8	4.0	2.6
Ta-N	2.3	1.3	1.6	3.0	1.3	1.4	0.9	0.9	0.2	1.3	0.8
La-N	69	45	43	52	62	45	10	47	77	37	41
Ce-N	120	86	79	83	120	82	23	133	67	80	182
Nd-N	91	79	36	--	81	--	--	--	70	28	46
Sm-N	8.2	6.4	5.8	4.5	7.6	4.0	1.0	3.8	8.3	4.9	0.6
Eu-N	0.89	1.50	1.40	1.41	1.50	0.51	0.29	1.23	1.48	1.56	0.66
Tb-N	1.30	1.50	1.30	0.68	1.50	0.79	0.47	1.69	0.46	1.05	0.91
Tm-N	2.9	3.2	2.8	2.7	3.0	1.0	0.9	2.0	1.4	2.2	2.2
Yb-N	3.90	3.00	2.30	1.80	2.70	1.00	1.02	2.65	2.46	1.03	1.03

Samples with field numbers beginning with EG are Eolus granite; TG are Trimble granite. Major oxide analyses by X-ray spectroscopy, U.S.G.S. (LOI - loss on ignition). Analysts: J. S. Wahlberg, J. Taggart, J. Baker. Trace element analyses by Nuclear Division, Union Carbide Corporation. Letter N after element indicates analysis by INAA. All other analyses by ICPS.

The negative Eu anomalies become more pronounced with increased silica. This indicates either increasing fractionation of feldspar, decreasing fractional crystallization of hornblende, or more likely, both. The hornblende-bearing granitoids have the smallest Eu anomalies, which were probably moderated by the fractionation of phases that tend to contribute to positive anomalies, such as hornblende and apatite. The strongest negative anomalies occur in hornblende-free rocks.

The steep slope of the REE curves and relatively low HREE contents suggests the presence of hornblende ± garnet in the residue of a mafic source. Such a source and hornblende-dominated fractionation are not characteristic of anorogenic magmatism (Collins and others, 1982).

Other trace-element data

Further information on the fractionation history of the Eolus batholith can be inferred from the behavior of trace elements such as Sr, Ba, and Rb. For example, Ba shows a linear increase with increasing K_2O in hornblende-bearing rocks, and a strong decrease in hornblende-free rocks (Fig. 6). The strong increase indicates that neither element was in equilibrium with a mineral

TABLE 1. CHEMICAL ANALYSES OF ROCKS OF THE EOLUS BATHOLITH (continued)

Field No.	TG-14	TG-15	EG-16	EG-17	EG-18	EG-19	EG-20	EG-21	EG-22	EG-23
					Oxides (wt. %)					
SiO_2	68.0	68.2	72.6	71.0	73.3	73.7	70.9	72.9	59.9	63.0
Al_2O_3	15.1	15.0	13.4	14.3	13.6	13.3	14.6	13.6	14.5	13.7
Fe_2O_3	3.39	2.81	2.41	2.22	2.46	1.93	2.51	1.89	10.5	8.79
MgO	0.96	0.75	0.55	0.63	0.52	0.42	0.70	0.46	1.87	1.64
CaO	1.32	1.71	0.60	0.60	0.79	0.99	1.40	0.72	4.25	2.97
Na_2O	3.60	3.70	3.24	3.24	3.24	3.61	3.74	3.24	3.58	3.15
K_2O	5.29	4.95	5.30	5.32	5.23	4.66	4.84	5.40	2.64	3.51
TiO_2	0.40	0.31	0.22	0.21	0.20	0.17	0.26	0.17	1.35	1.15
P_2O_5	0.13	0.10	0.07	0.07	0.08	0.06	0.08	0.06	0.34	0.28
MnO	0.03	0.04	0.02	<0.02	0.03	0.03	0.04	0.03	0.16	0.12
LOI	0.97	0.86	0.82	1.12	0.74	0.61	0.52	0.85	0.76	0.70
Total	99.19	98.43	99.24	98.73	100.3	99.48	99.59	99.32	99.85	99.01
					Trace Elements (ppm)					
U-N	3.7	5.2	7.7	8.2	15.0	11.0	8.5	16.9	3.4	3.5
Th-N	33	30	33	36	54	35	36	37	6.6	6.3
Zr-N	418	353	279	255	143	255	353	351	--	--
Ni	4	6	4	4	4	2	5	7	9	9
Ba	985	866	442	571	367	357	578	396	890	1288
Cr-N	148	267	255	229	209	276	148	469	165	142
Cu	10	11	3	5	4	3	13	4	17	15
Li	23	29	17	13	20	32	54	22	23	25
Nb	14	23	11	15	14	16	20	24	22	20
Pb	25	24	29	28	30	38	20	28	10	14
Sc-N	5.41	5.16	3.77	3.67	5.22	3.58	5.47	3.35	25.70	18.70
Sr	177	153	63	82	55	71	102	60	213	197
V	29	22	13	17	13	10	19	12	74	63
Y	15	27	23	16	27	27	36	33	41	34
Zn	62	54	45	46	46	46	53	42	123	113
Rb-N	251	322	289	270	332	297	291	307	86	100
Cs-N	3.5	10.0	7.0	11.0	7.7	4.5	4.7	5.8	4.5	4.4
Ta-N	0.7	2.6	0.9	1.4	1.7	2.0	3.0	2.4	2.1	1.4
La-N	125	101	63	75	72	49	94	55	48	44
Ce-N	214	179	124	134	138	95	149	101	98	83
Nd-N	--	100	--	114	107	94	89	140	--	--
Sm-N	10.1	7.8	0.5	1.3	8.4	0.9	8.6	7.9	0.7	7.3
Eu-N	0.75	0.59	0.61	0.46	0.55	0.44	0.61	0.51	1.69	1.54
Tb-N	1.01	2.10	2.00	1.01	1.51	2.00	1.38	1.50	2.00	1.50
Tm-N	1.5	2.7	2.3	1.7	2.7	3.1	3.6	3.2	4.0	4.6
Yb-N	1.35	2.94	2.62	1.79	4.39	2.75	3.71	4.43	3.95	3.14

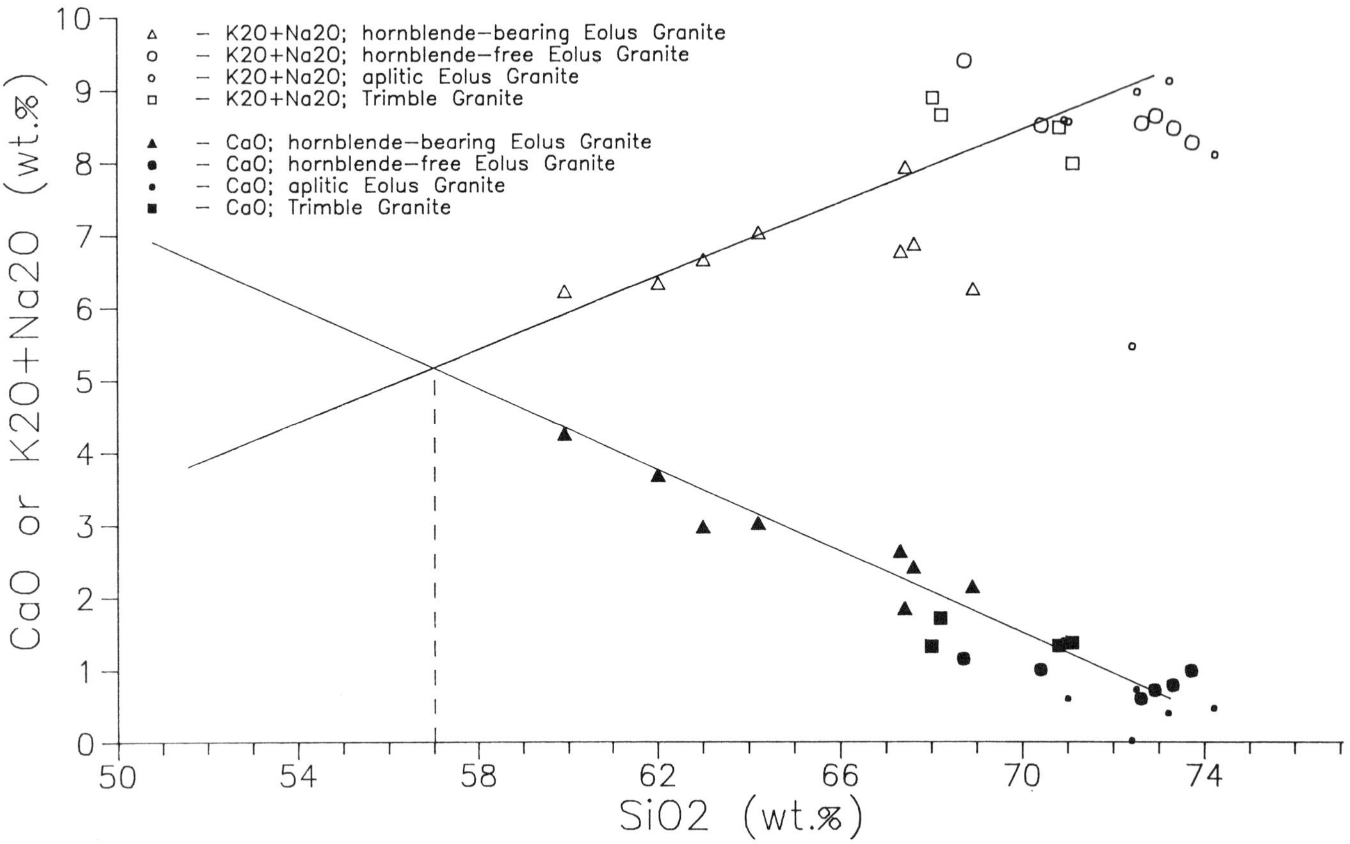

Figure 3. Plots of SiO_2 versus CaO and SiO_2 versus K_2O and Na_2O, showing their extrapolated intersection and the alkali-lime index (dashed line) of Peacock (1931).

having a high distribution coefficient (K_d) for these elements. The only minerals relevant to this suite having a high K_d for Ba are potassium feldspar and biotite (Arth, 1976), indicating that they were not early fractionating phases. The strong decrease of Ba with relatively constant K_2O in the hornblende-free rocks suggests that K-feldspar fractionated as this trend developed (high Rb contents preclude biotite as a fractionating phase).

It is apparent from the behavior of Rb versus a number of elements (Fig. 7) that relationships between samples are more complex than suggested by plots of major oxides. It appears that there are three trends present, although the number of analyses is too small to fully define them. These trends are also visually distinct in a number of other variation diagrams not included here. They indicate that all of the Eolus Granite may not be comagmatic. One sample of hornblende granodiorite, texturally and mineralogically identical to other hornblende-bearing varieties of Eolus Granite, falls on trend with samples of Trimble Granite. This discrepancy is not due to analytical error and suggests that some of the rocks that have been mapped as Eolus Granite may actually be of the same age and comagmatic with the Trimble Granite.

Although major-oxide behavior suggests strong genetic ties between all rocks of the Eolus batholith, the trends in Figure 7, and of course, the difference in age between the Eolus and Trimble Granites, indicate multiple episodes of partial melting. In order to maintain a geochemical coherence between different phases of the batholith, either: (1) a single source area was partially melted in successively smaller degrees of melting to explain the increase in silica from Eolus Granite to Trimble Granite, or (2) multiple source areas were melted. The latter possibility seems far more plausible, as repeated melting of a single source would produce magmas much drier than those that formed the Eolus batholith.

The three trends suggested in Figure 7 show that the Trimble Granite, and perhaps its more primitive cognates, is the most highly enriched in incompatible elements. The sharp decrease of such elements as strontium with increasing differentiation cannot be easily explained by partial melt fractionation. Melting experiments by Shaw (1970) indicate that bulk distribution coefficients between liquid and solid for elements such as strontium show little change with different degrees of partial melting. In contrast, Rayleigh fractionation models for fractional crystallization, as in Hanson (1978), adequately explain such rapid decreases. Thus, the rocks within a given trend are probably related by fractional crystallization rather than by partial melt fractionation.

One consequence of an apparently complex and extensive history of fractionation during the evolution of the Eolus batholith was a strong enrichment in late-stage differentiates of certain elements, particularly the large-ion lithophile (LIL) and high field strength (HFS) elements, and marked depletion in other

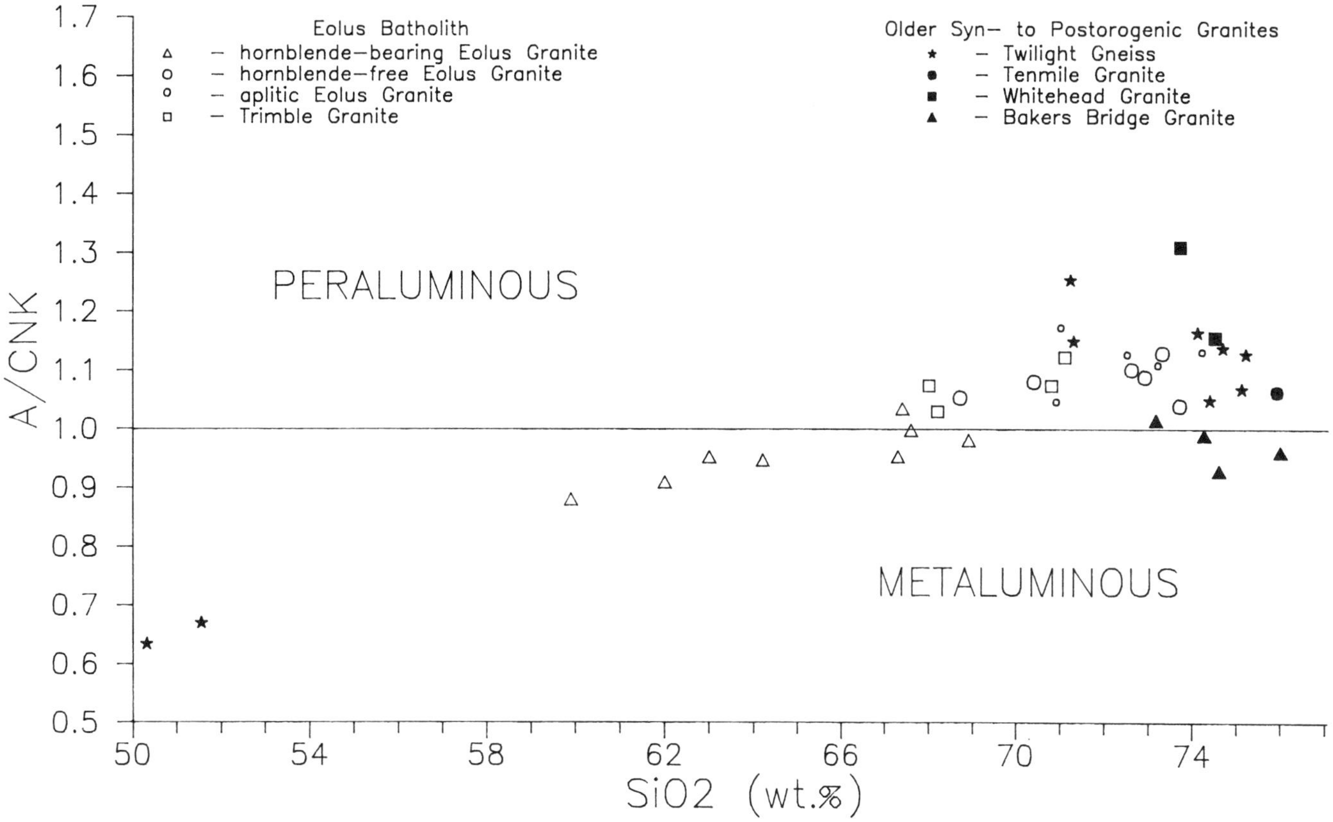

Figure 4. Plot of SiO_2 (wt. %) versus $Al_2O_3/(CaO + Na_2O + K_2O$ [mole %]), or A/CNK, showing alumina-saturation classification fields. Rocks with A/CNK >1 are peraluminous, those with A/CNK <1 are metaluminous. In addition to data for the Eolus batholith, older synorogenic and early postorogenic granitic rocks of the Needle Mountains are shown for comparison. Twilight Gneiss = 1,780 to 1,805 Ma; Whitehead Granite = 1,780 to 1,695 Ma; Tenmile Granite = 1,720 Ma; Bakers Bridge Granite = 1,720 to 1,692 Ma. Data for rocks other than the Trimble and Eolus Granites from Barker (1969).

elements, such as Sr and Ba. This extensive fractionation may also help explain a general chemical transition from metaluminous, calcalkaline traits to more subalkaline, peraluminous characteristics.

MINERALIZATION

Along with the strong enrichment of other LIL and HFS elements during fractionation, uranium shows a marked enrichment in late-stage differentiates. In fact, the Trimble Granite is host to a relatively large vein-type uranium deposit, with perhaps as much as 10,000 tonnes U_3O_8 (Collier, 1985). The age of the deposit is unknown, but it is probably considerably younger than the emplacement of the granites. It may have been the result of nearby Tertiary plutonism, or alternatively, it might have been due to hydrothermal circulation generated by high radioelement heat production from the Eolus batholith itself. Regardless of the mechanism of mineralization, it is likely that the granites of the Eolus batholith acted as the source rocks for the uranium.

Whether or not a granite may serve as a good source rock is a function not only of above-normal uranium concentrations, but also the location of uranium within labile or readily leachable phases. Granite fertility (a concept first introduced by Moreau and others, 1966) in the Eolus Granites was studied using fission-track radiographs and solid-state, natural alpha-track detectors.

Within the hornblende granodiorites of the Eolus Granite, which contain an average of only about 3.5 ppm uranium, uranium occurs mostly within euhedral zircon grains and within small, anhedral grains of allanite. Mass-balance calculations indicate that these refractory accessory minerals are capable of accounting for the entire uranium content of the rocks (Collier, 1985).

Uranium contents in the hornblende-free variants of the Eolus Granite are higher than the granodiorites, generally about 5 to 8 ppm, but range to as much as 16.9 ppm. Uranium detected in random thin sections occurs dominantly in nonlabile accessories such as zircon and allanite. In some samples, particularly in more felsic varieties, minor uranium is associated with iron oxides along cleavage planes in biotite, as well as within accessory oxides such as magnetite, hematite, and leucoxene. Uranium in these phases probably represents minor remobilization during deuteric alteration, although in general, uranium showed little evidence of major remobilization from accessory minerals.

Quantitatively, the biotite monzogranites contain less Zr, P_2O_5, and TiO_2 than the granodiorites, which is expressed by lower abundances of zircon and apatite and the near absence of sphene. These minerals are not nearly abundant enough to account for the higher uranium contents of the rocks. By examining radiographs of large slabs of rock, widely scattered grains (3 to 4 per 100 cm^2) of uranium-rich minerals were discovered. These were identified by scanning electron microscope as uranothorite, high-thorium uraninite, thorianite, and possibly chevkinite [(Ce, Y, Ca, U, Th)$_2$ (Ti, Fe, Mg)$_2$(Si, Al)$_2$O$_{11}$]. These more labile minerals were too widely scattered to be encountered by random thin sections, but despite their low abundances, can easily account for the bulk of the uranium in the monzogranites.

The average uranium content of the Trimble Granite is about 15 ppm. Uranium residence is very similar to the monzogranites of the Eolus Granite, although zircon and apatite grains are even less abundant and there is a somewhat larger volume percentage of uranium-bearing allanite. Again, the major contributors to the overall uranium content are widely scattered high-uranium phases such as uraninite and uranothorite.

The low uranium contents of the early hornblende-bearing differentiates do not suggest a source especially rich in radioelements. This is more characteristic of calcalkaline suites than of anorogenic granites, which commonly have uniformly high uranium contents and exhibit evidence of radioelement-enriched sources. The abundance of uranium in late-stage differentiates, particularly the Trimble Granite, and the fertility of these granites as uranium source rocks is probably due to two factors. The relatively high concentration of uranium in the Trimble Granite was the result of LIL concentration toward the end stages of the extensive fractionation history of the Eolus batholith. Part of this fractionation was an apparent unmixing of minimum melt and restite, resulting in strong decreases in TiO_2, P_2O_5, and Zr, thus reducing the abundance of refractory accessory minerals and limiting their ability to account for significant amounts of uranium in the rocks. As a result, relatively leachable uranium-rich minerals became more abundant with progressive differentiation, eventually producing rocks not only high in uranium, but with uranium contained in phases from which it could be easily leached and remobilized to form a deposit.

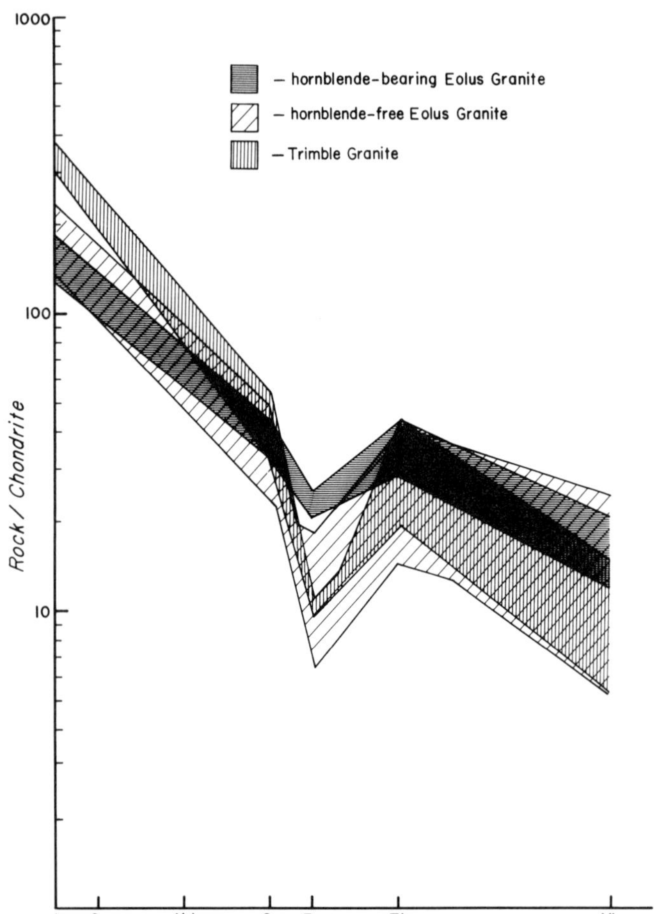

Figure 5. Chondrite-normalized rare-earth abundances for the Eolus and Trimble Granites.

PETROGENESIS AND CONCLUSIONS

Any model for the genesis of the Eolus batholith must take into account not only the chemical and mineralogic characteristics of the suite, but also the fact that the Eolus and Trimble Granites were formed during magmatic events that occurred throughout a large part of North America between 1.4 and 1.5 Ga and from 1.34 to 1.41 Ga. In particular, one must consider that the Eolus batholith contains a much wider range of compositions and is volumetrically more mafic than the typical potassic granites of anorogenic suites, yet must in some way be tectonically related to other rocks of the 1.5- to 1.4-Ga and 1.41- to 1.34-Ga provinces.

Source rocks

Collins and others (1982) proposed that melting of residue left after the production of previous granites is the probable mechanism for the genesis of the A-type granites of southeastern Australia. Anderson and Thomas (1985) call upon a peraluminous, metasedimentary source for the 1.49-Ga "anorogenic" Silver Plume and St. Vrain batholiths of Colorado. Unlike the rocks of southeastern Australia, these have chemical and mineralogical affinities to S-type, rather than I-type, granites. In either case, the proposed source was quartzofeldspathic and dehydrated by previous melting episodes, yielding dry, relatively homogeneous, alkaline magmas.

The chemistry and mineralogy of the Eolus batholith suggest a metaigneous source rock. Although older (ca. 1.7 Ga) synorogenic granites are present in the Needle Mountains, it is unlikely that melting of their residue could produce the Eolus batholith. Rocks of the Eolus batholith range in silica content from about 55 to 75 percent. This range is even further expanded if the Electra Lake Gabbro is genetically related to this suite, a relationship as yet unproved, but likely given the identical ages

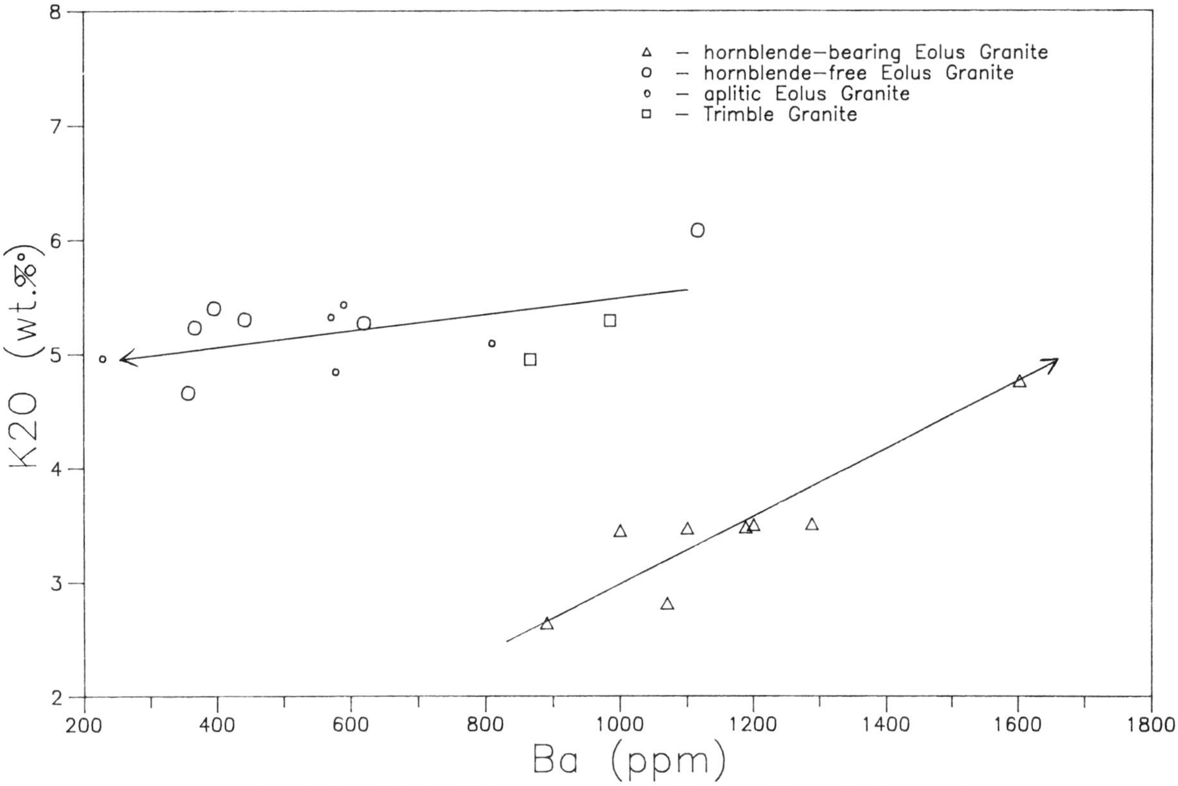

Figure 6. Variation of K_2O with Ba. Arrows point in direction of increasing fractionation.

and compositional continuity between the gabbros and the Eolus granodiorites. Such a range requires a different, more mafic, source than has been interpreted for typical A-type granites, including other 1.4-Ga, North American, "anorogenic" granites, most of which are relatively homogeneous and close to minimum-melt composition. In addition, the abundance of hornblende in the more mafic variants of the Eolus batholith and the trace-element evidence indicating that it was an early fractionating phase suggest hydrous melting conditions, again in contrast to anorogenic granite magmatism.

Most calcalkaline granitic suites with broad compositional ranges have trace-element characteristics consistent with hydrous melting of subducted oceanic crust or overlying underplated mantle wedge, with clinopyroxene+hornblende+garnet in the residue. A similar source is likely for the Eolus batholith. Perhaps in areas outside of the Needle Mountains, particularly in areas of thickened crust, mafic evolved-calcalkaline magmas could trigger melting higher in the crust and the generation of S- and A-type minimum-melt granite magmas more typical of the 1.5- to 1.4-Ga granitic province.

Tectonic setting

Plutonism from 1.4 to 1.5 Ga and 1.34 to 1.41 Ga in North America is generally referred to as anorogenic. Yet the Eolus batholith is spatially associated with a Cordilleran-type subduction zone that is believed to have been active only a few hundred kilometers to the southeast at the time the batholith was emplaced (Condie, 1982). It is intuitively difficult to rule out a tectonic association between "anorogenic" igneous activity and this orogenic belt, even if others have attempted to do so (Anderson, 1983). The Eolus batholith shows remarkable mineralogic and chemical similarities to batholiths found elsewhere in the world with clear ties to orogenic cycles, such as in the Andes and the northern Caledonides of Great Britain.

The calcalkaline Mesozoic-Cenozoic Coastal batholith of Peru is very similar to the Eolus batholith in rock types, range of SiO_2, K_2O/Na_2O ratios, LIL concentrations, and REE patterns (Atherton and Sanderson, 1985) These rocks were emplaced distal to an ocean-continent subduction zone and are associated with deep transcurrent shear systems, which are believed to have been preexisting lineaments reactivated at the close of an orogenic cycle (Pitcher and others, 1985). Besides its spatial association with a similar type of subduction zone, the Eolus batholith is located within the Colorado mineral belt, which is believed to represent a major Precambrian lineament (Warner, 1978; Tweto and Sims, 1963).

The Late Silurian–Early Devonian granites of the Scottish Caledonides are similar to those of the Andes in that they are associated with major lineaments, although compositions and tectonic styles differ. There are some small calcalkaline plutons similar to the Coastal and Eolus batholiths; however, the bulk of the

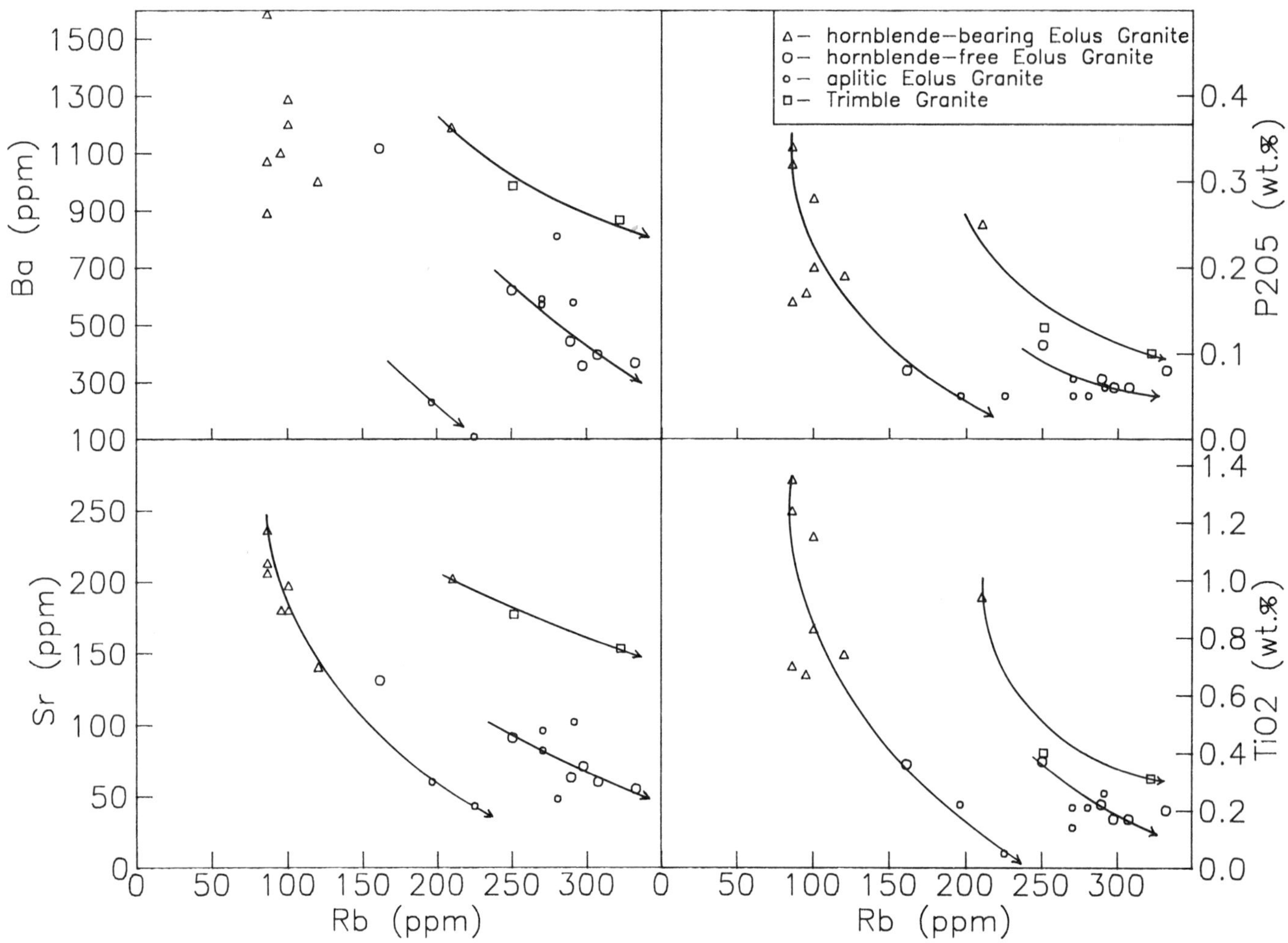

Figure 7. Variation diagrams of Ba, Sr, TiO_2, and P_2O_5 versus Rb.

granites are subalkaline S- and A-type plutons, similar in many ways to the "anorogenic" granites of North America. According to Watson (1984), the bulk of the granite magmatism in the Caledonides corresponded to a change in tectonic style from crustal thickening to deep faulting at the close of the Caledonian orogeny.

Thus the Peruvian Andes, the Scottish Caledonides, and the North American "anorogenic" province all share two common features: (1) they are spatially associated with subduction zones and, at least in the case of the Andes and Scotland, are associated with the waning stages of orogenic activity; and (2) they are situated on major lineaments. Significant magmatism along these lineaments was probably inhibited by compression during orogenic cycles, but as the cycles either ended or migrated away from the lineaments, one would expect a change from a dominantly ductile, compressive regime to one of brittle extension. Extension on deep faults would create low-pressure zones in the upper mantle or lower crust, which would be conducive to melting. The products of such melting could show a range in chemical characteristics, depending on a number of factors such as subduction-zone angles and the depth of melting, P_{H_2O} conditions and degrees of melting, mantle or crust heterogeneity, crustal contamination, etc. Any combination of these factors could explain the compositional differences between the Eolus batholith and other granites of the 1.5- to 1.4-Ga province.

It is difficult to completely divorce the Eolus batholith from the orogenic episodes that preceded it. Although the Eolus batholith falls within the time span of a proposed "anorogenic" province, its lithology and chemistry are more consistent with late- to postorogenic, rather than anorogenic, magmatism. The dissimilarities between the Eolus batholith and other rocks of the province clearly indicate the need for further investigations into plausible tectonic and petrogenetic models for the contemporaneous generation of both A-type and calcalkaline magmas.

REFERENCES

Anderson, J. L., 1983, Proterozoic anorogenic granite plutonism of North America, *in* Medaris, L. G., and others, eds., Proterozoic geology: Geological Society of America Memoir 161, p. 133–154.

Anderson, J. L., and Thomas, W. M., 1985, Proterozoic anorogenic two mica granites; Silver Plume and St. Vrain batholiths of Colorado: Geology, v. 13, p. 177–180.

Arth, J. G., 1976, Behavior of trace elements during magmatic processes; A summary of theoretical models and their applications: Journal of Research of the U.S. Geological Survey, v. 4, p. 41–47.

Atherton, M. P., and Sanderson, L. W., 1985, The chemical variation and evolution of the superunits of the segmented Coastal Batholith, *in* Pitcher and others, eds., Magmatism at a plate edge, the Peruvian Andes: Glasgow, Blackie Press and New York, Halstead Press, p. 208–227.

Barker, F., 1969, Precambrian geology of the Needle Mountains, southwestern Colorado: U.S. Geological Survey Professional Paper 644–A, 35 p.

Bickford, M. E., Barker, F., Wetherill, G. W., and Lee-Hu, C., 1968, Precambrian chronology in the Needle Mountains, southwestern Colorado; Pt. 2, Rb-Sr results [abs.]: Geological Society of America Special Paper 115, p. 14.

Chappell, B. W., and White, A.J.R., 1974, Two contrasting granite types: Pacific Geology, v. 8, p. 173–174.

Collier, J. D., 1985, Geology and uranium mineralization of the Florida Mountain area, Needle Mountains, southwestern Colorado: U.S. Geological Survey Open-File Report 84–336, 88 p.

Collins, W. J., Beams, S. D., White, A.J.R., and Chappell, B. W., 1982, Nature and origin of A-type granites with particular reference to southeastern Australia: Contributions to Mineralogy and Petrology, v. 80, p. 189–200.

Condie, K. C., 1982, Plate-tectonics model for Proterozoic continental accretion in the southwestern United States: Geology, v. 10, p. 37–42.

Gulson, B. L., and Krogh, T. E., 1975, Evidence of multiple intrusions, possible resetting of U-Pb ages, and new crystallization of zircons in the post-tectonic intrusions ("Rapakivi granites") and gneisses from Greenland: Geochimica Cosmochimica Acta, v. 39, p. 65–82.

Hanson, G. N., 1978, The application of trace elements to the petrogenesis of igneous rocks of granitic composition: Earth and Planetary Science Letters, v. 38, p. 26–43.

Kinnaird, J. A., Batchelor, R. A., Whitley, J. E., and MacKenzie, A. B., 1985, Geochemistry, mineralization, and hydrothermal alteration of the Nigerian high-heat producing granites, *in* Proceedings, High-heat production granites, hydrothermal circulation, and ore genesis, Cornwall, England: London, The Institution of Mining and Metallurgy, p. 169–195.

Larsen, E. S., and Cross, W., 1956, Geology and petrology of the San Juan region, southwestern Colorado: U.S. Geological Survey Professional Paper 258, 303 p.

Loiselle, M. C., and Wones, D. R., 1979, Characteristics and origin of anorogenic granites: Geological Society of America Abstracts with Programs, v. 11, p. 468.

Moreau, M., Poughon, A., Puybaraud, Y., and Sanselme, H., 1966, L'uranium et les granites: Chronique des Mines et de la Recherche miniere, v. 350, p. 47–50.

Peacock, M. A., 1931, Classification of igneous rocks: Journal of Geology, v. 39, p. 54–67.

Persson, L., 1978, The Revsund–Sorvik granites in the western parts of the province of Angermanland central Sweden: Sveriges Geologiska Undersokning, Series C, no. 741, 59 p.

Pitcher, W. S., Atherton, M. P., Cobbing, E. J., and Beckinsale, R. D., eds., 1985, Magmatism at a plate edge, the Peruvean Andes: Glasgow, Blackie Press and New York, Halstead Press, 328 p.

Plant, J. A., O'Brien, C., Tarney, J., and Hurdley, J., 1985, Geochemical criteria for the recognition of high heat production granites, *in* Proceedings, High heat production granites, hydrothermal circulation, and ore genesis, Cornwall, England: London, The Institution of Mining and Metallurgy, p. 263–285.

Shand, S. J., 1947, The eruptive rocks: New York, John Wiley & Sons, 488 p.

Shaw, D. M., 1970, Trace element fractionation during anatexis: Geochimica et Cosmochimica Acta, v. 34, p. 237–243.

Silver, L. T., and Barker, F., 1968, Geochronology of Precambrian rocks of the Needle Mountains, southwestern Colorado; Pt. 1, U-Pb zircon results [abs.]: Geological Society of America Special Paper 115, p. 204–205.

Tarney, J., and Saunders, A. D., 1979, Trace element constraints on the origin of Cordilleran batholiths, *in* Atherton, M. P., and Tarney, J., eds., Origin of granite batholiths; Geochemical evidence: Orpington, Kent, Shiva Publishing, p. 90–105.

Tweto, O., and Sims, P. K., 1963, Precambrian ancestry of the Colorado mineral belt: Geological Society of America Bulletin, v. 74, p. 991–1014.

Warner, L. A., 1978, The Colorado lineament; A middle Precambrian wrench fault system: Geological Society of America Bulletin, v. 89, p. 161–171.

Watson, J. V., 1984, The ending of the Caledonian orogeny in Scotland: Journal of the Geological Society of London, v. 141, p. 193–214.

Manuscript Accepted by the Society September 26, 1988

Printed in U.S.A.

Geological Society of America
Special Paper 235
1989

The role of crustal extension in the metamorphism of Proterozoic rocks in northern New Mexico

Jeffrey A. Grambling, Michael L. Williams,* Roger F. Smith,* and Christopher K. Mawer
Department of Geology, University of New Mexico, Albuquerque, New Mexico 87131

ABSTRACT

The Proterozoic metamorphic belt of northern and central New Mexico contains rocks of two distinctly different metamorphic grades, locally lying in direct contact. A large region exhibits coexisting kyanite, andalusite, and sillimanite. The triple-point assemblages resulted from peak metamorphic conditions falling near 500°C, 4 kb, and they occur across 75,000 km^2 in 14 separate mountain ranges. Triple-point metamorphic grade was attained after north-verging folding and ductile thrusting but before south-verging mylonitic shearing. Prograde and retrograde P-T paths were nearly isobaric.

However, two mountain ranges preserve a different metamorphic history. Granitic and migmatitic gneiss in the Santa Fe Range formed at 650° to 750°C. Sillimanite–K-feldspar gneiss in the northern Taos Range records peak conditions as high as 700° to 800°C, 9 to 11 kbar. In the latter area, high-grade metamorphism was followed by south-verging shearing, southeast-verging folding, and decompression, on a retrograde P-T path that passed through 500°C, 4 kb. Deformation produced a thick mylonitic shear zone that appears in scattered outcrops across the entire mountain range. The shear zone separates gneiss from structurally overlying kyanite–andalusite–sillimanite quartzite. Mineral textures indicate that the shearing occurred during and after the thermal peak in the gneissic rocks, and during the thermal peak in the overlying triple-point rocks.

We suggest that the kyanite–andalusite–sillimanite rocks are separated from the rocks of higher metamorphic grade by a synmetamorphic ductile detachment fault represented by the thick mylonitic shear zone. The detachment separated an actively extending lower crust from an upper crust that was not being thinned. We infer that the detachment extends in the subsurface beneath the entire belt where rocks show kyanite–andalusite–sillimanite metamorphic grade. Lower-crustal extension, with heat carried upward toward the detachment, seems to have caused the triple-point metamorphism.

INTRODUCTION

Mineral assemblages characteristic of pressure-temperature (P-T) conditions near the aluminum silicate triple point can be found within a 250 by 300 km area in northern and central New Mexico (Grambling, 1986b). The consistency in peak metamorphic conditions across so large a region is unusual.

Outcrops of high-grade gneiss lie within the triple-point belt and may be critical to understanding the cause of the triple-point metamorphism. The high-grade rocks are characterized by either migmatite or coexisting sillimanite and K feldspar. They appear in two separate mountain ranges 100 km apart, covering a combined surface area exceeding 1,500 km^2.

A contact between medium-grade and high-grade rocks is exposed on mountain summits north of Questa, New Mexico

*Present address: Williams, Department of Geology and Geography, University of Massachusetts, Amherst, Massachusetts 01003; Smith, Hall Southwest Water Consultants, 3303 Northland Drive, Suite 311, Austin, Texas 78731.

Grambling, J. A., Williams, M. L., Smith, R. F., and Mawer, C. K., 1989, The role of crustal extension in the metamorphism of Proterozoic rocks in northern New Mexico, *in* Grambling, J. A., and Tewksbury, B. J., Proterozoic geology of the southern Rocky Mountains: Boulder, Colorado, Geological Society of America Special Paper 235.

(Fig. 1). The contact is a subhorizontal mylonitic shear zone. Sillimanite–K feldspar gneiss lies beneath the shear zone, whereas triple-point rocks lie above it.

This study describes mineral assemblages, P-T paths, and deformation in the medium- and high-grade rocks. Its goals are to evaluate the relationship between the two metamorphic grades and to interpret the cause of the extensive Al_2SiO_5 triple-point metamorphism.

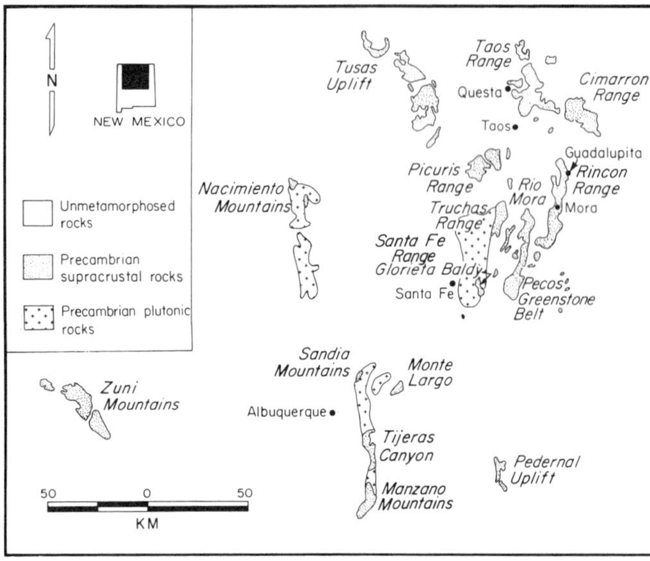

Figure 1. Distribution of Proterozoic supracrustal and plutonic rocks in northern and central New Mexico.

Regional Geology

Proterozoic rocks in the Taos Range north of Questa (Fig. 1) include amphibolite, quartzofeldspathic gneiss, and sillimanite–K feldspar gneiss. The Santa Fe (and Nacimiento?) Range exposes granitic gneiss, amphibolite, and migmatitic gneiss. Other mountain ranges contain assemblages characteristic of upper greenschist to lower amphibolite facies metamorphism. Mafic to felsic metavolcanic units, metapelite, meta-arkose, and quartzite are the most common rock types. Manganiferous schist with piemontite and Mn-andalusite is a rare but widespread lithology, found in a layer 1 to 50 m thick that separates felsic metavolcanic rocks (the Vadito Group) from an overlying, 1,000-m-thick pure quartzite (part of the Ortega Group). This manganiferous schist crops out in the Tusas, Taos, Picuris, Truchas, Rio Mora, Rincon, and Manzano Ranges (Fig. 1). Metaplutonic rocks, typically granitic, occur in places.

U-Pb zircon ages of the metaigneous rocks fall between 1,644 and 1,765 Ma (Bowring and Condie, 1982; Bowring and others, 1983, 1984; Reed, 1984). Locally, the Early Proterozoic rocks are intruded by granites with U-Pb zircon and Rb-Sr whole-rock ages around 1,450 Ma (Silver and others, 1977; Brookins, 1982; Anderson, 1983).

Methods

Mineral analyses were obtained on a JEOL 733 electron microprobe operated at 15 kV and 20 nA. Estimated analytical precision is $\pm$ 0.5 relative percent for all oxides except SiO_2, which has a precision of $\pm$1.5 relative percent.

Peak P-T conditions in medium-grade rocks were estimated from: Al_2SiO_5 equilibria, Holdaway (1971); Al_2SiO_5 solid-solution and andalusite–hematite–quartz thermobarometry, Grambling and Williams (1985); chloritoid + Al_2SiO_5 = staurolite + quartz, Rao and Johannes (1979); transition from greenschist to epidote-amphibolite facies in metabasalt, Apted and Liou (1983) and Moody and others (1983); pyrophyllite = kyanite + quartz, Haas and Holdaway (1973); and serpentine = forsterite + talc, Bowen and Tuttle (1949; not reversed). Garnet–biotite–plagioclase–muscovite barometry was based on the calibrations of Ghent and Stout (1981) and Hodges and Crowley (1985, equation R3). The Mg end-member reaction of Ghent and Stout (1981, p. 95, equation 1b, corrected with 16.675 in place of 16,675) was utilized in favor of their "1a" because pressures from "1a" are dependent on temperature whereas those from "1b" are not. Results were used for barometry only where garnet had X(Mn) below 0.25, X(Ca) below 0.10, and plagioclase had X(Ca) below 0.60. Garnet–biotite thermometry was taken from Williams (1987), modified from Ferry and Spear (1978) to account for Fe^{3+} in biotite and non-ideal mixing of Mn and Ca in garnet. The thermometer was used for garnet with X(Mn) below 0.40 and X(Ca) below 0.10. P-T paths were estimated using two independent approaches: aluminum silicate reaction textures, and quantitative thermobarometry. Calculations for the latter approach were based on garnet–biotite (Williams, 1987) and garnet–biotite–plagioclase–muscovite equilibria (Ghent and Stout, 1981, equation 1b). Paths were based on three points: inclusions of biotite, plagioclase, and muscovite in garnet; "peak" mineral compositions; and retrograde rims on garnet and plagioclase. The barometer of Hodges and Crowley (1985) was not used because it gives pressures that are affected by small errors in estimates of temperature.

Peak P-T conditions in high-grade gneiss were estimated using garnet-biotite thermometry (Indares and Martignole, 1985; Williams, 1987) and garnet–plagioclase–sillimanite–quartz barometry (Newton and Haselton, 1981) with compositions derived from garnet and plagioclase cores. Results were used only where garnet had X(Mn) below 0.25 and X(Ca) below 0.10. Retrograde P-T paths have been estimated by comparing peak to retrograde conditions. The latter were derived from garnet and plagioclase rims, retrograde biotite, and retrograde muscovite using Williams (1987) and Ghent and Stout (1981, equation 1b). Retrograde pressures from Newton and Haselton (1981) were not used because of their strong dependence on temperature.

The solid-solution thermobarometers have the following precision resulting from 2σ errors in microprobe analyses: aluminum silicate equilibria, less than $\pm$15°C, 0.3 kb; garnet–biotite–plagioclase–muscovite, ±25°C, 1 kb; and biotite–garnet–aluminum silicate–plagioclase–quartz, ±30°C, 1.5 kb. The text cites scatter that represents 2σ variations about the mean of all P-T estimates from a single locality, rather than estimated analytical precision. All P-T calculations assume that equilibrium was attained at each stage of metamorphism.

Structural data come from 1:12,000 geologic mapping supplemented with mesoscopic and microscopic structural analyses. The most detailed work has been done in the Tusas, northern Taos, northern Rincon, Truchas, and Rio Mora uplifts. Overprinting fold (F) and schistosity (S) relationships are described for individual map areas; a given fold generation (F_n) in one area may not correlate with folds having the same subscript in other areas.

DATA

Petrography

Medium-grade areas. Pyrophyllite has been recognized only in the northwestern corner of the Tusas Range (Fig. 1). It lies in the plane of foliation and is considered primary.

Garnet, biotite, staurolite, and chloritoid commonly form euhedral to subhedral porphyroblasts that cut across a schistosity formed by muscovite, chlorite, and FeTi oxides. Locally foliation deflects around the porphyroblasts, or staurolite shows "millipede" inclusion trails characteristic of syntectonic growth (Bell and Rubenach, 1983).

Kyanite forms prisms 0.5 to 2 cm long, elongate in the direction of mineral lineation. Most grains are kinked, but such grains coexist with a few others that cut across foliation or have overgrown microfolds of the foliation (cf., Grambling, 1981, Fig. 2). Most andalusite crystals (up to 10 cm across) overprint foliation, although some grains show sweeping extinction, subgrain development, and pull-apart cracks filled with muscovite. Sillimate, varying from fibrolite to prisms up to 1 cm in length, generally is randomly oriented. Exceptions occur in localized deformation zones within the Tusas, Truchas, Rio Mora, Sandia, and Monte Largo uplifts where sillimanite needles are folded or aligned (cf., Holdaway, 1978; Grambling, 1981, 1986a; Grambling and Williams, 1985; Vernon, 1987).

Plagioclase occurs as euhedral porphyroblasts in some rocks and smaller grains in the groundmass of others. Quartz typically forms coarse polygonal mosaics with 120° triple junctions.

Near mappable isograds, textures characteristic of mineral dissolution are seen. The most common breakdown textures are ragged, anhedral inclusions of kyanite within andalusite or mats of sillimanite; anhedral blebs of andalusite in mats of fibrous sillimanite or muscovite; coarse sillimanite needles that are epitaxial with and partially replace andalusite; and zincian relicts of staurolite partly to totally replaced by pseudomorphs of muscovite, biotite, or plagioclase.

Mineral textures differ markedly in medium-grade rocks north of Questa (Fig. 1). Quartzites preserve S/C fabrics except where local annealing has taken place. Muscovite, chlorite, and epidote are flattened and sheared. Kyanite, manganese-andalusite, sillimanite, and chloritoid are elongate in the direction of lineation, have pull-apart cracks normal to their length, and are asymmetric, recording the sense of shear. Aluminum silicate reaction textures are rare. Those observed include kyanite fringed by coarse to fibrolitic sillimanite; sillimanite forming trails off the ends of sheared, asymmetric kyanite; andalusite in pseudomorphs after kyanite; sillimanite fringing andalusite; and isoclinally folded andalusite, with some subgrain development but also with sweeping extinction, cut by nearly straight trails of sillimanite needles.

High-grade areas. Our work has concentrated on the northern Taos Range (Fig. 1). Garnet crystals in metapelites are up to 1 cm across, elongate in the plane of foliation, and partially surrounded by pseudomorphs of quartz, biotite, or plagioclase. One rock (RR79) contains highly flattened garnets with length-to-width ratios of 10:1 and sigmoidal shapes indicative of shearing. Subhedral to anhedral plagioclase occurs in three textural varieties. Most crystals are distributed evenly throughout the groundmass of the rocks. Others form partial pseusomorphs after garnet, and still others occur as porphyroblasts containing inclusions of sillimanite. Fibrolitic sillimanite is rare; sillimanite needles up to 2 cm long are more common. Most lie in the direction of lineation and are folded and sheared. In places the sillimanite needles cluster in pseudomorphs after kyanite.

Biotite grains up to 1 cm in length commonly have folded or sigmoidal shapes. Most are brown, but specimen RR79 contains fox red biotite. Brown biotite coexists with small shreds of green biotite in several rocks. The green grains form partial fringes around garnet and larger crystals of brown biotite.

Muscovite in gneiss of the Taos Range appears only as large porphyroblasts that cut across schistosity. Nearly all grains contain inclusions of sillimanite. Some sillimanite inclusions are folded where the host muscovites are not. Other muscovite grains are themselves sheared and folded. In contrast, muscovite appears as small interlocking grains that define foliation in migmatite and gneiss of the Santa Fe Range.

Peak metamorphic assemblages

Table 1 summarizes the mineral assemblages most useful for characterizing peak metamorphic conditions. All include quartz except those from ultramafic rocks in the Zuni Mountains. The regional distribution of assemblages is described below, presented by individual mountain range, in geographic order from north to south.

Mineral assemblages in the Tusas Mountains indicate an increase in metamorphic grade toward the southeast. Mappable isograds mark the first appearance of kyanite, andalusite, garnet, staurolite, and sillimanite, and the disappearance of staurolite (Fig. 2). The northern Taos Range (Fig. 3) exposes rocks of both medium and high metamorphic grade separated by a subhorizontal mylonitic shear zone. Medium-grade quartzite lies above the shear zone, and sillimanite–K feldspar gneiss lies beneath it, in both the Cedro Canyon (Fig. 4) and El Rito (Fig. 5) map areas.

TABLE 1. MINERAL ASSEMBLAGES USED TO HELP DEFINE PEAK METAMORPHIC PRESSURE AND TEMPERATURE

Location	Assemblages*	Notes
Tusas Range		
Northwest	ky-ppy; ky-andal; andal-hem; actin-chl-ep-plag	Ppy in plane of foliation, interpreted as primary
Central	ky-andal; ky-sill; st-gar-bi; hbl-chl-ep-plag	Ref: Williams, 1987; Grambling and Williams, 1985
Southeast	ky-sill; sill-gar-bi-plag-musc; hbl-plag	
Taos Range		
Cedro Canyon		
above shear zone	ky-sill-Mn andal-musc; gar-bi; Mn andal-Kfs; ctd-sill	All porphyroblasts fresh but intensely sheared
below shear zone	sill-Kfs-gar-bi-plag; porphyroblasts of retrograde musc	Ref: Smith, 1988
El Rito		
above shear zone	sill-musc; chl-musc-ep	
below shear zone	sill-Kfs-gar-bi; porphyroblasts of retrograde musc	
Questa	sill-Kfs-gar-bi-plag; porphyroblasts of retrograde musc	
Southern Taos Range		
station 85-144	sill; gar-bi-chl-plag-hbl-ilm-cc	Sill collected 1 km away from gar-bi rocks
San Cristobal	ky-andal-st-chl-musc; st-chl-bi-musc; gar-bi-chl-musc; hbl-plag-ep-chl-ilm	
Cerrito Colorado	ky-andal-st-bi-chl-musc; st-ctd-musc; ky-ctd-musc; hbl-plag-ep-chl-ilm	
Cimarron Range	ctd-sill-musc-hem; hbl-plag	Ref: Leyenberger, 1983
Picuris Range	ky-andal-sill; gar-bi-st; ctd-st-sill	Distribution of ky, andal, sill controlled by topography Ref: Holdaway, 1978; Grambling and Williams, 1985
Rincon Range		
North	ky-Mn andal; ky-st-gar-bi-musc; ky-andal-st-ctd-musc; hbl-plag-chl-ep	
Central and South	sill; andal; gar-bi-sill-musc-gahnite	Ref: Baltz and O'Neill, 1980a, b; this study
Truchas Range		
North	sill; ky-sill; ctd-st-sill-hem; ky-sill-chl-bi-st-gar-plag-hem-musc; hbl-plag-chl-ep-bi	All st is zincian, all gar is spessartine-rich; st commonly surrounded by plag or musc
Central	ky-andal-sill-bi-st-hem-musc-plag; hbl-plag-chl-ep	Ref: Grambling, 1981; Grambling and Williams, 1985
South	ky-andal-gar-bi-st-chl-musc; gar-bi-st-chl-musc-plag	
Rio Mora		
High elevations	ky; gar-bi-st-chl-musc-plag	Ref: Grambling and Williams, 1985
Low elevations	sill; gar-bi-st-sill-chl-musc-plag; hbl-plag-ep-chl	
Pecos greenstone belt	gar-bi-pl-musc; hbl-plag-chl-epid-bi; actin-plag-chl-ep	Ref: Klich, 1983; this study
Glorieta Baldy	gar-bi-st-sill-musc; gar-bi-chl-musc; hbl-plag-chl-bi	Ref: Renshaw, 1983
Zuni Mountains	hbl-plag-chl-ep; serp-talc	Ref: Lambert, 1983
Sandia Mountains		
Placitas	contact metamorphic aureole	Ref: Berkely and Callender, 1979
Monte Largo	andal-sill-musc; sill-ctd	Sill in pseudomorphs after ky; andal corroded
Tijeras Canyon	andal-sill; andal-st; ky	
Manzano Mountains		
Southern	gar-bi-st-chl-musc; gar-bi-st-chl-sill-musc; ky-andal-sill-ctd-musc	
Central	ctd-chl-bi-musc; chl-cc-albite-musc	Pseudomorphs of chl after st (rare)
Pedernal Hills	ky-musc; st-q; actin-chl-albite-bi-chinozoisite; ep-diopside-gar	Ref: Gonzales, 1968; Armstrong and Holcombe, 1982; Bauer and Williams, 1985

*Abbreviations:

actin = actinolite	ctd = chloritoid	ilm = ilmenite	ppy = pyrophyllite
andal = andalusite	ep = epidote	Kfs = K-feldspar	serp = serpentine
bi = biotite	hem = hematite	ky = kyanite	sill = sillimanite
cc = calcite	hbl = hornblende	musc = muscovite	st = staurolite
chl = chlorite	gar = garnet	plag = plagioclase	

All listed assemblages also contain quartz, except in the serpentine-talc rocks of the Zuni Mountains.

Medium-grade metamorphic rocks dominate the southern Taos Range (Fig. 6), the Cimarron Range, the Picuris Range (Holdaway, 1978; Grambling and Williams, 1985), the Rincon Range (Fig. 7), the Truchas Range (Grambling, 1981; Grambling and Williams, 1985), the Rio Mora uplift (Grambling and Williams, 1985), Glorieta Baldy, and the Pecos greenstone belt (Klich, 1983).

Migmatitic gneiss in the Santa Fe Range (Fig. 1) contains primary muscovite but lacks assemblages with sillimanite and K feldspar. All analyzed garnet grains are rich in MnO. The migmatites appear to have been derived from partial melting of supracrustal rocks (Metcalf, 1985, 1986) and are intruded by a suite of foliated granitic rocks.

Granitic plutons with ages of 1,440 to 1,470 Ma (Brookins, 1982; Bolton, 1976) have been examined in the Sandia and Manzano Mountains. A contact metamorphic aureole surrounds the northwestern side of the Sandia granite (Fig. 8). Berkley and Callender (1979) mapped zones of chlorite phyllite, andalusite schist, and sillimanite schist in this aureole. The sillimanite zone has relict andalusite except within 50 m of the intrusion, where sillimanite and K feldspar coexist. Kyanite has been found in two samples but is highly embayed and probably not a contact-metamorphic phase. The southeastern boundary of the granite, in Tijeras Canyon (Fig. 8), shows no contact aureole. The boundary is a ductile shear zone between granite and andalusite-sillimanite schist; samples with andalusite or kyanite as the only Al_2SiO_5 polymorph have been collected a few kilometers farther south. In the southern Manzano Mountains, the Priest pluton (Fig. 9) also lacks a contact aureole. It lies adjacent to pelitic schist and quartzite that contain the mineral assemblages kyanite–andalusite–sillimanite, sillimanite–chloritoid, and garnet–biotite–staurolite–chlorite–muscovite.

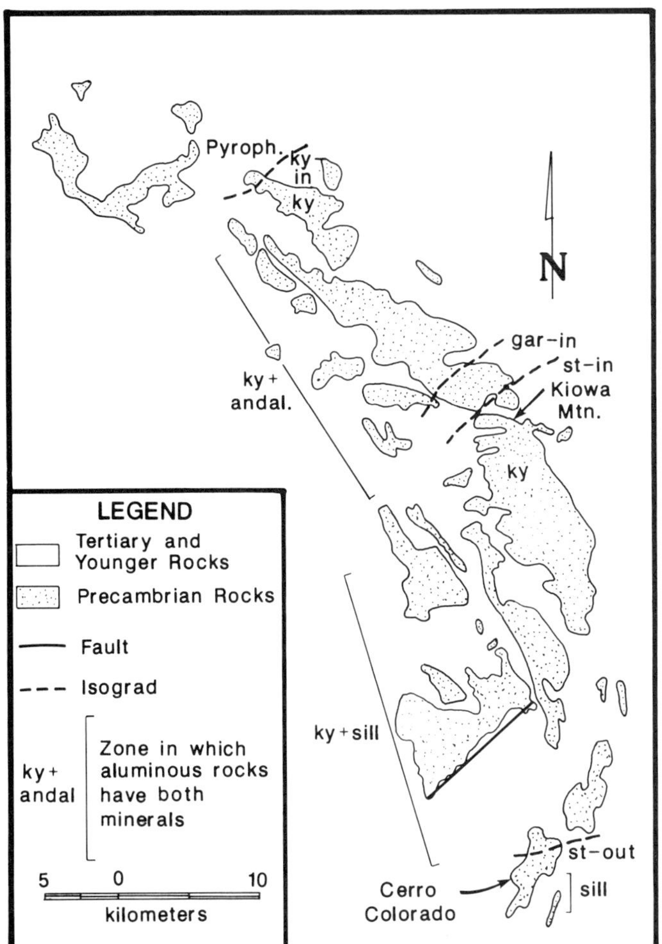

Figure 2. Metamorphic isograds in Proterozoic rocks of Tusas Mountains. Simplified from Williams (1987).

Mineral chemistry and zoning

Medium-grade areas. Euhedral garnets have Mn, Ca, and Fe/(Fe + Mg) falling from core toward rim, but these trends reverse in 5 to 250-μ-thick outer rims (Fig. 10). Positions of the zoning reversals vary by 20 μ or more for different cations. Reverse zoning is best developed where garnet lies in direct contact with biotite or chlorite. Garnets included within staurolite lack reverse-zoned rims (Fig. 11).

Ca in plagioclase typically decreases from cores to points 4 to 40 μ from rims, then varies erratically in thin and incomplete outer rims (Fig. 10). Plagioclase inclusions in staurolite lack the

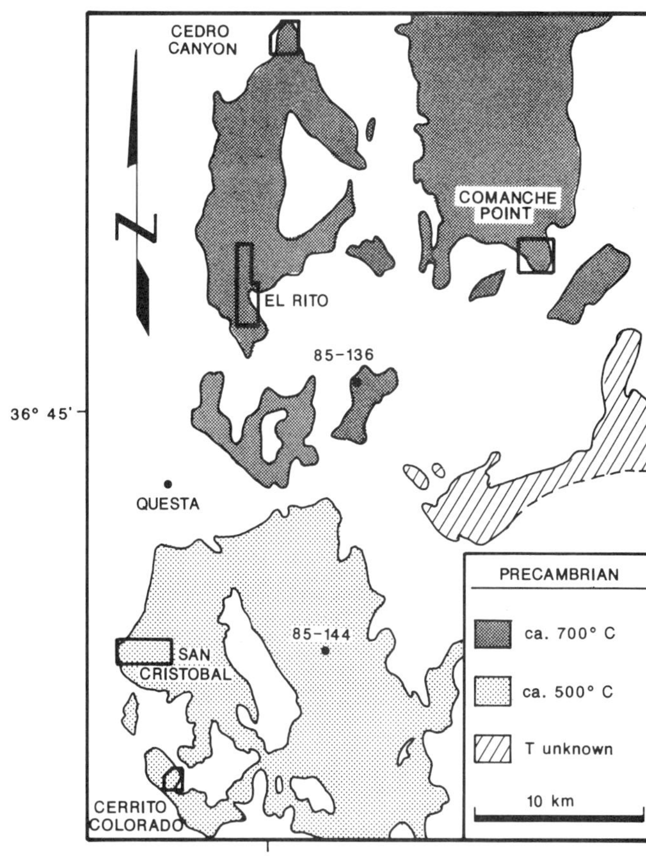

Figure 3. Simplified metamorphic map of Proterozoic rocks in Taos Range showing areas examined in detail.

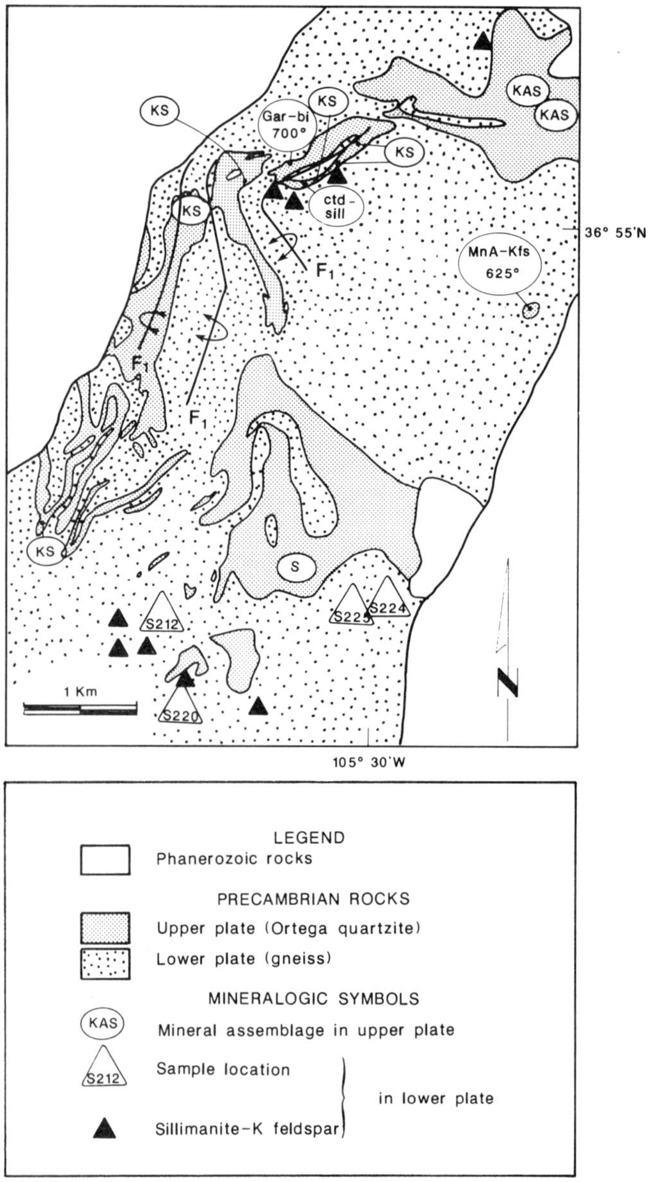

Figure 4. Geologic map of Cedro Canyon area, northern Taos Range, showing sample locations, axial traces of representative folds, and tectonic contact between medium- and high-grade Proterozoic rocks. Medium-grade Ortega quartzite is separated from high-grade rocks by a bedding-parallel shear zone. Simplified from Smith (1988). K = kyanite, A = andalusite, S = sillimanite; andalusite has variable Mn_2O_3.

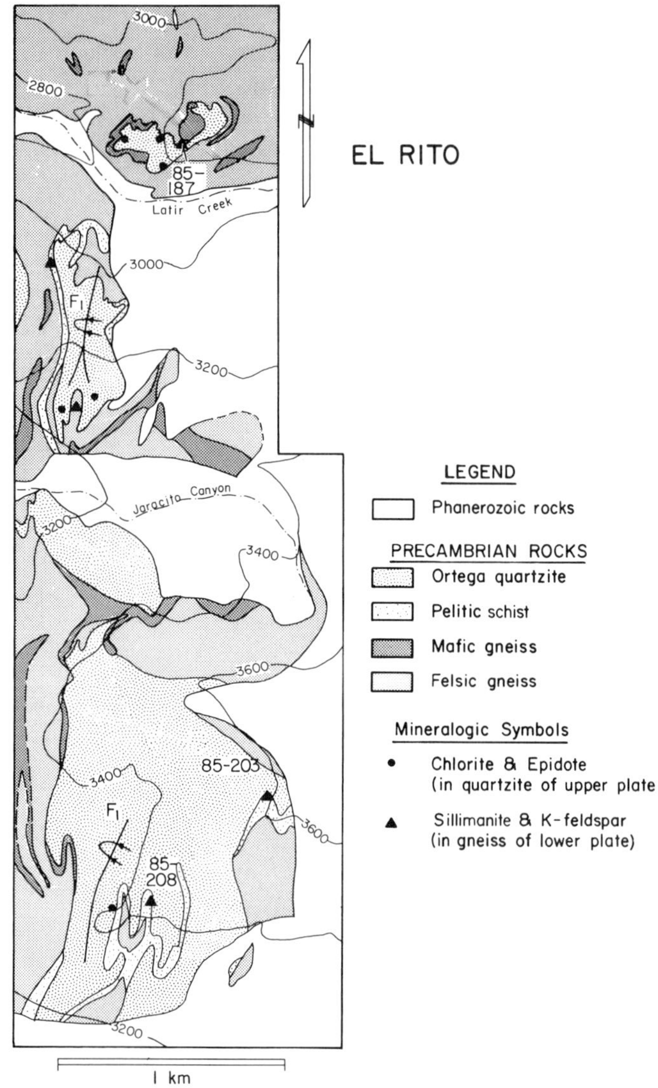

Figure 5. Geologic map of El Rito study area, northern Taos Range, showing sample locations, axial traces of major folds, and tectonic contact between medium- and high-grade rocks. Medium-grade Ortega quartzite is separated from high-grade rocks by a bedding-parallel shear zone. Topographic contours at 200-m intervals.

outer rims. Plagioclase inclusions in garnet have cores that are more sodic than any part of matrix plagioclase, but Ca increases toward rims of the inclusions and, at the contact with enclosing garnet, approaches that in the cores of matrix plagioclase grains. Patches of sericite 10 to 30 μ across occur inside plagioclase in a few rocks. Plagioclase is zoned to nearly pure albite against the sericite.

Matrix biotite is compositionally homogeneous within single thin sections. Biotite is modally abundant, garnet reverse-zoned rims are narrow, and volumetric ratios biotite:(garnet rims) exceed 100. Inclusions of biotite in garnet are common. The inclusions differ from matrix biotites in Ti, Fe, and Mg. Muscovite chemistry is constant on the scale of a thin section except where the mineral occurs as inclusions in other phases.

Garnet and plagioclase zoning is interpreted to be prograde except for rims, which are considered retrograde; "peak" compositions were taken just inside the rims. Sodic cores of plagioclase inclusions within garnet are interpreted to predate the growth of garnet. Zoning in plagioclase adjacent to sericite is assumed to result from post-metamorphic alteration. Muscovite and biotite inclusions in garnet or staurolite are interpreted to record prepeak composition, and matrix grains to preserve peak compositions. The high ratio of matrix biotite to garnet reverse-zoned rims

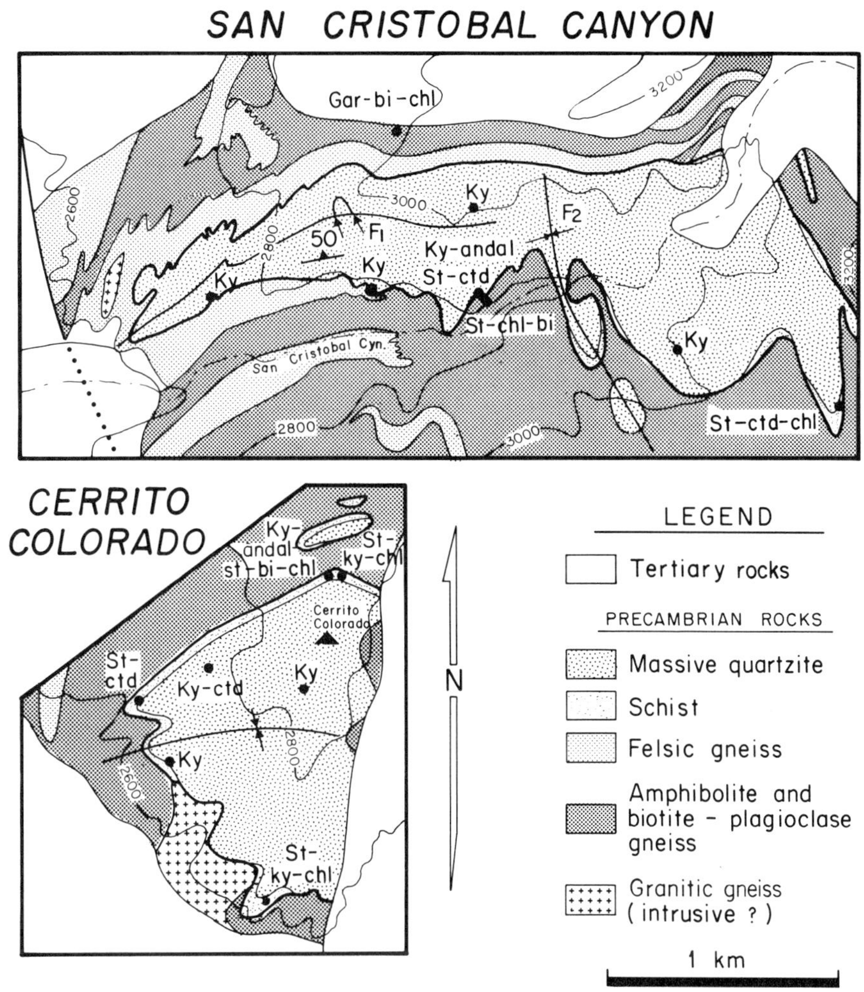

Figure 6. Geologic maps of Proterozoic rocks in San Cristobal and Cerrito Colorado areas, southern Taos Range. Ortega quartzite and minor schist are separated from underlying units by a bedding-parallel shear zone (heavy line). Metamorphic grade does not vary across this shear zone.

suggests that matrix biotite changed negligibly in composition during the development of reverse-zoning in garnet. Thus, matrix biotite is assumed to record both peak and retrograde composition.

Compositions of representative minerals from a number of areas in the triple-point belt are presented in Tables 2 and 3.

High-grade areas. Analytical data come from the northern Taos Range. Garnet cores are unzoned, but rims are zoned (Fig. 12). Rim zoning, best developed adjacent to biotite, shows increases in Fe and Mn at the expense of Mg. Plagioclase has irregular Ca-rich cores surrounded by incomplete, asymmetric, slightly more sodic rims (Fig. 12). Sillimanite crystals have variable Fe_2O_3. Smaller grains, and larger ones where bent in hinges of microfolds, have the lowest Fe_2O_3 in a given thin section.

Fresh biotite has stoichiometric K_2O, but TiO_2 varies by up to 3 weight percent in single thin sections. The lowest-Ti analyses show the highest MgO and Al_2O_3, and are associated with tiny rutile inclusions aligned parallel to 001. Some low-Ti analyses come from hinges of microfolds. Elsewhere, compositional variations in biotite can be seen on traverses normal to 001. In the rocks that contain two textural varieties of biotite, small green grains show low but consistent TiO_2, whereas coexisting porphyroblasts of brown biotite vary widely in TiO_2. Muscovite composition is constant on the scale of a thin section.

Garnet and plagioclase cores are considered to have been homogenized by volume diffusion near peak metamorphic conditions. Rims are interpreted to have formed during retrogression. High-Ti biotite is assumed to have been stable at or near the thermal peak; biotite of intermediate Ti, Al, and Mg to be partially retrograded; and biotite of lowest Ti to be the most thoroughly re-equilibrated. The variable composition of sillimanite suggests that it may have been stable at the thermal peak and during much of the retrogression. Based on its texture, muscovite is considered a retrograde phase.

Kyanite and andalusite occur in a few gneisses collected near Questa. They have grown across foliation and are considered retrograde in origin. Compositions of representative minerals from high-grade gneiss are presented in Table 4.

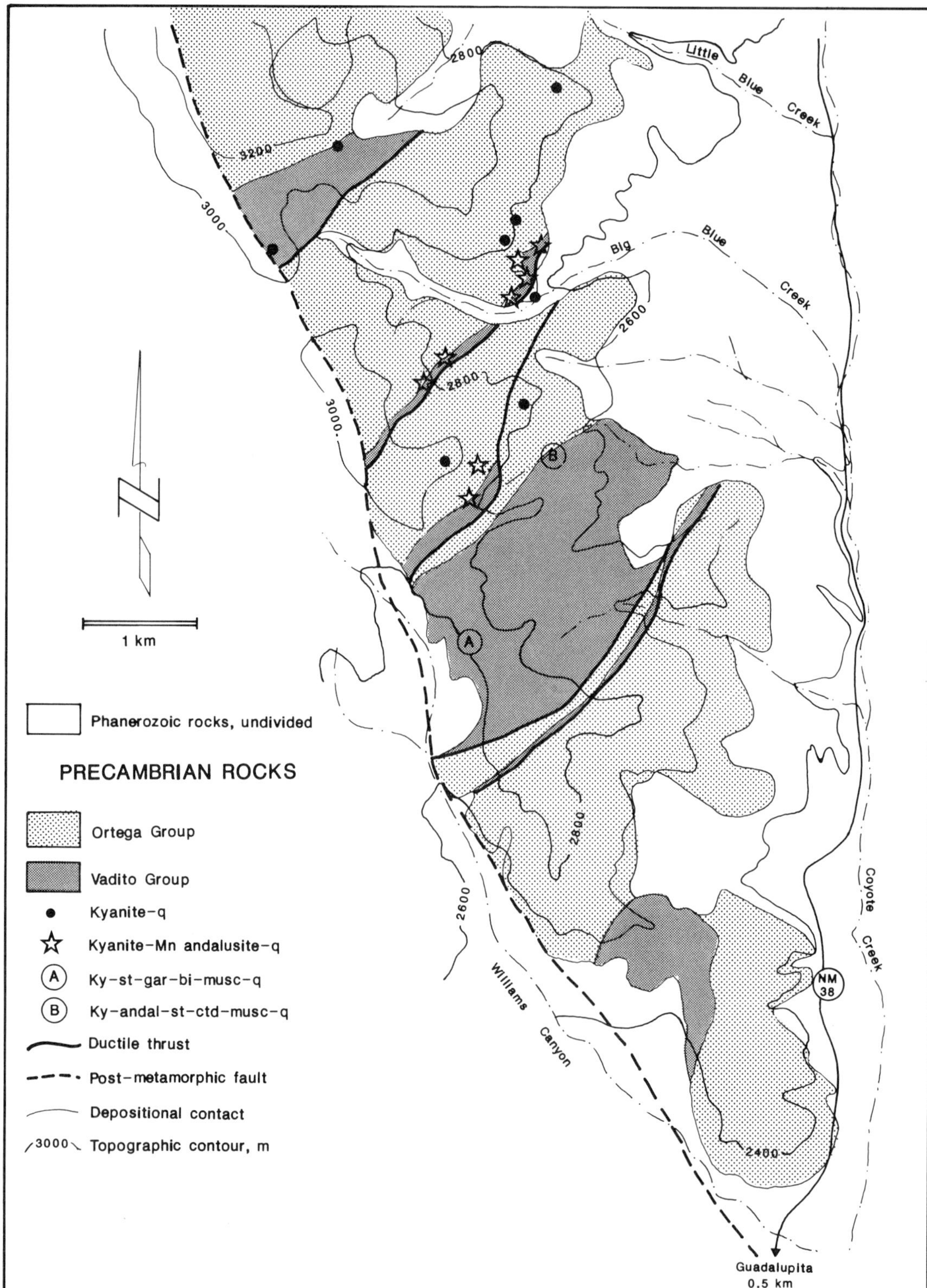

Figure 7. Simplified geologic map of Proterozoic rocks in the northern Rincon Range, showing distribution of selected mineral assemblages.

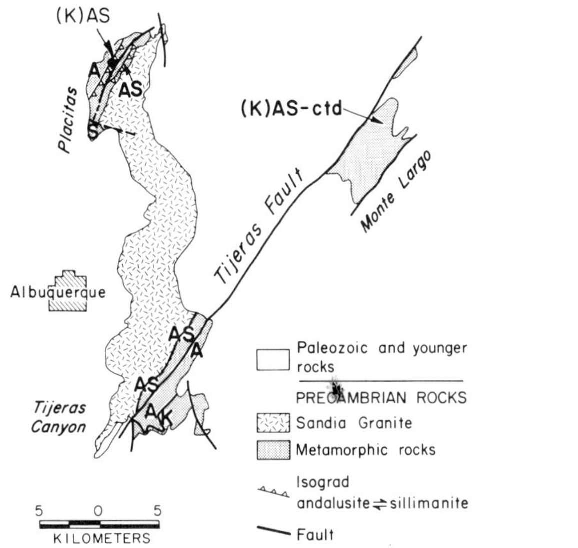

Figure 8. Metamorphic map of Sandia Mountains including Placitas, Monte Largo, and Tijeras Canyon study areas. K = kyanite, A = andalusite, S = sillimanite.

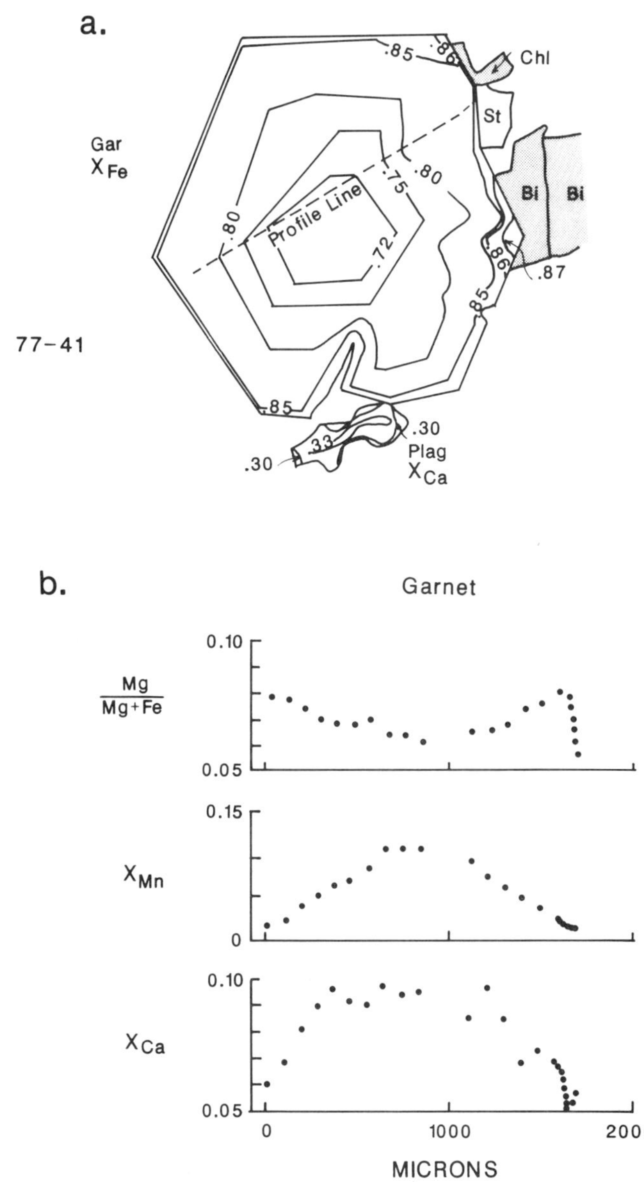

Figure 10. Characteristic chemical zoning maps (a) and profiles (b) for garnet and plagioclase, specimen 77-41, southern Truchas Range. Chlorite outlined in upper right is folded. Garnet profile line starts at upper right rim of garnet but does not reach left-hand rim, resulting in apparent asymmetry in (b).

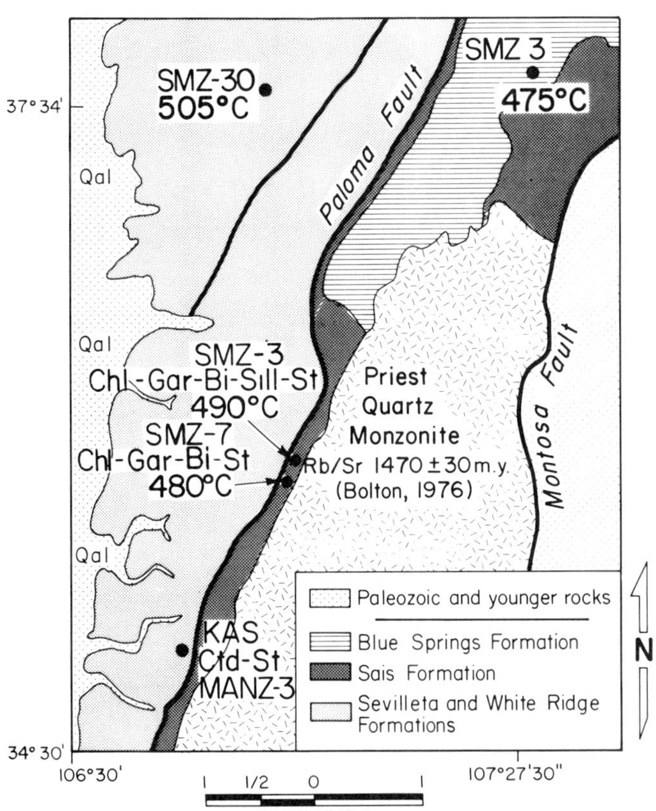

Figure 9. Simplified geologic map of part of the southern Manzano Mountains showing sample localities, mineral assemblages, and garnet–biotite temperatures. Geology from Bauer (1983).

TABLE 2. MINERAL COMPOSITIONS AND CALCULATED P-T VALUES FOR SAMPLES WITH ASSEMBLAGE GARNET-BIOTITE-PLAGIOCLASE-MUSCOVITE

Sample and Location	Garnet					Plagioclase		Biotite		Muscovite		$\ln K_{1b}$, from Ghent and Stout, 1981[1]	$\ln K_{R3}$, from Hodges and Crowley, 1985[1]	P,[5] Kb, K_{1b}	P,[5] Kb, K_{R3}
	X_{Fe}	X_{Mg}	X_{Mn}	X_{Ca}	t (μm)[2]	X_{Ca}	t (μm)[2]	Fe	Mg	X_K	X_{Al}^{VI}				
Southern Tusas Range															
W84-117a	0.662	0.088	0.162	0.088		0.55		0.447	0.367	—	—	10.16	5.87	3.4	1.9
near Cerro	0.663	0.104	0.130	0.104		0.46		0.432	0.385	0.88	0.90	8.79	4.39	4.7	5.1
Colorado	—	—	—	—		—		—	—	—	—	—	—	—	—
W85-253[6]	0.536	0.131	0.270	0.063		0.28		0.275	0.567	—	—	9.13	3.97	4.5[6]	3.4[6]
near Cerro	0.527	0.148	0.258	0.068		0.29		0.313	0.499	0.91	0.89	8.35	3.95	5.2[6]	5.5[6]
Colorado	—	—	—	—		—		—	—	—	—	—	—	—	—
Truchas Range															
77-39	0.785	0.053	0.080	0.082		0.34		0.408	0.319	0.81	0.98	9.88	3.87	3.9	3.1
near Pecos Baldy	0.857	0.076	0.023	0.043		0.20		0.471	0.312	0.76	0.96	9.18	4.27	4.4	3.5
	0.856	0.070	0.023	0.051	15	0.29	11	—	—	—	—	10.05	4.71	3.7	2.6
77-41	0.786	0.053	0.075	0.086		0.37		—	—	—	—	10.18	4.28	3.6	2.8
near Pecos Baldy	0.846	0.080	0.020	0.054		0.31		0.450	0.336	0.82	0.98	9.79	4.71	3.9	3.2
	0.854	0.073	0.019	0.054	230	0.30	31	—	—	—	—	9.98	4.68	3.8	2.8
77-42	0.817	0.068	0.079	0.037		0.22		0.487	0.303	—	—	10.12	5.35	3.7	1.9
near Pecos Baldy	0.836	0.074	0.041	0.049		0.20		0.483	0.299	0.78	0.98	8.69	3.97	4.8	3.8
	0.849	0.068	0.038	0.046	51	0.12	7	—	—	—	—	7.63	2.69	5.8	4.4
84-162[6]	—	—	—	—		—		—	—	—	—	—	—	—	—
near Pecos Baldy	0.473	0.127	0.318	0.083		0.25		0.268	0.586	0.85	0.90	8.27	3.09	5.2[6]	4.5[6]
	—	—	—	—		—		—	—	—	—	—	—	—	—
82-TRU2[6]	0.454	0.110	0.376	0.060		0.34		0.292	0.515	—	—	10.30	5.60	3.5[6]	1.8[6]
near Truchas Peak	0.491	0.153	0.313	0.043		0.30		0.285	0.530	0.80	0.89	10.06	5.57	3.6[6]	3.2[6]
	0.498	0.147	0.300	0.055	54	0.27	5	—	—	—	—	9.07	4.64	4.5[6]	2.9[6]
82-TPST[6]	—	—	—	—		—		—	—	—	—	—	—	—	—
near Truchas Peak	0.471	0.102	0.400	0.027		0.26		0.301	0.595	0.85	0.87	12.53	6.82	1.2[6]	2.1[6]
	0.452	0.083	0.426	0.039	63	0.19	12	—	—	—	—	11.10	5.21	2.8[6]	2.4[6]
82-231[6,7]	—	—	—	—		—		—	—	—	—	—	—	—	—
near Truchas Peak	0.490	0.127	0.269	0.114		0.84		0.322	0.483	0.97	0.91	10.22	5.93	3.4[6,7]	3.0[6,7]
	0.480	0.115	0.273	0.132	90	0.80	22	—	—	—	—	9.93	5.63	3.8[6,7]	2.2[6,7]
Rio Mora Uplift															
80-18a	—	—	—	—		—		—	—	—	—	—	—	—	—
along Rio Mora	0.816	0.093	0.039	0.052		0.29		0.437	0.381	0.80	0.95	9.77	4.36	3.8	4.0
	—	—	—	—		—		—	—	—	—	—	—	—	—
80-178a	0.845	0.088	0.044	0.027		0.17		0.415	0.387	—	—	10.31	5.00	3.5	2.2
along Rio Mora	0.842	0.095	0.037	0.026		0.15		0.458	0.330	0.77	0.96	9.34	4.81	4.3	3.5
	0.857	0.074	0.040	0.029	36	0.13	4	—	—	—	—	9.33	4.17	4.3	3.2

TABLE 2. MINERAL COMPOSITIONS AND CALCULATED P-T VALUES FOR SAMPLES WITH ASSEMBLAGE GARNET-BIOTITE-PLAGIOCLASE-MUSCOVITE (continued)

Sample and Location	Garnet X_{Fe}	Garnet X_{Mg}	Garnet X_{Mn}	Garnet X_{Ca}	Garnet t (μm)[2]	Plagioclase X_{Ca}	Plagioclase t (μm)[2]	Biotite Fe	Biotite Mg	Muscovite X_K	Muscovite X_{Al}^{VI}	lnK_{1b}, from Ghent and Stout, 1981[1]	lnK_{R3}, from Hodges and Crowley, 1985[1]	P,[5] Kb, K_{1b}	P,[5] Kb, K_{R3}
Rio Mora Uplift (continued)															
81-71a[7]	0.530	0.054	0.240	0.177		0.94		—	—	0.98	0.84	11.96	5.20	2.2[6,7]	2.1[6,7]
near Gascon,	0.541	0.102	0.183	0.174		0.93		0.319	0.522	0.98	0.95	10.06	4.59	3.5[7]	4.2[7]
east of Rio Mora	0.554	0.104	0.199	0.143	63	0.90	14	—	—	—	—	10.47	5.05	3.3[7]	2.5[7]
81-71b[6,7]	0.482	0.062	0.288	0.168		0.93		—	—	—	—	11.92	5.60	2.2[6,7]	1.8[6,7]
near Gascon,	0.524	0.096	0.171	0.209		0.93		0.319	0.528	0.98	0.84	9.94	4.30	3.7[7]	6.4[7]
east of Rio Mora	0.525	0.100	0.194	0.181	33	0.93		—	—	—	—	10.27	4.86	3.5[7]	2.7[7]
81-234S[7]	0.716	0.212	0.036	0.70		0.79		0.245	0.639	—	—	10.80	5.44	3.1[7]	2.1[7]
near Pacheco Lake,	0.689	0.217	0.030	0.064		0.51		0.316	0.507	0.94	0.93	8.99	5.01	4.6	3.7
east of Rio Mora	0.675	0.174	0.087	0.064	50	0.35	7	—	—	—	—	8.51	4.16	5.0	3.3
82-12[6]	0.371	0.176	0.413	0.041		0.23		—	—	—	—	9.50	4.52	4.2[6]	2.5[6]
near Pyramid Peak,	0.364	0.184	0.411	0.041		0.23		0.184	0.637	0.79	0.90	9.35	4.35	4.2[6]	4.3[6]
north of Rio Mora	0.350	0.137	0.467	0.046	35	0.16	40	—	—	—	—	8.77	3.32	4.8[6]	3.9[6]
82-62[6]	0.376	0.132	0.403	0.090		0.38		—	—	—	—	9.39	4.33	4.3[6]	2.7[6]
near Pyramid Peak,	0.378	0.164	0.385	0.074		0.35		0.218	0.626	0.86	0.89	9.09	4.35	4.5[6]	4.3[6]
north of Rio Mora	0.382	0.166	0.386	0.066	90	0.38	38	—	—	—	—	9.62	5.07	4.1[6]	2.5[6]
Northern Pecos Greenstone Belt															
C81-181c	0.837	0.085	0.030	0.045		0.36		0.442	0.354	0.91	0.89	10.85	5.87	3.1	1.9
south of Sparks	0.789	0.133	0.024	0.054		0.34		0.437	0.358	0.90	0.90	8.81	5.06	4.7	3.7
Creek	0.812	0.117	0.028	0.043	47	0.31	12	—	—	—	—	9.60	5.55	4.1	2.2
C81-105	0.726	0.067	0.146	0.060		0.29		0.482	0.286	—	—	9.41	5.14	3.7	2.1
along Rio Valdez	0.735	0.076	0.113	0.077		0.30		0.470	0.308	0.87	0.91	8.61	4.08	4.5	4.6
	0.709	0.068	0.148	0.076	20	0.31	19	—	—	—	—	9.08	4.54	4.1	3.0
C81-310[6]	—	—	—	—	—	—		—	—	—	—	—	—	—	—
Villa San Pio	0.584	0.132	0.240	0.044		0.27		0.329	0.457	0.85	0.90	9.54	5.06	4.0	3.7
	0.571	0.125	0.261	0.044	59	0.21	21	—	—	—	—	8.95	4.61	4.0[6]	2.9[6]
C81-244[6]	—	—	—	—	—	—		—	—	—	—	—	—	—	—
north of Maestas	0.386	0.162	0.408	0.045		0.27		0.199	0.633	0.80	0.91	8.88	4.72	3.7[6]	4.0[6]
Creek	0.387	0.149	0.420	0.045	27	0.27		—	—	—	—	10.10	4.91	3.5[6]	2.7[6]

[1]Data presented in order: mineral core, "peak," retrograde rim (where available).
[2]Thickness of retrograde rim.
[3]Fe, Mg as cations per 11/3 oxygens.
[4]Data presented in order: inclusion in garnet, matrix.
[5]Pressures calculated from Ghent and Stout (1981, equation 1b) and Hodges and Crowley (1985, equation R3). Pressures calculated from R3 are sensitive to errors in T; those calculated from 1b are not.
[6]X(Mn, gar) above 0.25.
[7]X(Ca, gar) above 0.10, and X(Ca, plag) above 0.60.

TABLE 3. MINERAL COMPOSITIONS IN OTHER AREAS OF TRIPLE-POINT METAMORPHISM

Sample	85-144					ROC-16		ROC-48a		8-2195		A911-6ab		B911-6ab		10175		49ccf		886ncf	
Location	Taos Range west of Gold Hill					Upper Rociada, 15 km SW of Mora						Glorieta Baldy				Pecos greenstone belt					
																Elk Mtn.		Spring Mtn.			
	Gar	Bi	Hbl	Plag	Calc	Gar	Bi	Gar	Bi	Gar	Bi	Gar	Bi	Gar	Bi	Gar	Bi	Gar	Bi	Gar	Bi
FeO	29.30	25.29	23.79	0.05	1.38	24.23	26.04	32.34	20.02	34.29	20.79	25.00	21.03	32.89	21.03	35.28	20.44	20.39	26.23	26.40	21.02
MgO	0.65	6.49	3.62	—	0.34	0.72	5.07	2.30	9.02	2.19	9.63	1.58	9.76	2.57	9.77	2.43	9.38	0.92	6.22	1.87	10.12
MnO	5.63	0.12	0.26	—	0.60	9.19	0.55	7.33	0.16	1.60	0.06	2.23	0.21	1.24	0.24	4.50	0.08	4.93	0.22	11.00	0.32
CaO	6.70	0.06	11.13	6.23	53.60	8.24	0.01	1.20	0.00	4.77	0.00	11.69	0.02	6.02	0.00	1.40	0.00	9.47	0.08	4.02	0.00
TiO_2	0.00	1.09	0.26	—	—	0.11	2.65	0.01	2.05	0.06	1.46	0.09	1.79	0.05	1.91	0.01	1.69	0.01	2.82	0.01	1.10
Al_2O_3	21.13	19.50	18.26	25.67	—	20.57	17.00	20.89	20.23	21.16	19.79	22.55	17.66	21.23	17.49	21.52	19.69	20.30	17.33	20.96	18.51
SiO_2	37.18	34.24	39.34	59.11	—	37.01	34.35	36.40	36.61	36.47	34.67	36.98	34.90	37.45	35.78	36.12	35.54	36.07	33.51	36.93	35.37
K_2O	—	9.77	0.54	0.09	—	—	9.74	—	8.76	—	8.72	—	8.59	—	9.00	—	8.39	—	8.63	—	8.80
Na_2O	—	0.04	1.29	8.19	—	—	0.05	—	0.20	—	0.21	—	0.12	—	0.17	—	0.42	—	0.16	—	0.13
Total	100.59	96.57	98.52	99.40	55.92	100.07	65.46	100.47	97.05	100.54	95.33	100.12	94.08	101.45	95.39	101.26	95.63	99.09	95.20	101.19	95.38

Sample	TLL-011b	SMZ-3					SMZ-7				SMZ-30		SMZ-37		Manz-3			
Location	Cimarron Mtns.						Southern Manzano Mtns.											
	Ctd	Gar	Bi	St	Sill	Chl	Gar	Bi	St	Chl	Gar	Bi	Gar	Bi	Ky	An	Sill	Ctd
FeO	23.66	38.63	22.36	13.72	0.42	29.40	38.33	22.03	14.06	29.44	37.34	23.69	38.12	22.93	0.52	1.16	0.39	26.83
MgO	1.33	1.82	7.98	0.93	0.03	11.80	1.92	8.57	0.97	12.16	1.34	6.16	1.56	7.33	0.00	0.00	0.00	0.95
MnO	3.45	0.11	0.00	0.07	0.00	0.05	0.55	0.00	0.00	0.01	1.73	0.04	1.44	0.00	0.00	0.97	0.00	0.44
CaO	—	2.47	0.00	—	0.00	—	2.28	0.00	—	—	2.38	0.03	1.35	0.04	—	—	—	—
TiO_2	0.10	0.01	1.52	0.47	0.00	0.08	0.01	1.41	0.36	0.08	0.02	1.69	0.00	1.88	0.00	0.00	0.00	0.03
ZnO				0.28					0.18									
Al_2O_3	40.32	21.19	20.82	55.30	63.00	23.84	21.08	20.40	55.78	24.70	20.91	20.14	21.05	19.48	63.74	63.07	62.94	41.19
SiO_2	23.62	37.09	33.66	27.63	36.00	23.60	37.41	34.85	27.44	23.54	37.32	34.40	36.65	33.86	36.44	35.57	36.65	23.40
K_2O	—	—	8.29	—	—	—	—	8.86	—	—	—	8.97	—	9.18	—	—	—	—
Na_2O	—	—	0.15	—	—	—	—	0.14	—	—	—	0.24	—	0.16	—	—	—	—
Total	92.81	101.32	94.78	98.40	99.45	88.82	101.58	96.26	98.80	90.02	101.04	95.36	100.17	94.88	100.70	100.77	99.98	92.84

TABLE 4. MINERAL COMPOSITIONS FROM HIGH-GRADE GNEISSES OF THE NORTHERN TAOS RANGE

Sample and Location		Garnet				Plag	Biotite			Muscovite			K_D Gar-Bi	T°C, Williams (1987)	T°C, Indares and Martignole (1985)	P,kb[1] Newton and Haselton (1981)	P,kb[2] Ghent and Stout (1981)
		X_{Fe}	X_{Mg}	X_{Mn}	X_{Ca}	X_{Ca}	X_{Fe}	X_{Mg}	X_{Ti}	X_{Al}^{VI}	X_K	X_{Al}^{VI}					
Cedro Canyon																	
S212	Peak	0.832	0.047	0.080	0.041	0.214	0.651	0.127	0.090	0.131	—	—	0.289	785	680	7-9.5	—
	Retro.	0.838	0.047	0.082	0.033	0.113	0.620	0.165	0.075	0.139	0.950	0.914	0.211	630	—	7.8	5.8
S220	Peak	0.807	0.129	0.030	0.034	0.161	0.557	0.262	0.052	0.128	—	—	0.340	830	795	10.6-11.1	—
	Retro.	0.847	0.089	0.042	0.022	0.089	0.464	0.342	0.004	0.191	0.952	0.882	0.127	500	—	5.5	4.5
S224	Peak	0.781	0.084	0.060	0.077	—	0.552	0.247	0.075	0.123	—	—	0.240	735	655	—	—
	Retro.	0.762	0.046	0.094	0.098	—	0.540	0.271	0.055	0.133	—	—	0.119	495	—	—	—
S225[3]	Peak	0.724	0.082	0.058	0.137	0.470	0.566	0.232	0.071	0.131	—	—	0.276	795[3]	770[3]	10.7-11.3[3]	—
	Retro.	0.742	0.063	0.079	0.119	0.138	0.462	0.361	0.004	0.171	0.950	0.904	0.109	475[3]	—	8.4[3]	7.2[3]
El Rito																	
85-187a	Peak	0.704	0.210	0.066	0.021	—	0.370	0.447	0.040	0.141	—	—	0.247	665	660	—	—
	Retro.	0.747	0.112	0.103	0.038	—	0.371	0.467	0.027	0.136	—	—	0.119	460	—	—	—
85-203	Peak	0.616	0.083	0.231	0.070	—	0.462	0.270	0.065	0.125	—	—	0.231	705	740	—	—
Questa																	
85-136	Peak	0.722	0.156	0.082	0.040	0.250	0.473	0.254	0.058	0.148	—	—	0.226	660	620	5.3-6.1	—
ENE of Questa	Retro.	0.825	0.055	0.086	0.034	0.230	0.474	0.264	0.041	0.167	0.920	0.950	0.120	500	—	2.4	3.1
RR79	Peak	0.865	0.057	0.053	0.026	—	0.562	0.203	0.075	0.159	—	—	0.182	580	—	—	—
N of Questa	Retro	0.872	0.042	0.063	0.023	—	0.557	0.200	0.062	0.180	—	—	0.133	480	—	—	—

[1]Range in P is derived by combining T from Williams (1987) and T from Indares and Martignole (1985) with geobarometer.
[2]Peak P not listed because muscovite is not part of peak assemblage.
[3]X_{Ca} in garnet above 0.10.

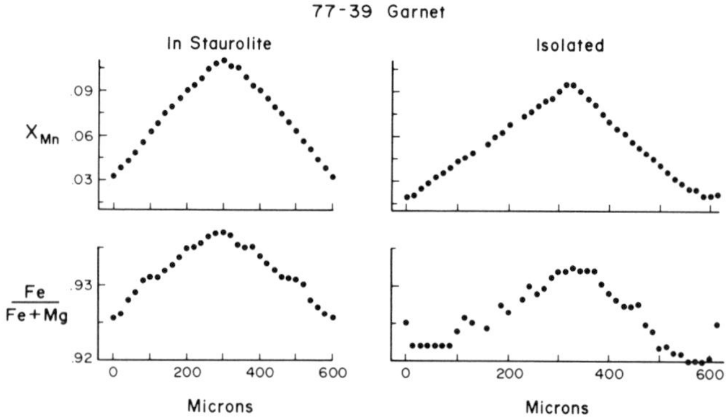

Figure 11. Chemical zoning profiles across two garnets in specimen 77-39, southern Truchas Range. Left-hand profile from garnet included in staurolite; right-hand profile from garnet in matrix of rock. Note absence of rim zoning in inclusion.

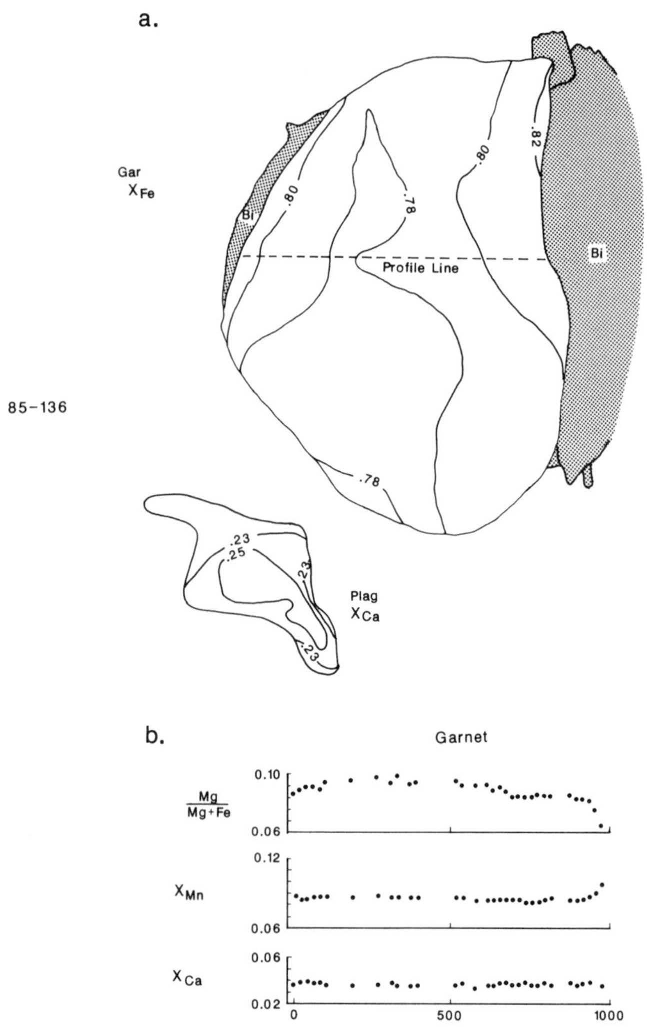

Figure 12. Chemical zoning map (a) and profile (b) across garnet and plagioclase, specimen 85-136, northeast of Questa, Taos Range. Sample is from high-grade gneiss. Note subdued zoning in garnet core, edge zoning, and general asymmetry to zoning map.

Structures

Areas outside Taos Range. Granitic rocks that give dates near 1,450 Ma are not intensely deformed except in local zones of high shear strain.

Other medium-grade Precambrian rocks are intensely deformed. The earliest fabric (S_1) is a bedding-parallel schistosity defined by muscovite, chlorite, and ilmenite in pelitic schist; hornblende and chlorite in metabasalt; and kyanite, muscovite, and hematite in aluminous schist and quartzite. Associated mineral lineations (kyanite, hornblende, aggregates of muscovite, chlorite, or quartz) plunge north or south in the plane of S_1.

Mesoscopic layer-parallel isoclinal folds, abundant in some areas, have axial planes parallel to the S_1 fabric. Most of these F_1 minor folds are rootless antiforms that have easterly trending hinges and verge to the north (Fig. 13). Cross-beds show that the folds are anticlinal as well as antiformal. Kyanite and muscovite fabrics are axial-planar to them.

Map-scale folds associated with S_1 are rare but present. Most are isoclinal synclines, inclined to the north, with amplitudes of several kilometers and easterly trending axial traces. South-dipping ductile shear zones truncate the upper limbs of nearly all synclines. Cross beds in rocks immediately below the shear zones are overturned; in rocks above the shear zones they are upright. Several such sheared folds crop out in the Rio Mora uplift (one is shown in Fig. 14), one in the northern Truchas Range (Grambling, in preparation), one in the northern Picuris Range (Holcombe and others, 1986; Bauer, 1987), and one in the Tusas Range (Williams, 1987, Fig. 2-13).

The early structures are overprinted by later deformation, which is variably developed. Effects of this later deformation include the microfolding of S_1, an intense crenulation cleavage in micaceous rocks, and local rotation of F_1 hinges (cf., Klich, 1983, p. 72–77 and Plate 1). In some regions S_1 is overprinted only by one younger fabric, S_2, that strikes east to northeast and dips 50

Figure 13. Recumbent, rootless isoclinal fold in Ortega quartzite, South Fork Rio de la Casa, northeastern part of Rio Mora uplift. View is to west; figure for scale. Kyanite lies on foliation axial-planar to fold; cross beds suggest that fold, outlined by kyanite-free layer, is anticlinal.

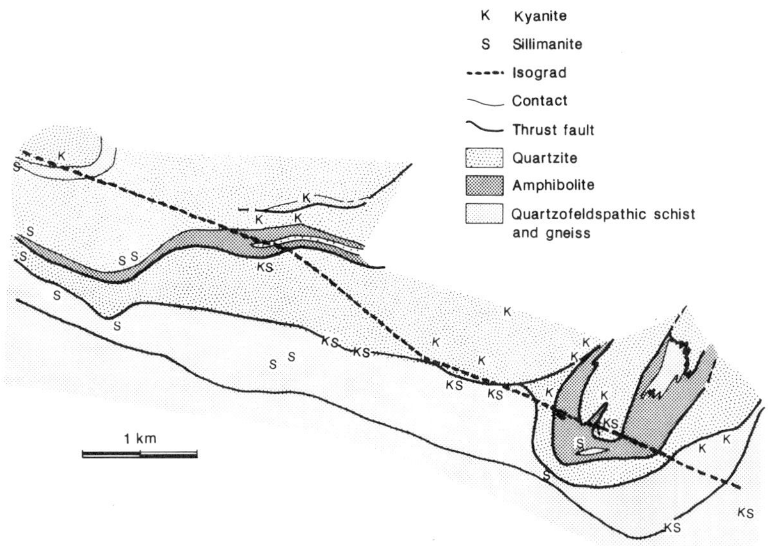

Figure 14. Cross section of Proterozoic rocks in eastern Rio Mora uplift showing rock types, structures, and kyanite–sillimanite isograd. Ductile thrusts have emplaced amphibolite (Vadito Group) above quartzite of younger Ortega Group. Note that isograd cuts across all folds and ductile thrusts. Cross section is equivalent to a downplunge projection because rocks are exposed in nearly vertical cliffs, 1,000 m high, perpendicular to plunge of folds. Plane of section is N30°W,65°NE; northwest to right.

to 80° to the south. In other areas S_2 is overprinted by an S_3 foliation that also strikes east and dips south. Where rocks show multiple foliations, the younger ones have progressively steeper southward dips, but all share a common intersection lineation that plunges gently to the west (cf., Grambling and Codding, 1982; Williams, 1987). Extension lineations associated with the younger foliations, defined by elongate aggregates of micas, plunge steeply down-dip.

Mesoscopic F_2 folds are common. They are open to tight, with axial surfaces that dip steeply south and strike nearly parallel to the traces of F_1 folds. Map-scale F_2 folds have amplitudes less than 1 km. They occur in antiform-synform sets that terminate against ductile shear zones (Grambling and others, 1988, Fig. 2). Whereas rocks on one side of the shear zone are multiply foliated and folded, those on the other side commonly lack fabrics overprinting S_1. These observations suggest that shearing and multiple folding were synchronous.

Small, upright, north-trending crenulations overprint the earlier foliations in the Picuris (Nielsen and Scott, 1979) and Truchas Ranges (Grambling, 1979), but are rare elsewhere. They seem to be associated with the steeply dipping, north-trending, nearly vertical Picuris-Pecos fault, which may record significant dextral-strike-slip motion (Miller and others, 1963).

Kinematic indicators. Pods of kyanite-quartz rock up to several hundred meters long, flattened in S_1 and elongate in the direction of mineral lineation, are found locally. The pods, surrounded by muscovitic schist, contain complex internal foliations defined by concentrations of aligned kyanite and less abundant muscovite. Early planar fabrics are truncated by subparallel later planar fabrics, and in places asymmetric quartz veins have developed along the truncations (Fig. 15). The surrounding schist shows multiple subparallel foliations with no consistent crosscutting relationships. Fabrics are reminiscent of mylonites, although quartz grains show thorough annealing. Asymmetric structures preserved in quartz veins and in anastomosing kyanite-quartz layers suggest a northward sense of shear.

Stratigraphic units are stacked across south-dipping S_1 shear zones in the Tusas (Williams, 1987), Rincon (Fig. 16), Truchas (Grambling, in preparation), and Rio Mora (Fig. 14) study areas. Asymmetric fabrics, as well as mapped offsets along shear zones of small displacement, show north-directed material transport.

A few outcrops contain south-verging mylonites in which

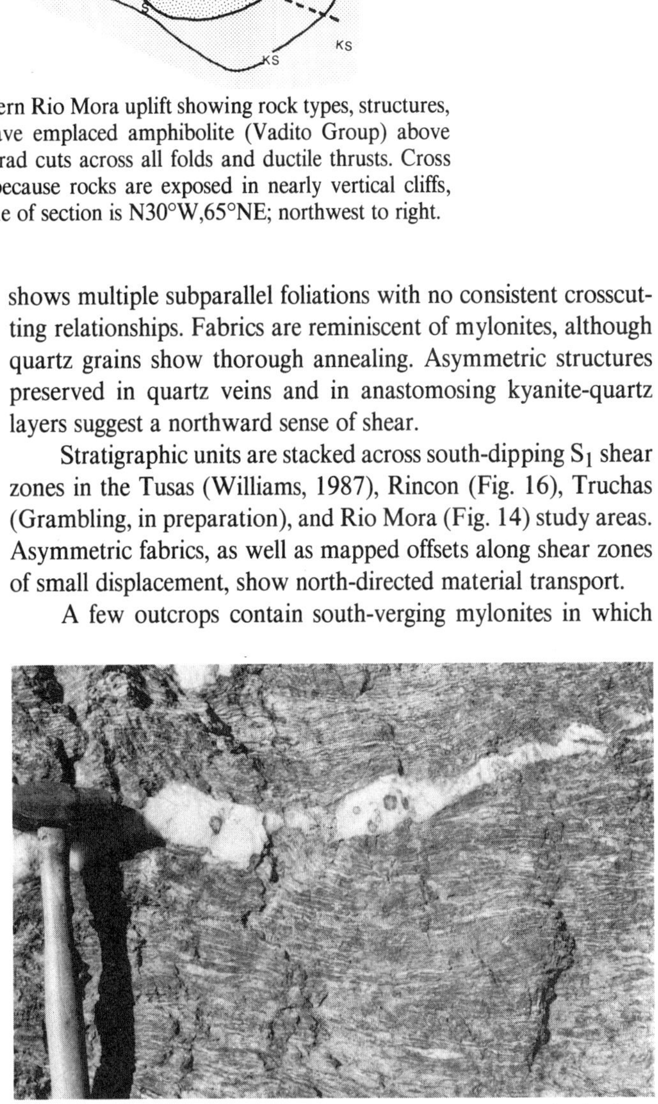

Figure 15. Truncation of early stage of S_1 (below quartz vein) by later stage of S_1 (above quartz vein) in sheared kyanite–quartz gneiss. View is to west, on nearly vertical exposure parallel to kyanite lineation. Offset of quartz vein, and asymmetries in kyanite fabric (not shown), indicate dextral shear (top to the north). Cliffs 100 m southeast of Enchanted Lake, northeastern part of Rio Mora uplift.

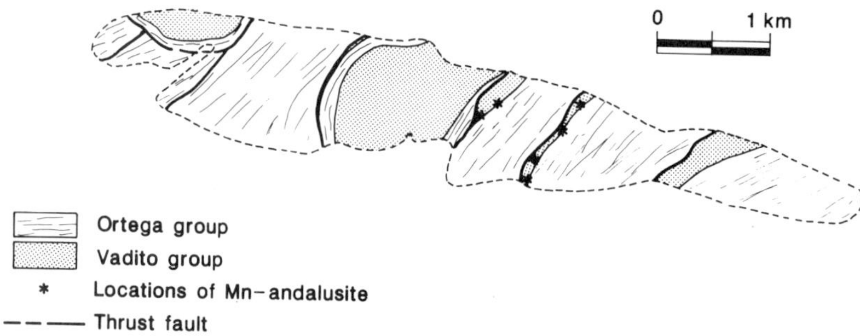

Figure 16. Cross section of Proterozoic rocks in northern Rincon Range (cf. Fig. 7) showing major map units and ductile thrusts. Cross section approximates downplunge projection because rocks are exposed in steep mountain faces with topographic relief exceeding 1,000 m. Plane of projection is N50°W,45°NE; northwest to right.

quartz ribbons, S/C fabrics, and grain-size reductions are well preserved. Such mylonites have been seen in the Tusas, Picuris, Pedernal, and Manzano Ranges (Bauer and Williams, 1985; Bauer, 1987; Williams, 1987; this study). Their poor annealing stands in marked contrast to the thorough annealing of the north-verging fabrics. No direct overprinting between north-verging and south-verging fabrics has been seen.

Southern Taos Range. A bedding-parallel schistosity (S_m) is defined by the flattening of muscovite and chlorite. Mineral lineations formed by kyanite and amphibole plunge north or south in the plane of schistosity. Mica-free quartzite has a mylonitic fabric parallel to S_m. This fabric is defined by flattened quartz grains that show S/C relations and well-preserved ribbons. Oriented samples have not been examined.

Mylonitic quartzite at San Cristobal Canyon and Cerrito Colorado (Fig. 6) structurally overlies mafic and felsic metavolcanic rocks. Map units in the metavolcanic sequence are truncated at the base of the quartzite. There are no apparent differences in metamorphic grade across the truncations.

San Cristobal Canyon exposes east-trending, map-scale F_1 folds that are isoclinal and inclined to the south. Crossbedding shows that the major fold outlined by quartzite (Fig. 6) is a syncline with minimum amplitude of 1.5 km. We do not know if the S_m fabric is related to F_1 folding.

F_2 folds, open to tight, overprint F_1 isoclines in San Cristobal Canyon. The F_2 folds have amplitudes less than 500 m, with upright axial planes that strike north. Interference between F_1 and F_2 folds has created the basin-and-dome map pattern shown in Figure 6.

Cerrito Colorado is dominated by east-trending folds. They have amplitudes of a few hundred meters, are open, upright, and fold an earlier schistosity (S_m?). North-trending folds are not present; the apparent dome-and-basin interference patterns in this area (Fig. 6) are an artifact of topography.

Northern Taos Range. A pervasive bedding-parallel mylonitic fabric (S_m) dominates quartzite and structurally underlying gneiss in both the Cedro Canyon (Fig. 4) and El Rito (Fig. 5) map areas. Intrafolial isoclines, but no mappable folds, are associated with S_m.

Gneiss, quartzite, and the shear zone that separates them are overprinted by map-scale folds. South-inclined F_1 folds with amplitudes and wavelengths of about 0.5 and 1 km, respectively, are the dominant fold structures (Figs. 4 and 5). They are tight to isoclinal, with a well-developed axial planar schistosity. The F_1 folds are highly non-coaxial, giving the map pattern the appearance of polydeformation. However, no post-F_1 structures have been seen.

Highly non-coaxial folds repeat sheared contacts between medium-grade quartzite and high-grade gneiss across much of the northern Taos Range. The sheared contacts appear repeatedly at similar elevations, indicating that enveloping surfaces to the folds are subhorizontal. The mylonitic contact between medium-grade and high-grade rocks is thought to be a single folded surface that extends across the entire northern Taos Range.

Kinematic indicators. Quartzite from Cedro Canyon shows asymmetric porphyroclasts and locally preserved S/C fabrics (Fig. 17). The underlying gneiss has asymmetric pegmatite veins and apparent coarsened S/C textures. Each fabric indicates southerly movement of quartzite with respect to gneiss (Smith, 1988).

One specimen of sillimanite gneiss from Cedro Canyon shows poorly preserved microstructures that may record northward shearing (Smith, 1988). Two types of hornblende fabric have been observed in outcrops of mafic gneiss at El Rito. The earlier is a lineation that wraps around rare north-verging, recumbent, mesoscopic, isoclinal folds; the later is a foliation axial-planar to the folds. Overprinting relations between the rare north-verging structures and the pervasive south-verging fabrics have not been seen.

DISCUSSION

Peak P-T conditions

Medium-grade areas. The best controls on metamorphic conditions come from assemblages with kyanite–andalusite–sillimanite. Mineral compositions in the triple-point assemblages, using the thermobarometer of Grambling and Williams (1985),

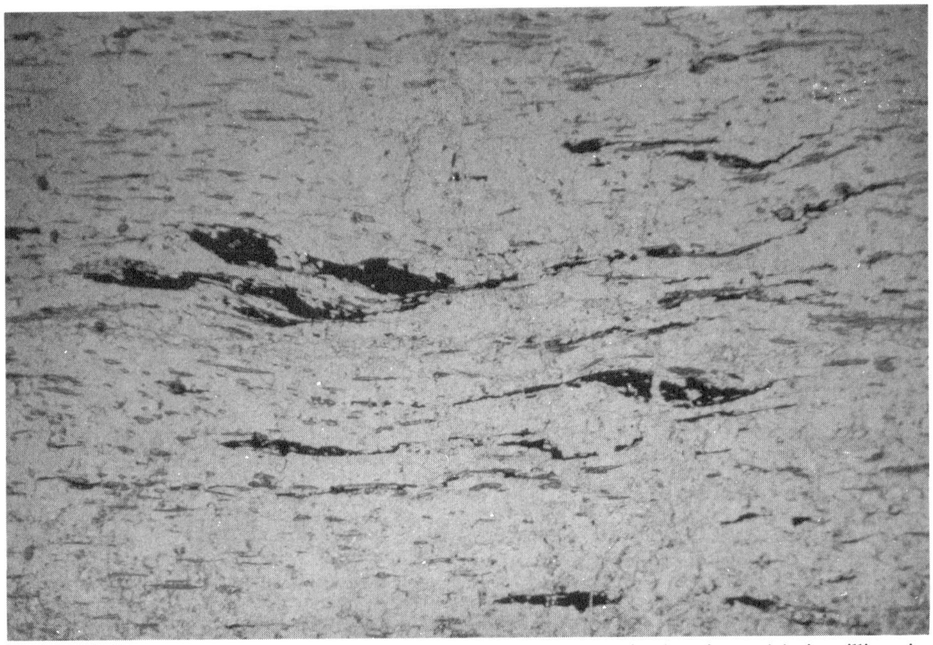

Figure 17. Photomicrograph of asymmetric hematite "fish" in quartzite, kyanite–andalusite–sillimanite zone, Cedro Canyon, northern Taos Range. View is toward the west; asymmetric structures record sinistral shear (top to the south). Base of photo is 6.3 mm across; plane-polarized light.

indicate the following: 505°C, 3.8 kb in the Picuris Range; 520°C, 4 kb in the southern Manzano Range; 520°C, 4.5 kb in Cedro Canyon (in quartzite); 530°C, 4.3 kb in the Truchas Range; and 540°C, 4.5 kb at Rio Mora. The three aluminum silicates occur separately in the Tusas Range and Tijeras Canyon, suggestive of similar P-T conditions.

Three aluminum silicates coexist across an area so large that one must question whether they formed near chemical equilibrium. Grambling and Williams (1985) presented chemical and geologic evidence supporting the hypothesis of equilibrium. The chemical evidence is that minor elements (Fe^{3+}, Mn^{3+}) are distributed regularly among coexisting polymorphs.

The geologic evidence takes two forms. The first is that the spatial distribution of polymorphs is consistent with the attainment of equilibrium. Individual mountain ranges have topographic relief exceeding 1,000 m, allowing isograds to be mapped in three dimensions. Sillimanite ± kyanite occurs at elevations below 2,400 m in the northern Picuris Range. Rocks contain three aluminum silicates at 2,500 to 2,550 m, kyanite-andalusite at 2,600 to 2,900 m, and only kyanite at higher levels. The northern Truchas Range has sillimanite as the only Al_2SiO_5 polymorph. Kyanite first appears near the 3,993 m summit of Truchas Peak, overlying the topographically lower sillimanite-bearing rocks. The planar kyanite-sillimanite isograd dips gently south. In the southernmost Truchas Range, kyanite occurs on mountain summits and andalusite at lower elevations. The kyanite-andalusite isograd is subhorizontal, 200 m below the summits of Pecos Baldy and East Pecos Baldy. Triple-point assemblages are common in the central Truchas Range where the kyanite-andalusite and kyanite-sillimanite isograds intersect (Grambling, 1981). Aluminum silicate isograds in the Rio Mora Uplift are best exposed along a cliff east of Rio Mora where elevation falls 1,000 m in slightly over 2 km. Kyanite occurs along the crest, and sillimanite appears at lower elevations. The two polymorphs coexist across a 50- to 180-m vertical interval, below which kyanite disappears. The narrow zone of coexistence can be traced laterally on the cliff for 15 km (Grambling and Williams, 1985). Green andalusite enriched in Mn_2O_3 (0.3 to 12 weight percent) has been found in a stratiform layer at the Vadito-Ortega contact. Rocks are tightly folded but the kyanite-sillimanite isograd is not. Wherever the Mn-rich layer crosses the isograd, kyanite–sillimanite–manganese andalusite has been found. Each of these geometric arrangements supports the hypothesis that equilibrium was closely approached during metamorphism.

The second form of geologic evidence supporting equilibrium among kyanite, andalusite, and sillimanite is that mineral assemblages and solid-solution thermobarometers yield P-T estimates near Holdaway's (1971) experimentally determined triple point. Chloritoid-sillimanite, which requires pressure and temperature near the triple point, has been collected in seven map areas (Table 1). Andalusite–hematite–quartz temperatures in triple-point areas are 500°C in the central Tusas Range, 500°C in the Picuris Range, 515°C in quartzite of Cedro Canyon, and 540°C at Rio Mora.

Garnet-biotite temperatures in rocks where garnet has low Mn and Ca (Tables 2 and 3) are as follows: 10 rocks across the Picuris Range, 500 to 525°C; 1 specimen from the southern

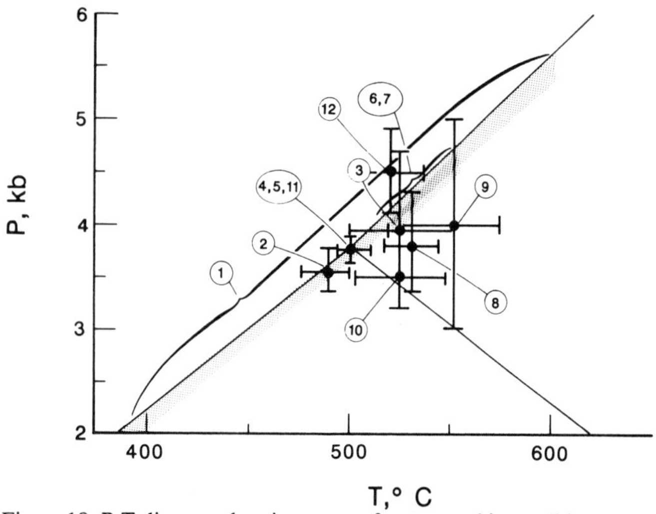

Figure 18. P-T diagram showing range of metamorphic conditions preserved in medium-grade rocks of northern and central New Mexico. Al_2SiO_5 equilibria from Holdaway (1971). 1 = Tusas Range, 2 = southern Taos Range, 3 = Cimarron Range, 4 = Rincon Range, 5 = Picuris Range, 6 = Truchas Range, 7 = Rio Mora, 8 = Glorieta Baldy, 9 = Pecos greenstone belt, 10 = Monte Largo, 11 = southern Manzano Range, 12 = Cedro Canyon, quartzite above shear zone.

Rincon Range, 535°C; 23 samples near the kyanite-andalusite isograd in the southern Truchas Range, 520 ± 20°C, and 15 from the kyanite-sillimanite isograd at Truchas Peak, 550 ± 35°C; 20 rocks near the kyanite-sillimanite isograd at Rio Mora, 550 ± 30°C, and 10 from higher elevations in the same mountain range, 500 to 520°C; 1 specimen from Glorieta Baldy, 530°C; 1 sample from the central Pecos greenstone belt, 520°C, and 4 from the northern part of the belt, 550 ± 60°C; and 4 samples from the southern Manzano Mountains, 490 ± 25°C. No aluminum silicates have been found in the central Pecos greenstone belt. Sillimanite, found in every other area, requires pressures close to the triple point at temperatures below 550°C.

Microprobe analyses have been obtained on assemblages with garnet–biotite–plagioclase–muscovite in several areas (Table 2). The barometers of Ghent and Stout (1981) and Hodges and Crowley (1985), respectively, applied only to specimens where garnet has low Mn and Ca, and plagioclase has low Ca, give the following results: 3 rocks from the southern Truchas Range, 4.4 ± 0.9 kb or 3.5 ± 0.6 kb; 3 from Rio Mora, 4.2 ± 0.8 or 3.7 ± 0.5 kb; and 4 from the northern Pecos greenstone belt, 4.3 ± 1 or 4.0 ± 0.8 kb.

Kyanite–andalusite–staurolite and kyanite-chloritoid without sillimanite have been found at Cerrito Colorado and San Cristobal Canyon in the southern Taos Range, and north of Guadalupita in the Rincon Range; the lack of sillimanite but the presence of staurolite requires pressure and temperature not far below the triple point. Assemblages in quartzite above the ductile shear zone at El Rito, although not particularly restrictive of pressure and temperature, do not contradict the interpretation that temperature peaked near 500°C. The regional occurrence of mafic assemblages transitional from upper greenschist to lower amphibolite facies is consistent with temperatures near the Al_2SiO_5 triple point.

Medium-grade metamorphic rocks *not* indicative of P-T conditions near the Al_2SiO_5 triple point have been identified in only three places:

(1) the central Manzano Range, where assemblages suggest upper greenschist facies conditions;

(2) Cedro Canyon, in quartzite collected at the gneiss-quartzite contact, where one andalusite–hematite–quartz rock with K feldspar has given 625 ± 25 °C, and one garnet-biotite rock has given 700 ± 50°C (Smith, 1988); and

(3) the Tusas Range (Fig. 2), where rocks record a significant gradient in peak pressure and temperature. In the northwestern part of the range, mineral assemblages imply conditions near 400°C, 2 to 3 kb; andalusite–hematite–quartz thermometry (6 specimens) gives 410°C. Four garnet–biotite pairs 4 km north of Kiowa Mountain give 480°C. Two garnet-biotite pairs at Kiowa Mountain yield 500°C, with temperature rising to 525°C 8 km farther south (8 specimens). Pelitic schists with kyanite and sillimanite exposed near Cerro Colorado have given garnet-biotite temperatures of 585 to 605°C (25 samples). Comparison of this temperature to the kyanite-sillimanite boundary suggests pressure of 5 to 6 kb. Pressures calculated for two samples from garnet–biotite–plagioclase–muscovite are 4.7 to 5.5 kb, and from garnet–plagioclase–sillimanite–quartz are 5.7 to 6.2 kb (Table 1).

Figure 18 summarizes the range of peak P-T conditions shown by all data from the 250 km by 300 km medium-grade metamorphic belt. The triple-point belt covers a surface area comparable to the combined areas of southern and central New England (Fig. 19).

Manganese and geobarometry. Although garnet–biotite–plagioclase–muscovite assemblages used to calculate the pressures reported above had low Mn and Ca in garnet, other samples have higher Mn. Analytical data are sufficient to examine the effects of Mn and Ca in garnet, and Ca in plagioclase, on calculated pressures. In continuous outcrops between the northern Pecos greenstone belt and the Rio Mora uplift (Fig. 1), six rocks have been analyzed with X(Mn, garnet) below 0.25, X(Ca, garnet) below 0.10, and X(Ca, plagioclase) below 0.60. Five additional rocks have mineral compositions outside these ranges. Peak pressures calculated using the barometric calibrations of Ghent and Stout (1981) and Hodges and Crowley (1985) are listed in Table 2 and summarized in Figure 20. They show no changes with mineral composition.

High-grade areas. Migmatite in the Santa Fe Range, in which partial melting seems to have occurred in the stability field of muscovite, requires temperatures of 650 to 750°C, at pressure higher than 4 kb (Metcalf, 1985, 1986).

Mineral assemblages in the Taos Range restrict both pressure and temperature (Table 4). Four specimens of garnet–biotite gneiss from Cedro Canyon give peak temperatures of 785 ± 75°C using Williams (1987), and 725 ± 135°C using Indares and Martignole (1985). Two of the rocks also contain plagioclase, sillimanite, and quartz; the barometer of Newton and Haselton (1981) gives 9.5 to 11 kb at the higher temperature or 7 to 10.6 kb at the lower temperature. Two garnet-biotite specimens from El Rito were analyzed. The first, with sillimanite but no K feldspar or muscovite, gave 665°C with both garnet-biotite calibrations. The second, with retrograde muscovite (not analyzed) but no K feldspar, yielded 705 to 740°C. Neither rock contains plagioclase. Specimen 85-136, collected northeast of Questa, yielded 5.3 to 6.1 kb, 620 to 660°C. Specimen RR79, a garnet-biotite rock from just north of Questa, gave peak temperature of 580°C using Williams (1987); the Indares and Martignole (1985) calibration yielded 475°C. This is the sillimanite gneiss in which garnet has a sigmoidal, sheared shape, and the low temperature may reflect volume diffusion in garnet during deformation.

Peak P-T conditions in high-grade rocks of the Taos Range are summarized in Figure 21. Peak pressures decrease systematically from north to south, a trend that can be interpreted in two ways. The northern Taos Range may expose a deeper structural level than the southern Taos Range. Alternatively, cooling rates may have been slower in the south than north, allowing for more extensive retrograde re-equilibration.

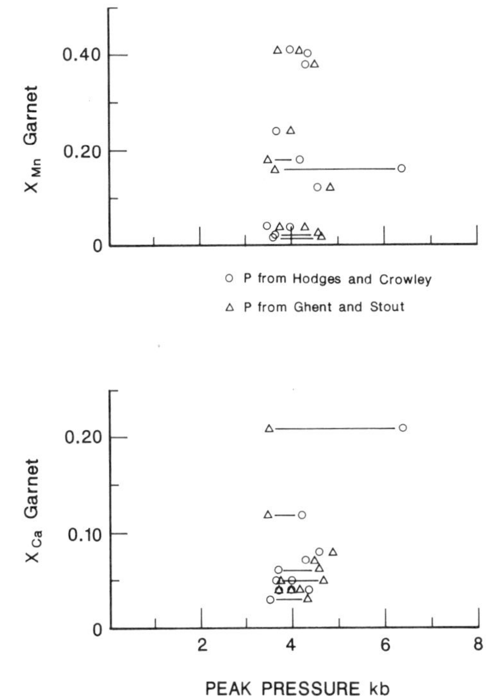

Figure 20. Relationship between Mn and Ca in garnet and pressures calculated from two calibrations of garnet–biotite–plagioclase–muscovite barometer (Ghent and Stout, 1981; Hodges and Crowley, 1985). Ca in plagioclase varies with Ca in garnet (Table 2).

P-T paths

P-T paths have been estimated in seven mountain ranges showing medium-grade metamorphism and in two areas of high metamorphic grade.

Medium-grade areas. The prograde P-T path of six study areas (Cedro Canyon, quartzite above the shear zone; Picuris, Truchas, Monte Largo, and Manzano Ranges) can be derived from aluminum silicate reaction textures. Each area shows a paragenetic sequence kyanite to andalusite to sillimanite, requiring heating at constant or slightly increasing pressure.

Garnet–biotite–plagioclase–muscovite thermobarometry has been applied to inclusion, peak, and retrograde assemblages in four areas, using mineral compositions in Table 2. Only pressures derived from garnet samples with low Mn and Ca are considered here. Two samples from the southern Tusas Range (W84-117a, W85-253) suggest that pressure increased from 4 ± 1.6 kb to 5 ± 0.7 kb as temperature rose 75°C. Three rocks collected in the southern Truchas Range (77-39, 77-41, 77-42) have given inclusion, peak, and retrograde metamorphic conditions of 3.7 ± 0.3 kb, 450°C; 4.4 ± 0.9 kb, 520°C; and 4.4 ± 2 kb, 475°C. Inclusions in one rock from Rio Mora (80-178a) give P-T conditions near 475°C, 3.5 kb. It and another rock (81-234s) give peak P-T conditions of 4.5 ± 0.4 kb, 540 ± 35°C; and retrograde P-T conditions of 4.6 ± 1 kb, 475°C. Three specimens from the northern Pecos greenstone belt (C81-181c, C81-105 and C81-310) give inclusion pressures of 3.4 ± 0.8 kb at 475 to 525°C, peak pressures of 4.4 ± 0.7 kb at 550°C, and retrograde pressures of 3.9 ± 0.6 kb at 475 to 500°C.

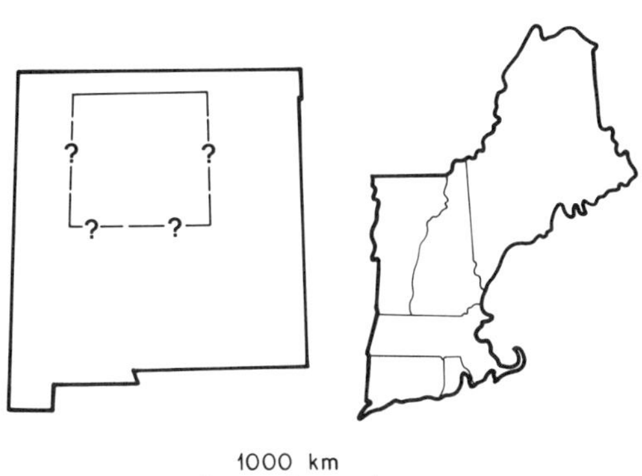

Figure 19. Comparison of size of Al_2SiO_5 triple-point zone in New Mexico (left) with size of New England (right). Queries on New Mexico map outline limits of Proterozoic exposures, not limits of triple-point conditions.

Figure 22 summarizes the P-T paths derived from garnet–biotite–plagioclase–muscovite equilibria. They coincide with prograde paths interpreted from aluminum silicate reaction textures but provide the additional information that retrograde cooling may have been isobaric to ca. 475°C, the lowest temperature preserved by retrograde mineral zoning.

High-grade areas (Taos Range). Mineral textures (pseudomorphs of sillimanite after kyanite, plagioclase after garnet) imply a P-T path showing decompression. This is supported by mineral zoning (Table 4). At the northern end of the Taos Range, specimen S212 gives peak and retrograde conditions of 8 kb, 725°C and 5.8 kb, 630°C; specimen S220 gives peak and retrograde conditions of 10.8 kb, 815°C, and 4.5 kb, 500°C. Specimen

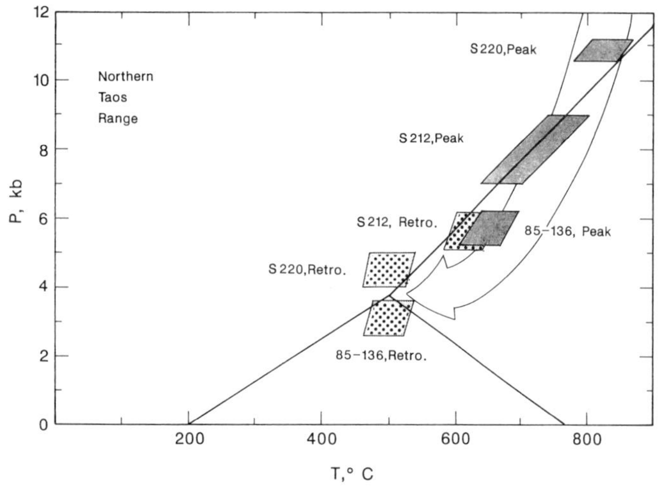

Figure 21. P-T diagram for gneissic rocks of northern Taos Range. Dark pattern shows peak metamorphic conditions (garnet and plagioclase cores) calculated from Williams (1987) and Indares and Martignole (1985) with Newton and Haselton (1981). Light pattern shows retrograde conditions (garnet and plagioclase rims, retrograde muscovite) calculated from Williams (1987) with Ghent and Stout (1981). Retrograde pressures derived from Newton and Haselton (1981) are similar (Table 4).

85-136, collected near Questa, gives peak and retrograde conditions of 5.7 kb, 640°C and 3.1 kb, 500°C. Results are summarized in Figure 21.

Peak P-T estimates lie along a single trajectory, a metamorphic field gradient, that overlaps with retrograde P-T paths calculated from single specimens. Retrograde conditions recorded by two of the three rocks are, within analytical error, those of the aluminum silicate triple point. These conditions are consistent with the late-stage andalusite and kyanite found locally near Questa.

Relative timing of metamorphism and deformation

Medium-grade areas outside the Taos Range. Kyanite, stable in quartzite during the north-verging deformation, requires pressure and temperature greater than 2 kb and 380°C. Assuming that the inferred P-T paths are correct, pressures during the initial growth of kyanite may have been closer to 3 to 4 kb. Quartz fabrics are thoroughly annealed, and metamorphic isograds cut across north-verging folds and ductile thrusts (Fig. 14; see also Grambling, 1986a, Fig. 2; Williams, 1987, p. 88). Apparently the thermal peak was established after north-directed deformation. Textures shown by staurolite, and by sillimanite in local shear zones, suggest that some deformation continued at temperatures close to the thermal peak. The sense of vergence of this deformation is not known. Because the scattered south-verging mylonites that cut the triple-point terrane are much less annealed than the north-directed tectonite fabrics, a reasonable interpretation is that the south-directed shearing occurred at temperatures lower than the northward shearing, after both the northward deformation and the thermal peak.

The north-verging deformation may have caused crustal thickening (Grambling, 1986b). However, any such thickening either was less than the errors associated with thermobarometry (cf., Fig. 22), or occurred prior to the growth of garnet. The observation that P-T paths do not show decompression can be interpreted in two ways. Either extreme crustal thickening was not achieved by north-directed tectonism, or such thickening was relieved by processes that did not affect the P-T path.

Quartzite in Cedro Canyon. We have found no evidence of north-verging deformation in this quartzite, although its record may have been erased by the pervasive south-verging shearing. The alignment and sigmoidal shapes shown by chlorite, epidote, chloritoid, muscovite, and all three aluminum silicates, and the S/C fabrics recorded in quartz (locally annealed), argue that shearing either accompanied or postdated the thermal peak. If shearing postdated the thermal peak, one might expect to see retrogression of peak minerals or reversals in the kyanite to andalusite to sillimanite paragenetic sequence. These features are not seen, indicating that the south-verging shear accompanied the thermal peak.

The pervasive nature of the south-verging shear fabrics in Cedro Canyon contrasts with fabrics in other areas of triple-point metamorphic grade, where such fabrics are rare. The rare fabrics elsewhere may be genetically related to the mylonites of Cedro Canyon.

High-grade areas. Observations in mafic gneiss at El Rito suggest that north-verging deformation did occur in the high-grade rocks, within the stability field of hornblende. Other evidence linking north-directed deformation to metamorphism has not been found. The rarity of north-verging shear fabrics, com-

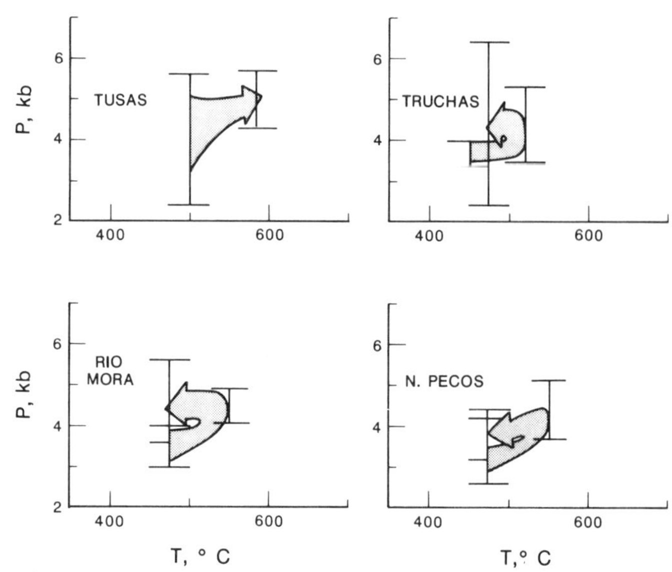

Figure 22. Prograde and retrograde P-T paths for Tusas, Truchas, Rio Mora, and northern Pecos greenstone belt study areas. Derived from garnet–biotite–plagioclase–muscovite equilibria using Williams (1987) and Ghent and Stout (1981).

pared to the pervasiveness of south-verging shear fabrics, supports the interpretation that southward shearing postdated northward shearing.

The asymmetric shape of most peak metamorphic minerals requires that south-verging deformation occurred during or after the thermal peak. Because sillimanite grains and intergrown K feldspar lie in the shear foliation, with no sign of retrogression in many rocks, it is likely that some of the shearing occurred near the thermal peak. Porphyroblasts of retrograde muscovite in some samples have overgrown relicts of sheared and folded sillimanite, but the muscovite is not itself deformed. Other retrograde muscovite porphyroblasts are themselves sheared and folded. The textures are consistent with the interpretation that retrogression and deformation occurred together.

Migmatitic leucosomes in gneiss from the Santa Fe Range commonly are folded. Muscovite defines a foliation parallel to the leucosome layering. Deformation either accompanied or postdated partial melting.

Absolute timing of metamorphism

Deformed, metamorphosed supracrustal and plutonic rocks have U-Pb zircon dates as young as 1,650 Ma across the study area (in the Taos Range, Reed, 1984; east of the Manzano Range, Bowring and others, 1983). Late-stage granitic rocks are ca. 1,450 Ma in age, but their relationship to regional metamorphism is not well known. Many Rb-Sr whole-rock dates from supracrustal rocks fall near 1,400 Ma (Ward, 1986), $^{40}Ar/^{39}Ar$ ages on muscovite are 1,300 to 1,400 Ma (J. Irwin, personal communication to Grambling, 1987; P. Copeland, personal communication to Williams, 1987), and K-Ar and Rb-Sr mineral dates are 1,250 to 1,350 Ma (Ward, 1986). These dates may pertain to metamorphism or uplift of the metamorphic belt.

Triple-point assemblages generally are found within 500 m of the overlying Proterozoic-Paleozoic unconformity. This implies that the metamorphic belt experienced negligible tilting between the peak of metamorphism and 350 Ma, the approximate age of the oldest rocks overlying the unconformity (Baltz and O'Neill, 1980a, b). The present-day topographic relief in the southern Rocky Mountains results mainly from Cenozoic tectonism (Manley, 1984), which disrupted Proterozoic isobaric surfaces in such a way that the triple-point rocks now lie at elevations from 1,800 to 3,650 m above sea level.

INTERPRETATIONS

If retrograde P-T conditions determined from high-grade gneiss in the Taos Range are valid, their coincidence with the triple point can be interpreted in two ways. The gneiss may have experienced two unrelated metamorphic events, one near 700° and one near 500°C. The latter independent thermal event may have occurred regionally at pressures near 4 kb. An alternate view is that the gneiss records one metamorphic event associated with simultaneous decompression and cooling, and that such an event was linked directly to the triple-point regional metamorphism. The remainder of this chapter evaluates these two hypotheses.

Constraints on the hypotheses

The following observations constrain the two hypotheses.

(a) In the northern Taos Range, sillimanite–K feldspar gneiss directly underlies kyanite–andalusite–sillimanite quartzite.

(b) The two metamorphic grades are separated by a mylonitic shear zone.

(c) Shearing occurred during and after the thermal peak in gneiss, and during the thermal peak in quartzite.

(d) Inferred P-T paths from gneiss and quartzite join near the aluminum silicate triple point.

(e) Triple-point P-T conditions are exposed widely outside the northern Taos Range.

(f) Triple-point rocks in the Taos Range show the same prograde P-T path as those elsewhere, heating at constant or slightly increasing pressure; we infer that all triple-point rocks record the same retrograde path.

(g) Rocks in the northern Taos Range preserve rare evidence of synmetamorphic, north-verging deformation (in the gneiss; possibly earlier than south-verging shearing), and abundant evidence of southward shearing (in all rocks).

(h) Rocks outside the Taos Range show abundant evidence of north-directed deformation, probably overprinted by rare south-verging shear zones.

(i) Quartzite in Cedro Canyon seems to show local thermal highs near the contact with underlying gneiss.

(j) Migmatitic gneiss of the Santa Fe Range (Fig. 1) lies in the center of the triple-point belt, opening the possibility that high-grade gneisses directly underlie the entire belt of medium metamorphic grade.

Evaluating the hypotheses

Hypothesis 1: Two metamorphic events. Medium- and high-grade rocks in the Taos Range may be separated by an unconformity or a shear zone unrelated to the triple-point metamorphic event. Gneissic rocks, which in this hypothesis lie below the stratigraphic or structural discontinuity, saw an early M_1 metamorphism at high temperature. This was followed by one of two sequences of events: (1) exposure of the gneiss, deposition of younger rocks, shearing of the unconformity, and then M_2 metamorphism at Al_2SiO_5 triple-point conditions; or (2) tectonic emplacement of unmetamorphosed rocks on top of gneiss, M_2 metamorphism of both tectonic units, and further shearing of the contact during the M_2 thermal peak.

This hypothesis directly explains features (a) and (b). It can account for most other observations but requires special circumstances to do so. Feature (c) can be a result of south-verging M_1 shearing in gneiss, near the M_1 thermal peak, overprinted by later south-verging M_2 shearing in gneiss and quartzite. Item (d) could be a composite of the M_1 and M_2 events, (g) and (h) may have no particular tectonic significance, and (i) can have resulted from shear heating during M_2. High-grade rocks in the Santa Fe Range (feature j) could lie below the same discontinuity that is hypothesized in the Taos Range, if it were regional in its extent.

Hypothesis 1 provides no mechanism to generate the widespread triple-point P-T conditions (feature e) or to produce nearly isobaric P-T paths in the medium-grade rocks (feature f). Its failure to explain these features, and the special circumstances needed to explain other observations, make this hypothesis not particularly attractive.

Hypothesis 2: One metamorphic event. Synmetamorphic decompression typically is attributed to surficial erosion, following the pioneering work of England and Richardson (1977). Erosion cannot have created the P-T path shown by high-grade gneiss of the Taos Range, if metamorphism of the gneiss was related to that of the structurally overlying triple-point rocks, because P-T paths in the two sequences were different. If the data can be interpreted in terms of a single metamorphic event, a mechanism is required to allow simultaneous decompression and cooling in the high-grade rocks, and isobaric heating in the triple-point rocks. Extensional deformation provides a plausible mechanism.

Consider a major detachment fault separating an actively extending lower crust from a more rigid upper crust. Wernicke (1985, p. 109–111) has suggested that rocks below a detachment undergo substantial thinning during the early phases of extensional deformation, at a time when imbricate normal faults have not yet developed in rocks above the detachment. Selverstone (1988) has presented data supporting this interpretation. A detachment fault may be represented by the mylonitic shear zone that separates medium-grade rocks from high-grade rocks in the northern Taos Range. The detachment may extend southward in the subsurface beneath the entire triple-point metamorphic belt.

This hypothesis provides a direct and simple interpretation for all observations. Items (a) through (d) follow directly from the hypothesis. The widespread consistency in upper-plate peak temperatures (feature e) resulted from regionally uniform transfer of heat across the décollement. Isobaric heating of rocks above the décollement (feature f) developed because the upper plate was not being thinned, but it was being heated by upward transfer of heat from actively thinning underlying units. Extension should result in the eventual cooling and decompression of all rocks. Mineral zoning in the upper plate does not record decompression because imbricate normal faults did not develop above the detachment until the rocks had cooled below the temperature at which cation interdiffusion was effective in garnet. Van den Eeckhout and Zwart (1988) have observed features similar to (g) and (h) in the Pyrenees. They explained them as resulting from extension: rocks above a detachment preserve early tectonite fabrics that have been largely erased by extensional shearing near and below the detachment. The upper-plate rocks in the northern Taos Range, where only south-verging shear is preserved, are less than 100 m thick; in this hypothesis it is inferred that north-verging structures should be found there at higher structural levels. Feature (i) resulted from either shear heating or local anomalies in the rate of extension. High-grade gneisses in the Santa Fe Range (feature j) may represent another exposure of rocks beneath the décollement.

The poorly annealed south-verging mylonites that are seen locally in the triple-point rocks developed, in this hypothesis, when extension moved into the upper plate. Continued extension aided in the eventual exposure of the metamorphic belt.

CONCLUSIONS

Although the two hypotheses need further testing, we favor the second because it provides an elegant way to integrate our geological, structural, and metamorphic observations. Thus, we interpret the high-grade metamorphism, the medium-grade metamorphism, and the tectonic juxtaposition of the two metamorphic grades to have resulted from a single geologic process. We infer that this process was extensional deformation. We suggest that the early phases of extension were concentrated below a major décollement located at a crustal depth near 15 km (4.5 kb). Thinning below the detachment brought hot rocks toward it, transferring heat into an upper crust that was not being thinned. This heat was the cause of the triple-point metamorphic event. As extension continued, rocks began to cool; extension then began thinning the upper tectonic plate. Continued crustal extension is interpreted to have caused the eventual exposure of the metamorphic terrane, probably around 1,250 to 1,350 Ma.

If our conclusions are correct, they raise an interesting possibility. The P-T path of gneissic rocks in the northern Taos Range is characteristic of those interpreted to result from rapid uplift and erosion (England and Richardson, 1977; Hollister, 1982; England and Thompson, 1984). However, surficial erosion alone did not form the decompression P-T path shown by gneiss of the northern Taos Range. It is conceivable that P-T paths in other areas showing decompression were generated by extensional deformation, rather than weathering and erosion, of a tectonically thickened crust. Distinguishing between the two mechanisms may prove difficult.

ACKNOWLEDGMENTS

Research was supported by the New Mexico Bureau of Mines and Mineral Resources and by NSF Grants EAR-8309503, EAR-8406572, and EAR-8707565. NSF Grant EAR-8201211 partially supported purchase of the electron microprobe. Additional support came from Anaconda Minerals, Inc. We also thank T. H. Bell, W. Cavin, K. Condie, J. Connolly, H. W. Day, R. J. Holcombe, S. P. Phipps, J. Reed, Jr., J. M. Robertson, and L. A. Woodward for contributing samples or discussions. The manuscript benefited greatly from critical reviews by R. Bateman, F. S. Spear, R. Vidale-Buden and R. P. Wintsch.

REFERENCES CITED

Anderson, J. L., 1983, Proterozoic anorogenic granite plutonism of North America, *in* Medaris, L. G. Jr., and others, eds., Proterozoic geology: Geological Society of America Memoir 161, p. 133–154.

Apted, M. J., and Liou, J. G., 1983, Phase relations among greenschist, epidote amphibolite, and amphibolite in a basaltic system: American Journal of Science, v. 283-A, p. 328–354.

Armstrong, D. G., and Holcombe, R. J., 1982, Precambrian rocks of a portion of the Pedernal Highlands, Torrance County, New Mexico: New Mexico Geological Society Guidebook 33, p. 203–210.

Baltz, E. H., and O'Neill, J. M., 1980a, Preliminary geologic map of the Mora River area, Sangre de Cristo Mountains, Mora County, New Mexico: U.S. Geological Survey Open-File Report 80–374, scale 1:24,000.

—— , 1980b, Preliminary geologic map of the Sapello River area, Sangre de Cristo Mountains, Mora and San Miguel Counties, New Mexico: U.S. Geological Survey Open-File Report 80–398, scale 1:24,000.

Bauer, P., 1983, Geology of Precambrian rocks of the southern Manzano Mountains, New Mexico [M.S. thesis]: Albuquerque, University of New Mexico, 133 p.

—— , 1987, Precambrian geology of the Picuris Range, north-central New Mexico [Ph.D. thesis]: Socorro, New Mexico Institute of Mining and Technology, 259 p.

Bauer, P., and Williams, M. L., 1985, Structural relationships and mylonites in Proterozoic rocks of the northern Pedernal Hills, central New Mexico, *in* Lucas, S. G., ed., Santa Rosa–Tucumcari region: New Mexico Geological Society Guidebook 36, p. 141–145.

Bell, T. H., and Rubenach, M. J., 1983, Sequential porphyroblast growth and crenulation cleavage development during progressive deformation: Tectonophysics, v. 92, p. 171–194.

Berkley, J. L., and Callender, J. F., 1979, Precambrian metamorphism in the Placitas–Juan Tabo area, northwestern Sandia Mountains, New Mexico, *in* Ingersoll, R. V., and Woodward, L. A., eds., Guidebook of Santa Fe country: New Mexico Geological Society Guidebook 30, p. 181–188.

Bolton, W. R., 1976, Precambrian geochronology of the Sevilleta metarhyolite and the Los Pinos, Sepultura, and Priest plutons of the southern Manzano uplift, New Mexico [M.S. thesis]: Socorro, New Mexico Institute of Mining and Technology, 45 p.

Bowen, N. L., and Tuttle, O. F., 1949, The system $MgO\text{-}SiO_2\text{-}H_2O$: Geological Society of America Bulletin, v. 60, p. 439–460.

Bowring, S. A., and Condie, K. C., 1982, U-Pb zircon ages from northern and central New Mexico: Geological Society of America Abstracts with Programs, v. 14, p. 304.

Bowring, S. A., Kent, S. C., and Sumner, W., 1983, Geology and U-Pb geochronology of Proterozoic rocks in the vicinity of Socorro, New Mexico, *in* Chapin, C. E., and Callender, J. F., eds., Socorro region II: New Mexico Geological Society Guidebook 34, p. 137–142.

Bowring, S. A., Reed, J. C., Jr., and Condie, K. C., 1984, U-Pb geochronology of Proterozoic volcanic and plutonic rocks, Sangre de Cristo Mountains, New Mexico: Geological Society of America Abstracts with Programs, v. 16, p. 216.

Brookins, D. G., 1982, Radiometric age of Precambrian rocks from central New Mexico, *in* Grambling, J. A., and Wells, S. G., eds., Albuquerque country II: New Mexico Geological Society Guidebook 33, p. 187–190.

England, P. C., and Richardson, S. W., 1977, The influence of erosion upon mineral facies of rocks from different metamorphic environments: Geological Society of London Journal, v. 134, p. 201–213.

England, P. C., and Thompson, A. B., 1984, Pressure-temperature-time paths of regional metamorphism; I, Heat transfer during the evolution of regions of thickened continental crust: Journal of Petrology, v. 25, p. 894–928.

Ferry, J. M., and Spear, F. S., 1978, Experimental calibration of the partitioning of Fe and Mg between garnet and biotite: Contributions to Mineralogy and Petrology, v. 66, p. 113–117.

Ghent, E. D., and Stout, M. Z., 1981, Geobarometry and geothermometry of plagioclase-biotite-garnet-muscovite assemblages: Contributions to Mineralogy and Petrology, v. 76, p. 92–97.

Gonzales, R. A., 1968, Petrography and structure of the Pedernal Hills, Torrance County, New Mexico [M.S. thesis]: Albuquerque, University of New Mexico, 78 p.

Grambling, J. A., 1979, Precambrian geology of the Truchas Peaks region, north-central New Mexico, and some regional implications, *in* Ingersoll, R. V., and Woodward, L. A., eds., Santa Fe country: New Mexico Geological Society Guidebook 30, p. 135–143.

—— , 1981, Kyanite, andalusite, sillimanite, and related mineral assemblages in the Truchas Peaks region, New Mexico: American Mineralogist, v. 66, p. 702–722.

—— , 1986a, A regional gradient in the composition of metamorphic fluids in pelitic schist, Pecos Baldy, New Mexico: Contributions to Mineralogy and Petrology, v. 94, p. 149–164.

—— , 1986b, Crustal thickening during Proterozoic metamorphism and deformation in New Mexico: Geology, v. 14, p. 149–152.

Grambling, J. A., and Codding, D. B., 1982, Stratigraphic and structural relationships of multiply deformed Precambrian metamorphic rocks in the Rio Mora area, New Mexico: Geological Society of America Bulletin, v. 93, p. 127–137.

Grambling, J. A., and Williams, M. L., 1985, The effects of Fe^{3+} and Mn^{3+} on aluminum silicate phase relations in north-central New Mexico, U.S.A.: Journal of Petrology, v. 26, p. 324–354.

Grambling, J. A., Williams, M. L., and Mawer, C. K., 1988, The Proterozoic tectonic assembly of New Mexico: Geology, v. 16, p. 724–727.

Hodges, K. V., and Crowley, P. D., 1985, Error estimation and empirical geothermobarometry for pelitic systems: American Mineralogist, v. 70, p. 702–709.

Holcombe, R. J., Bauer, P. W., and Callender, J. F., 1986, A kinematic model for polyphase deformation of Proterozoic rocks, Picuris Range, northern New Mexico, U.S.A., *in* Aldrich, M. J. and Laughlin, A. W., eds., Proceedings of the Sixth International Conference on Basement Tectonics, p. 199.

Holdaway, M. J., 1971, Stability of andalusite and the aluminum silicate phase diagram: American Journal of Science, v. 271, p. 97–131.

—— , 1978, Significance of chloritoid-bearing and staurolite-bearing rocks in the Picuris Range, New Mexico: Geological Society of America Bulletin, v. 89, p. 1404–1414.

Hollister, L. S., 1982, Metamorphic evidence for rapid (2 mm/yr) uplift of a portion of the Central Gneiss Complex, Coast Mountains, B.C.: Canadian Mineralogist, v. 20, p. 319–322.

Indares, A., and Martignole, J., 1985, Biotite-garnet geothermometry in the granulite facies; The influence of Al and Ti in biotite: American Mineralogist, v. 70, p. 272–278.

Klich, I., 1983, Precambrian geology of the Elk Mountain–Spring Mountain area, San Miguel County, New Mexico [M.S. thesis]: Socorro, New Mexico Institute of Mining and Technology, 147 p.

Lambert, E. E., 1983, Geology and petrochemistry of ultramafic and orbicular rocks, Zuni Mountains, Cibola County, New Mexico [M.S. thesis]: Albuquerque, University of New Mexico, 166 p.

Leyenberger, T. L., 1983, Precambrian geology of Cimarron Canyon, Colfax County, New Mexico [M.S. thesis]: Albuquerque, University of New Mexico, 93 p.

Manley, K., 1984, Brief summary of the Tertiary geologic history of the Rio Grande rift in northern New Mexico, *in* Baldridge, W. S., and others, eds., Rio Grande rift; Northern New Mexico: New Mexico Geological Society Guidebook 35, p. 63–66.

Metcalf, R. V., 1985, Formation of layered and schlieren migmatites by partial melting at Aspen Basin, New Mexico: Geological Society of America Abstracts with Programs, v. 17, p. 662.

—— , 1986, Production of muscovite trondhjemite and granite by anatexis at temperatures below the second sillimanite isograd: Geological Society of

America Abstracts with Programs, v. 18, p. 694.

Miller, J. P., Montgomery, A., and Sutherland, P. K., 1963, Geology of part of the southern Sangre de Cristo Mountains, New Mexico: New Mexico Bureau of Mines and Mineral Resources Memoir 11, 106 p.

Moody, J. B., Meyer, D., and Jenkins, J. E., 1983, Experimental characterization of the greenschist/amphibolite boundary in mafic systems: American Journal of Science, v. 283, p. 48–96.

Newton, R. C., and Haselton, H. T., 1981, Thermodynamics of garnet-plagioclase-Al_2SiO_5 geobarometer, *in* Newton, R. C., and others, eds., Thermodynamics of minerals and melts: Advances in Physical Geochemistry, v. 1, p. 131–148.

Nielsen, K. C., and Scott, T. E., Jr., 1979, Precambrian deformational history of the Picuris Mountains, New Mexico, *in* Ingersoll, R. V. and Woodward, L. A., eds., Santa Fe country: New Mexico Geological Society Guidebook 30, p. 113–120.

Rao, B. B., and Johannes, W., 1979, Further data on the stability of staurolite + quartz and related assemblages: Neues Jahrbuch für Mineralogie Monatshefte, p. 437–447.

Reed, J. C., Jr., 1984, Proterozoic rocks of the Taos Range, Sangre de Cristo Mountains, New Mexico, *in* Baldridge, W. S., and et al., eds., Rio Grande rift; Northern New Mexico: New Mexico Geological Society Guidebook 35, p. 179–185.

Renshaw, J. L., 1984, Precambrian geology of the Thompson Peak area, Santa Fe County, New Mexico [M.S. thesis]: Albuquerque, University of New Mexico, 197 p.

Selverstone, J., 1988, Evidence for east-west crustal extension in the eastern Alps; Implications for the unroofing history of the Tauern window: Tectonics, v. 7, p. 87–105.

Silver, L. T., Bickford, M. E., Van Schmus, W. R., Anderson, J. L., and Medaris, L. G., 1977, The 1.4–1.5 b.y. transcontinental anorogenic plutonic perforation of North America: Geological Society of America Abstracts with Programs, v. 9, p. 1176–1177.

Smith, R. F., 1988, Structural and metamorphic evolution of Proterozoic rocks in the northern Taos Range, Taos County, New Mexico [M.S. thesis]: Albuquerque, University of New Mexico, 85 p.

Van den Eeckhout, B., and Zwart, H. J., 1988, Hercynian crustal-scale extensional shear zone in the Pyrenees: Geology, v. 16, p. 135–138.

Vernon, R. H., 1987, Oriented growth of sillimanite in andalusite, Placitas–Juan Tabo area, New Mexico: Canadian Journal of Earth Sciences, v. 24, p. 580–590.

Ward, D. B., 1986, Rb-Sr dating techniques applied to a metamorphosed Proterozoic terrane in the southern Sangre de Cristo Mountains, north-central New Mexico [M.S. thesis]: Albuquerque, University of New Mexico, 152 p.

Wernicke, B., 1985, Uniform-sense normal simple shear of the continental lithosphere: Canadian Journal of Earth Sciences, v. 22, p. 108–125.

Williams, M. L., 1987, Stratigraphic, structural, and metamorphic relationships in Proterozoic rocks from northern New Mexico [Ph.D. thesis]: Albuquerque, University of New Mexico, 138 p.

Manuscript Accepted by the Society September 27, 1988

Printed in U.S.A.

Geological Society of America
Special Paper 235
1989

Proterozoic supracrustal rocks and plutons of the Cimarron Canyon area, north-central New Mexico

Reinhard A. Wobus
Department of Geology, Williams College, Williamstown, Massachusetts 01267

ABSTRACT

Precambrian supracrustal rocks of the Cimarron Canyon area consist of quartzite, amphibolite, and a previously undescribed bimodal metavolcanic assemblage with associated metasedimentary rocks. These layered rocks, and the plutons that intrude them, differ on opposite sides of the Fowler Pass fault, a northwest-trending reverse fault of presumed Laramide age. The rocks east of the fault consist of a weakly metamorphosed sequence of felsic and mafic volcanic rocks and interlayered volcaniclastic sediments (now phyllite, chlorite schist and metasiltstone). The felsic rocks, which locally contain well-preserved eutaxitic fabric and bipyramidal quartz phenocrysts, are anomalously low in K_2O and may represent highly altered tuffs. The metabasalts are tholeiitic and locally show amygdular texture. The metavolcanic rocks were intruded sequentially by small plutons that range in composition from gabbro to granodiorite and granite but do not appear to be comagmatic with granitic rocks west of the fault, the former being appreciably lower in potash and higher in lime and soda. West of the fault, the weakly to strongly foliated granitic rocks contain elongate roof pendants of quartzite (feldspathic and muscovitic near contacts) and smaller pendants of amphibolite of uncertain origin.

Stratigraphic relations between the quartzites and the metavolcanic rocks, which lie on opposite sides of the Fowler Pass fault, are not determinable in this area. Lithologic similarity of these rocks to radiometrically dated supracrustal rocks in the Sangre de Cristo Mountains and Tusas Mountains to the west suggests a late early Proterozoic age for the deposition and metamorphism of the Precambrian framework of the Cimarron Mountains.

INTRODUCTION

The Cimarron Mountains are a south-southeast–trending spur of the Sangre de Cristo uplift of north-central New Mexico (Fig. 1). The heavily forested ridges and peaks of the range rise more than 1,800 m above the Great Plains to the east, to elevations of more than 3,600 m. On the west, the Moreno Valley separates the Cimarron Mountains from the Taos Range, the highest section of the Sangre de Cristo Mountains in New Mexico.

The canyon of the Cimarron River provides the best exposures of the Precambrian rocks, which are widely exposed in the Cimarron Mountains from the canyon southward along the range crest. Two tributary drainage basins, Tolby Creek and Clear Creek, also dissect the Precambrian terrane in the canyon area and provide access to the higher parts of the range.

A large portion of the Cimarron Mountains is now within the Philmont Ranch, owned by the Boy Scouts of America. The present study considers the Precambrian rocks outside of the ranch boundaries but within the Colin Neblett Wildlife Management Area of the New Mexico Game and Fish Department, which adjoins Philmont on the west. The wildlife area boundaries serve as an arbitrary boundary for the map area of this report.

The Precambrian rocks in the eastern part of Cimarron Canyon have been previously mapped and described in a U.S. Geological Survey Professional Paper on the Philmont Ranch

Wobus, R. A., 1989, Proterozoic supracrustal rocks and plutons of the Cimarron Canyon area, north-central New Mexico, 1989, *in* Grambling, J. A., and Tewksbury, B. J., Proterozoic geology of the southern Rocky Mountains: Boulder, Colorado, Geological Society of America Special Paper 235.

(Robinson and others, 1964). That report was intended for a general audience and did not attempt to describe the Precambrian rocks in detail; only two Precambrian map units ("granodiorite" and "gneiss, schist, and quartzite") were defined. No mention was made in that report of the bimodal metavolcanic rocks of the canyon area; in fact, some of the quartzites described are actually felsic metavolcanic rocks.

The westernmost part of Cimarron Canyon was mapped by Clark and Read (1972). Again, only two map units were defined for the Precambrian rocks in the canyon ("granitic rocks" and "mafic gneiss"), with no account of the quartzite and metavolcanic rocks that also occur there. The Cenozoic structural geology of the central part of the Cimarron Mountains, including Cimarron Canyon, was discussed by Goodknight (1976), who did not subdivide the Precambrian rocks in his report.

The present study of the Precambrian rocks of the central

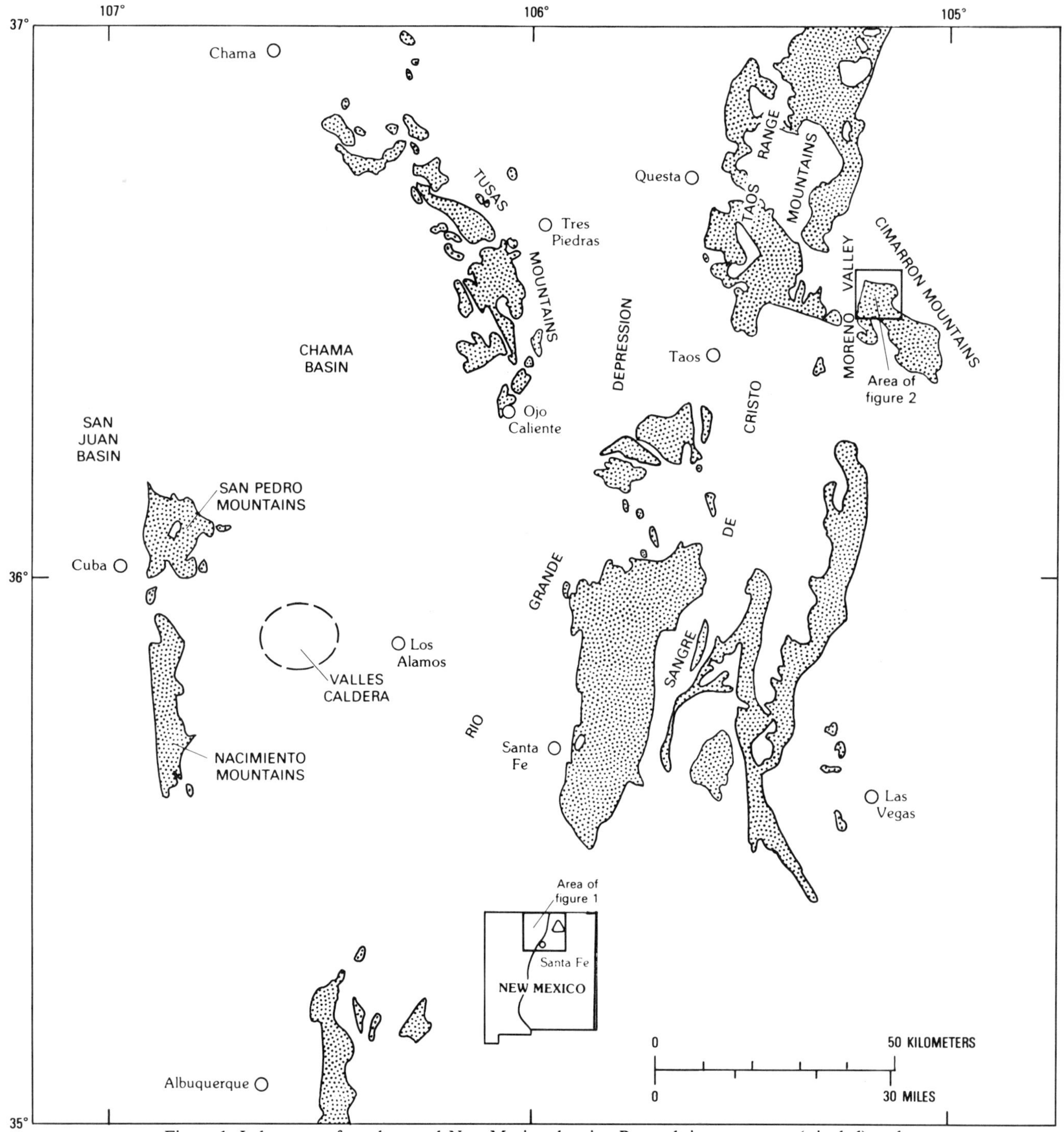

Figure 1. Index map of north-central New Mexico showing Precambrian exposures (stippled) and location of Cimarron Mountains.

part of the Cimarron Mountains took place during the fall of 1982 and late spring of 1983 as part of a U.S. Geological Survey project to map the Wheeler Peak Quadrangle (scale 1:100,000).

GENERAL GEOLOGY

The Precambrian rocks of Cimarron Canyon consist of two different sequences that are juxtaposed along the Fowler Pass fault, a southwest-dipping high-angle reverse fault of presumed Laramide age (Fig. 2). Displacement of several thousand feet has been reported across the southern portion of this fault southeast of the present study area (Goodknight, 1976); its throw diminishes toward the northwest. East of the fault in Cimarron Canyon, the Precambrian rocks are bimodal metavolcanic and associated metasedimentary rocks of the upper greenschist facies, all of which have been intruded by small calc-alkaline to calcic stocks of Precambrian age. West of the fault, in the upthrown block, elongate roof pendants and screens of quartzite and amphibolite are surrounded by fine- to medium-grained, weakly to strongly foliated Precambrian granite.

The Precambrian rocks of the canyon area have all been cut by basaltic dikes, too small to map at this scale, with a general northwest trend. The age of these dikes is uncertain, but they and their Precambrian host rocks are unconformably overlain by sedimentary rocks of the Dockum Group (Triassic) and are in fault contact with other Mesozoic and Cenozoic sedimentary units. Dikes of similar composition and trend in the Taos Range to the west are cut by the unconformity at the base of Mississippian strata (Reed, 1984).

In the western part of the study area, the Precambrian rocks are cut by the Lost Cabin fault, a normal fault with the downthrown side to the west. The Lost Cabin fault brings Precambrian rocks into contact with Tertiary basalts just south of the map area (Robinson and others, 1964). The unconformity and the major fault zones are commonly occupied by dikes and thick sills of Tertiary rhyodacite. The sills commonly weather along vertical joints and form spectacular pillared outcrops like those at the Palisades in Cimarron Canyon.

PRECAMBRIAN ROCK UNITS

The Fowler Pass fault separates the Precambrian rocks of the Cimarron Canyon area into two terranes whose relative ages cannot be ascertained. These two terranes are described separately in this report.

Precambrian rocks east of the Fowler Pass fault

Felsic metavolcanic rocks. Felsic metavolcanic rocks form jagged resistant outcrops in Cimarron Canyon that could be mistaken for quartzite from a distance. Most layers within this rock unit are light colored, fine grained, and porphyritic, containing quartz phenocrysts (some bipyramidal) as much as 6 mm long. No phenocrysts of feldspar or of mafic minerals were observed. Some of the felsic metavolcanic rock layers show relict eutaxitic

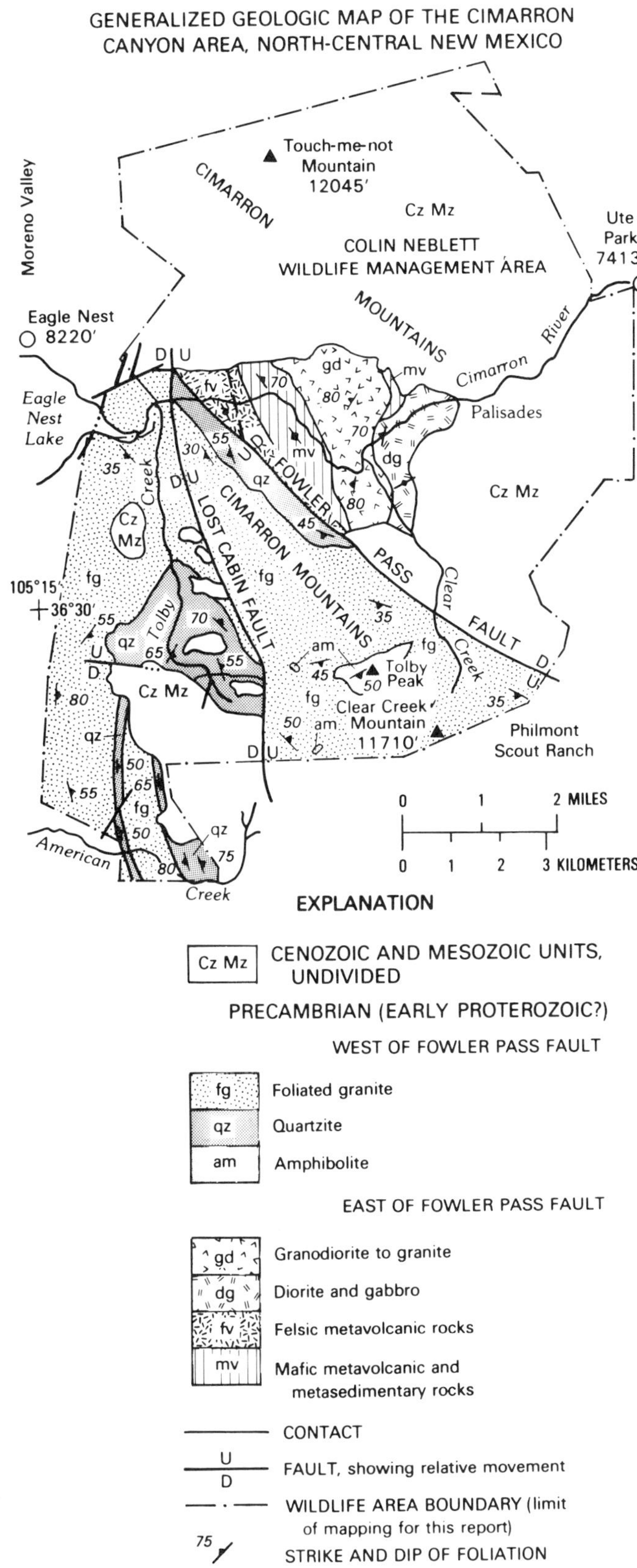

Figure 2. Generalized geologic map of Cimarron Canyon and vicinity, emphasizing Precambrian rock types.

TABLE 1. CHEMICAL ANALYSIS OF PROTEROZOIC METAVOLCANIC ROCKS AND PLUTONS FROM CIMARRON CANYON AREA*

METAVOLCANIC ROCKS EAST OF FOWLER PASS FAULT

	Mafic metavolcanic rocks				Felsic metavolcanic rocks			
Field No.	CM-39A	CM-50	CM-51	CM-109	CM-7	CM-32	CM-100B	CM-112
Major elements (wt. %)								
SiO_2	46.0	48.2	44.0	50.8	73.3	72.8	76.7	74.8
Al_2O_3	16.0	13.4	18.0	14.3	11.7	13.2	11.5	12.5
$FeTO_3$	12.5	11.9	14.3	13.0	4.60	3.34	3.45	3.75
MgO	7.00	5.32	7.14	4.16	3.62	0.84	1.24	0.51
CaO	13.2	8.32	9.29	6.59	1.41	4.81	1.63	1.55
Na_2O	1.84	1.98	1.62	2.88	2.58	1.09	2.59	4.24
K_2O	0.23	0.38	1.12	0.26	0.23	1.88	0.58	0.59
TiO_2	0.40	0.30	0.53	0.86	0.23	0.15	0.19	0.20
L.O.I.	1.93	9.44	3.21	5.59	2.21	1.32	1.31	1.02

PLUTONS EAST OF FOWLER PASS FAULT

	Metagabbro		Metadiorite		Granodiorite-granite	
Field No.	CM-5	CM-115	CM-34	CM-114	CM-4	CM-34A
Major elements (wt. %)						
SiO_2	49.5	50.0	60.7	53.6	73.2	73.7
Al_2O_3	14.5	14.4	15.3	16.6	13.4	14.4
$FeTO_3$	14.3	13.5	8.30	8.11	2.32	1.78
MgO	6.37	6.58	3.00	6.31	0.56	0.40
CaO	10.2	10.3	6.45	9.19	2.16	2.59
Na_2O	2.47	2.41	2.39	2.42	3.39	4.23
K_2O	0.26	0.28	1.15	0.75	2.33	1.17
TiO_2	1.82	1.61	0.39	0.67	0.16	0.12
L.O.I.	0.22	0.40	1.34	1.71	1.32	1.03

PRECAMBRIAN ROCKS WEST OF FOWLER PASS FAULT

		Foliated granite			Amphibolite
Field No.	CM-30	CM-42	CM-111	CM-117	CM-21
Major elements (wt. %)					
SiO_2	76.6	78.0	77.7	74.8	50.4
Al_2O_3	12.0	11.5	11.5	11.9	16.4
$FeTO_3$	1.19	1.35	1.23	4.73	9.73
MgO	<0.10	<0.10	<0.10	0.22	7.81
CaO	0.35	0.24	0.20	0.16	9.57
Na_2O	2.80	2.73	2.41	0.83	2.73
K_2O	5.65	4.97	5.42	4.78	0.90
TiO_2	0.16	0.11	0.09	0.50	0.77
L.O.I.	0.36	0.65	0.55	1.74	1.01

*Total iron as $FeTO_3$; L.O.I., lost on ignition at 900°C.
X-ray spectroscopic analyses by U.S. Geological Survey, Branch of Analytical Chemistry, Lakewood, Colorado, A. J. Bartel, K. Stewart, and J. Taggart, analysts.

fabric and may have been ash flows, whereas others contain flattened fragments of a variety of rock types and appear to be metamorphosed lithic tuffs.

The quartz phenocrysts are commonly polycrystalline, strained, and fractured. The extremely fine grained groundmass is highly sericitic, with only quartz and chlorite as other major constituents. The altered nature of these rocks, not apparent in hand specimen, is confirmed by major-element chemical analysis, which shows high SiO_2 (72 to 76 percent) but very low K_2O (<2 percent) and high volatiles ("lost on ignition"; Table 1). The compositions of the felsic metavolcanic rocks are compared with those of lithologically similar rocks in north-central New Mexico in an AFM diagram (Fig. 3).

Mafic metavolcanic and associated metasedimentary rocks. Metabasalt and associated metasedimentary rocks (chlorite schist, phyllite, and metasiltstone) are locally interlayered with the felsic metavolcanic rocks and form the dominant layered rock unit in the easternmost part of the Precambrian exposures in Cimarron Canyon. Foliations in all the layered rocks strike northwest and dip steeply northeast. No pillow structures were observed, but porphyritic, fragmental, and amygdaloidal varieties of dark green to black metabasalt are common.

In thin section, the metabasalt shows a relict trachytic fabric in its altered groundmass, which is composed of chlorite, epidote, sodic andesine, calcite, and opaque minerals. Phenocrysts consist of subhedral plagioclase, faintly zoned and poorly twinned. Lenticular zones surrounded by chlorite and containing epidote, calcite, and clay minerals probably represent flattened amygdules or vesicles. Chemically the metabasalt ranges from 44 to 51 percent SiO_2 (Table 1) and shows a tholeiitic affinity on an AFM diagram (Fig. 3). No metavolcanic rocks of intermediate composition (andesite, dacite) were observed.

Calc-alkaline plutonic rocks. The metavolcanic and interlayered metasedimentary rocks have been intruded by small stocks ranging from gabbro and diorite to granodiorite and granite. All of these rock types are weakly to moderately foliated, although mineral lineations are generally lacking. All are cut by numerous thin shear zones (linear zones of streaking and alteration up to several meters wide) and diabase dikes with a general northwest trend. There is, however, no evidence to verify that the largest stock (granodiorite to granite) is in fault contact with the adjacent metavolcanic wall rocks as shown in Robinson and others (1964). Instead, inclusions, non-linear contacts, and cross-cutting relations within and around the granodiorite and other plutons are common enough to establish relative ages of intrusion from more mafic to more felsic rock types.

Metagabbro crops out only in the area immediately west of the Palisades (Fig. 2), where it forms dark outcrops that are cut by thin dikes of pink pegmatite. The metagabbro is a medium- to coarse-grained rock consisting of minor olivine (approximately Fo_{80}) surrounded by abundant clinopyroxene; both minerals are partly altered to hornblende and chlorite. The plagioclase is well-zoned labradorite that is subophitically enclosed by the larger pyroxene grains.

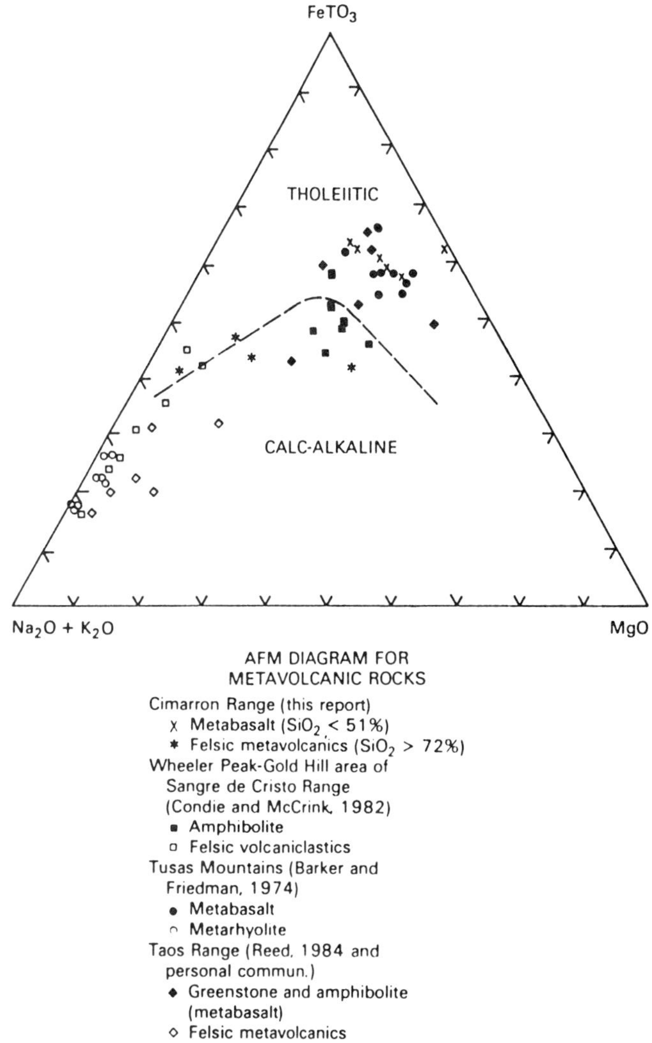

Figure 3. AFM diagram showing compositions of felsic and mafic metavolcanic rocks of Cimarron Canyon and nearby areas of north-central New Mexico. Compositions of Proterozoic volcanic and subvolcanic rocks from Wheeler Peak–Gold Hill parts of Sangre de Cristo Mountains (Condie and McCrink, 1982), Taos Range (Reed, 1984), and Tusas Mountains (Barker and Friedman, 1974) are shown for comparison. Dashed line separating fields of tholeiitic and calc-alkaline rocks is from Irvine and Baragar, 1971.

Metadiorite nearly surrounds the metagabbro outcrops and contains numerous smaller inclusions of metagabbro and of mafic metavolcanic rocks. The metadiorite is speckled black and white, medium grained, equigranular, and weakly foliated. It is composed of nearly equal amounts of blue-green hornblende (locally associated with clusters of biotite grains) and altered, faintly zoned andesine, with approximately 10 percent quartz and minor opaque minerals. All grains are anhedral, and the quartz is highly strained. Alignment of grains is not apparent in thin section.

Pink, medium- to coarse-grained biotite granodiorite to granite forms the largest pluton east of the Fowler Pass fault in Cimarron Canyon. This weakly foliated rock contains inclusions of all the older rock units in the eastern part of the canyon and forms some of the most rugged outcrops within the canyon. The rock is distinguished by bluish gray to purple quartz lenses as much as 1 to 3 cm long, which are composed of strained and fractured quartz grains. The feldspars are highly altered, with perthitic microcline about half as abundant as separate grains of poorly twinned albite. Biotite occurs as irregular grains, which are altered to yellow-brown iron oxides.

Major-element chemical analyses of all of the plutonic rocks of the Cimarron Canyon area (both east and west of Fowler Pass fault) are shown in Table 1, and the results are plotted on a Harker-type variation diagram (Fig. 4). The Harker diagram confirms the calc-alkaline trend suggested by the field relationships of the plutons east of the Fowler Pass fault, a trend that does not appear continuous through the compositions of the granitic rocks west of the fault.

Precambrian rocks west of the Fowler Pass fault

The Precambrian rocks underlying the highest parts of the Cimarron Mountains south of the Cimarron River and west of the Fowler Pass fault are very different from those to the east

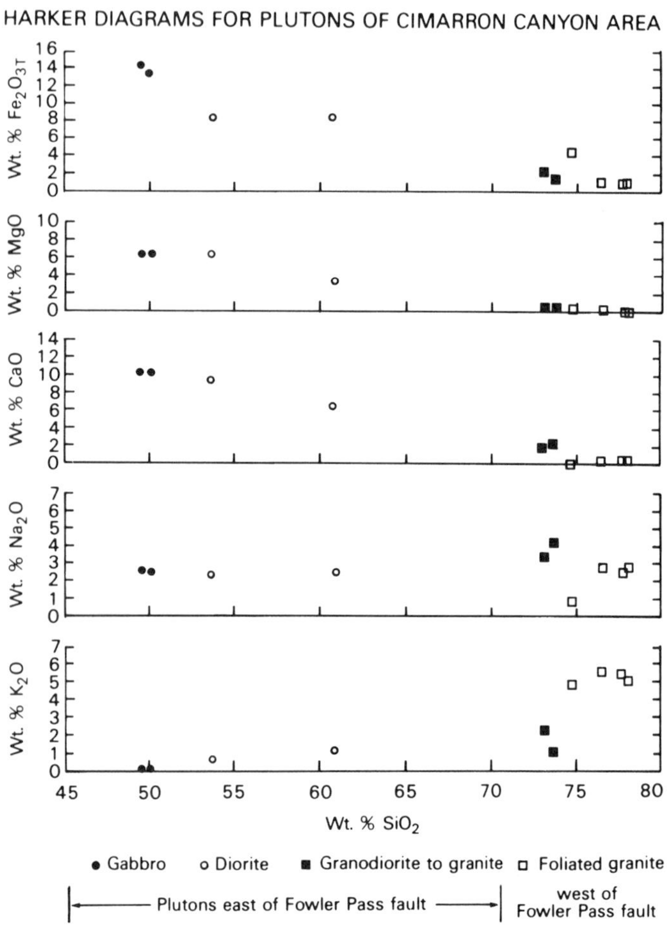

Figure 4. Harker-type variation diagram for plutons of Cimarron Canyon area. Gabbro, diorite, and granodiorite to granite plutons occur east of Fowler Pass fault; foliated granite occurs west of fault. Total iron expressed as Fe_2O_3.

(Fig. 2). West of the fault, irregularly shaped to elongate roof pendants and screens of quartzite, and smaller pendants of amphibolite, are surrounded by fine- to medium-grained, weakly to strongly foliated granite.

Quartzite. White, gray, and reddish vitreous quartzite forms prominent outcrops along the high ridge east of Tolby Creek and immediately south of the Cimarron River. The quartzite is in fault contact with felsic metavolcanic rocks along the east side of this ridge. Quartzite also occurs in and west of the Tolby Creek drainage as a complexly folded sheet-like roof pendant in granite; near the southern edge of the map area this body of quartzite appears to be approximately 90 m thick.

Although bedding is clearly visible in most quartzite outcrops, no other primary depositional structures were observed, and no sense of stratigraphic orientation could be ascertained. Especially near contacts with granite, the quartzite is highly muscovitic and even has developed distinct grains of potassium feldspar; pegmatite stringers about 5 cm wide in quartzite are also common near the granite.

Mineralogically the quartzite is typically 90 to 95 percent quartz, often highly strained or even granulated. Muscovite is commonly present in small amounts throughout the extent of the quartzite, and locally clots or clusters of muscovite grains may represent altered kyanite or sillimanite.

Amphibolite. Large xenoliths or roof pendants of amphibolite in granite occur primarily along the ridge northwest of Clear Creek Mountain, with the largest of these underlying much of Tolby Peak. The amphibolite is weakly foliated, medium grained, and mottled black and white. It is commonly cut by thin dikes and stringers of granite and pegmatite, especially near contacts with the granite. Petrographically the amphibolite is a granoblastic arrangement of hornblende and andesine, with minor secondary biotite and no quartz. Its fabric is more highly recrystallized than that of any mafic rocks east of the Fowler Pass fault, where relict igneous textures prevail. The amphibolite is also distinct chemically, showing appreciably higher alumina and lower iron than the metabasalts to the east (Table 1).

Granite. Pink to pale red granite underlies the highest parts of the Cimarron Mountains south of Cimarron Canyon and is the most widespread rock type west of the Fowler Pass fault. The granite is of two textural varieties: (1) a medium-grained, strongly foliated type cut by the other; (2) a fine-grained, moderately foliated granite (Fig. 5). Both types contain minor biotite, and each is locally muscovitic, with the muscovite occurring in irregular clusters of grains. Microcline is the dominant mineral in both textural varieties of granite, followed by quartz (anhedral, strained) and plagioclase (intermediate oligoclase).

Chemically these two textural varieties of granite are virtually identical (Table 1). They are peraluminous, very low in calcium, and contain considerably higher potash and lower soda than the calc-alkaline plutons east of the Fowler Pass fault. On a Harker-type variation diagram (Fig. 4), they do not appear continuous with the calc-alkaline trend of the plutons to the east, especially in terms of Na_2O and K_2O.

STRATIGRAPHIC RELATIONS, AGE, AND CORRELATION

Because the quartzite and metavolcanic parts of the Precambrian terrane of Cimarron Canyon were only observed in fault contact with one another, no stratigraphic sequence can be derived for the layered metamorphic rocks of this area. In other associations of Precambrian quartzite and bimodal metavolcanic rocks in north-central New Mexico, quartzite has often been shown to overlie the metavolcanics stratigraphically, as in the Taos Range 25 km to the west (Reed, 1984), the Picuris Range 55 km to the southwest (Gresens and Stensrud, 1974; Bauer, 1984), and the Tusas Mountains 80 km to the west (Gresens and Stensrud, 1974; Wobus, 1984).

None of the Precambrian rocks of the Cimarron Mountains has yet been dated radiometrically. However, zircon from lithologically similar metarhyolite in the Taos Range to the west yields a concordia age of about 1,750 Ma, and foliated granitic plutons that cut metavolcanic rocks in the same area range in age from 1,730 to 1,675 Ma (Bowring and others, 1984). These ages are comparable to those for felsic metavolcanic rocks and granitic plutons throughout northern New Mexico (Silver, 1984), and it seems likely that the Precambrian rocks of Cimarron Canyon are also of late early Proterozoic age.

Figure 5. Inclusion of foliated medium-grained granite in fine-grained, weakly foliated granite about 0.5 km west of Clear Creek Mountain. Pen for scale is 12 cm long.

ACKNOWLEDGMENTS

I wish to acknowledge the assistance of Julie Bryant, Mary Anne Duncan, Gary Gianniny, and Kirsten Pauly, then senior geology majors at Colorado College, during the field work in Fall 1982, and of Pat Murphy who assisted in the field during May 1983. I have benefited from discussions of the regional geology, both in the field and office, with Jack Reed and Chuck Pilmore of the U.S. Geological Survey, who also reviewed the manuscript for the Survey. Thanks are also due to Jeff Grambling, Mike O'Neill, and Howard Stensrud for their comments and suggestions during GSA review. Portions of this project were funded by a Faculty Research Grant from Williams College, whose support is gratefully acknowledged.

REFERENCES

Barker, F., and Friedman, I., 1974, Precambrian metavolcanic rocks of the Tusas Mountains, New Mexico: New Mexico Geological Society Guidebook 25, p. 115–117.

Bauer, P. W., 1984, Stratigraphic summary and structural problems of Precambrian rocks, Picuris Range, New Mexico: New Mexico Geological Society Guidebook 35, p. 199–204.

Bowring, S. A., Reed, J. C., Jr., and Condie, K. C., 1984, U-Pb geochronology of Proterozoic volcanic and plutonic rocks, Sangre de Cristo Mountains, New Mexico: Geological Society of America, Abstracts with Programs, v. 16, no. 4, p. 216.

Clark, K. F., and Read, C. B., 1972, Geology and ore deposits of the Eagle Nest area, New Mexico: New Mexico Bureau of Mines and Mineral Resources Bulletin 94, 152 p.

Condie, K. C., and McCrink, T. P., 1982, Geochemistry of Proterozoic volcanic and granitic rocks from the Gold Hill–Wheeler Peak area, northern New Mexico: Precambrian Research, v. 19, p. 141–166.

Goodknight, C. S., 1976, Cenozoic structural geology of the central Cimarron Range, New Mexico: New Mexico Geological Society Guidebook 27, p. 137–140.

Gresens, R. L., and Stensrud, H. L., 1974, Recognition of more metarhyolite occurrences in New Mexico and a possible Precambrian stratigraphy: Mountain Geologist, v. 11, p. 109–204.

Irvine, T. N., and Baragar, W.R.A., 1971, A guide to the chemical classification of common volcanic rocks: Canadian Journal of Earth Sciences, v. 8, p. 523–548.

Reed, J. C., Jr., 1984, Proterozoic rocks of the Taos Range, Sangre de Cristo

Mountains, New Mexico: New Mexico Geological Society Guidebook 35, p. 179–185.

Robinson, G. D., Wanek, A. A., Hays, W. H., and McCallum, M. E., 1964, Philmont country; The rocks and landscape of a famous New Mexico ranch: U.S. Geological Survey Professional Paper 505, 152 p.

Silver, L. T., 1984, Observations on the Precambrian evolution of northern New Mexico and adjacent regions: Geological Society of America Abstracts with Programs, v. 16, p. 256.

Wobus, R. A., 1984, An overview of the Precambrian geology of the Tusas Range, north-central New Mexico: New Mexico Geological Society Guidebook 35, p. 193–198.

Manuscript Accepted by the Society September 29, 1988

Printed in U.S.A.

Geological Society of America
Special Paper 235
1989

Geology and geochemistry of early Proterozoic volcanic and subvolcanic rocks of the Pecos greenstone belt, Sangre de Cristo Mountains, New Mexico

James M. Robertson
New Mexico Bureau of Mines and Mineral Resources, New Mexico Institute of Mining and Technology, Campus Station, Socorro, New Mexico 87801
Kent C. Condie
Geoscience Department, New Mexico Institute of Mining and Technology, Campus Station, Socorro, New Mexico 87801

ABSTRACT

The Pecos greenstone belt, dated at ca. 1,720 Ma, consists of metamorphosed subaqueous basalts together with locally important felsic volcanic rocks, iron-formation, and metasedimentary rocks, some of volcanic provenance. Volcanic rocks define a compositionally bimodal suite of basalt (80 percent) and dacite-rhyolite (20 percent). Mafic volcanics are mostly fine-grained, massive to well-foliated amphibolite that locally display relict amygdules, pillows, and pillow breccia. Felsic volcanic rocks are mainly porphyritic flows and crystal-rich volcaniclastic units that contain variable amounts of quartz, K-feldspar, and plagioclase phenocrysts. The greenstone belt also includes a compositionally bimodal subvolcanic complex that intrudes, and is locally overlain by, portions of the volcano-sedimentary pile. The subvolcanic complex comprises concordant to discordant, hypabyssal intrusions of tonalite-trondhjemite (65 percent) and diabase-gabbro (30 percent). Ultramafic and mafic rocks with possible ophiolitic affinities constitute the remaining 5 percent of the subvolcanic complex.

Greenstone-belt rocks have undergone regional metamorphism of upper greenschist to lower amphibolite grade and show the effects of at least three periods of deformation. The metamorphic sequence is cut by pre- to syn-orogenic granites and quartz porphyries dated at ca. 1,650 Ma and by syn- to post-orogenic granitic rocks dated at ca. 1,500 to 1,450 Ma.

Mafic volcanic and subvolcanic rocks define four distinct chemical populations: (1) ultramafic, (2) high-Mg tholeiite, (3) tholeiite, and (4) calc-alkaline basalt (CAB). Geochemically, the high-Mg tholeiites are similar to Archean high-Mg tholeiites, whereas the tholeiites and calc-alkaline basalts are similar to their counterparts in modern arc systems. The ultramafic rocks and various basalt groups cannot be related to each other by closed-system fractional crystallization or batch melting of a single mantle source. Their incompatible-element ratios seem to demand at least three different mantle sources (one depleted). Felsic volcanics can be divided into three groups (one rhyolitic and two dacitic) on the basis of immobile-element abundances. The rhyolites and one dacite group can be produced by fractional crystallization of CAB. The other dacite group (Doctor Creek dacite) is not obviously or easily related, chemically or genetically, to the rest of the Pecos volcanic and subvolcanic rocks. It instead may belong to a younger (ca.

Robertson, J. M., and Condie, K. C., 1989, Geology and geochemistry of Early Proterozoic volcanic and subvolcanic rocks of the Pecos greenstone belt, Sangre de Cristo Mountains, New Mexico, *in* Grambling, J. A., and Tewksbury, B. J., Proterozoic geology of the southern Rocky Mountains: Boulder, Colorado, Geological Society of America Special Paper 235.

1,700 Ma) felsic-dominated volcanic succession that is widespread in northern New Mexico.

Rocks of the Pecos greenstone belt may represent a remnant of a back-arc basin that opened far enough to form oceanic crust and tap a depleted mantle source.

INTRODUCTION

Recent mapping in the southern Sangre de Cristo Mountains of north-central New Mexico has delineated an extensive volcano-sedimentary terrane of early Proterozoic age (ca. 1,720 Ma) that has been called the Pecos greenstone belt (Robertson and Moench, 1979). The greenstone belt consists of a closely interrelated package of metamorphosed subaqueous basalts and locally important felsic volcanic rocks, iron-formation, and metasedimentary rocks, most of volcanic provenance. A subvolcanic complex of tonalite-trondhjemite and diabase-gabbro, coeval with and chemically equivalent to the volcanic rocks, constitutes a hypabyssal plutonic portion of the greenstone belt. This investigation describes the general geology, petrography, and geochemistry of the volcanic and subvolcanic portions of the Pecos greenstone belt, presents 45 new major and trace-element analyses, and considers possible origins, interrelationships, and tectonic settings for these rocks.

The Pecos greenstone belt occupies the southern end of a broad north-south–trending band of 1,800 to 1,700 Ma metamorphosed Precambrian supracrustal rocks that extends from the Wyoming-Colorado border southward into north-central New Mexico (Condie, 1981b, 1986; Fig. 1). In southern Colorado and northern New Mexico, a common supracrustal assemblage in this band is compositionally bimodal volcanic and volcaniclastic rocks along with lesser amounts of related sedimentary rocks. Bimodal volcano-sedimentary belts, similar to but slightly older (ca. 1,770 to 1,730 Ma) than the Pecos greenstone belt, occur in the Tusas Mountains (Williams, 1987) and Taos Range (Reed, 1984) in northern New Mexico, and in the Needle Mountains northeast of Durango (Barker, 1969), the Gunnison area (Condie and Nuter, 1981; Bennett, 1984; Knoper and Condie, 1988), and the Salida area (Boardman, 1986) of southern Colorado.

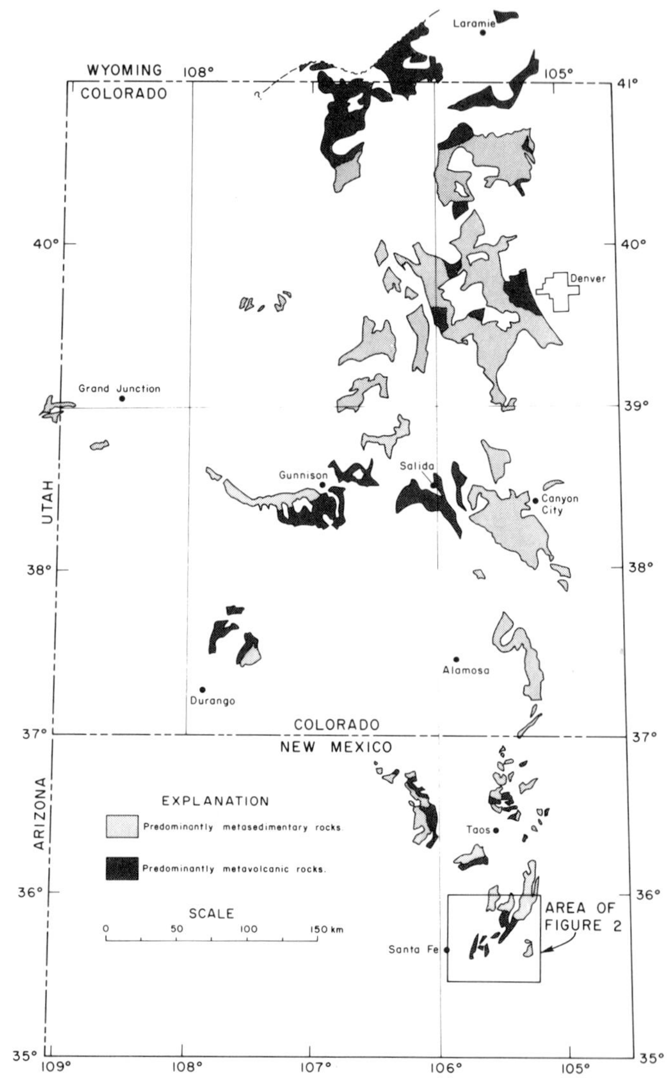

Figure 1. Geologic map of 1,800 to 1,700 Ma supracrustal rocks in Colorado and northern New Mexico. Modified from Tweto (1979), Condie (1981b), and unpublished mapping of J. M. Robertson and K. C. Condie.

The Precambrian geology of the southern Sangre de Cristo Mountains is summarized by Moench and Robertson (1980) and Moench and others (1988). The general geology of the Pecos greenstone belt is presented in Robertson and Moench (1979). The Pecos greenstone belt occupies an area of some 650 km^2, mainly in the headwaters of the Pecos River, northeast of Santa Fe (Fig. 2). It is bounded on the west by a major north-south fault zone, the Picuris-Pecos fault. To the north, greenstone-belt rocks have been thrust northward over younger Precambrian rocks of the Vadito and Ortega Groups (Grambling and Ward, 1987). To the south and east, the margins of the belt are marked by voluminous intrusions of younger Precambrian granitic rocks.

Precambrian rocks are exposed over approximately one-third of the area known to contain the Pecos greenstone belt, and are overlain by as much as 600 m of upper Paleozoic carbonate and clastic rocks in the remaining two-thirds. Four assemblages of Precambrian rock types (see Table 1) are present within the greenstone belt area: (1) metavolcanic rocks, with minor interbedded volcaniclastic rocks and iron-formation; (2) metasedimentary rocks, including feldspathic sandstone, aluminous shale or mudstone, conglomerate, small amounts of volcaniclastic graywacke, thin carbonate horizons, and possibly distal clastic facies of iron-formation; (3) intrusive felsic and mafic rocks of the subvolcanic complex; and (4) several generations of younger granitic rocks (Fig. 2). The greenstones are multiply deformed, variably metamorphosed to the upper greenschist or lower amphib-

TABLE 1. RELATIVE ABUNDANCES OF PRECAMBRIAN ROCKS IN THE PECOS REGION, SOUTHERN SANGRE DE CRISTO MOUNTAINS

Volcanic rocks 25%		Sedimentary rocks 25%	Plutonic rocks 50%	
Felsic	20%		Granites*	35%
Mafic	80%		Subvolcanic complex	65%
			tonalite and trondhjemite (65%)	
			gabbro and diabase (30%)	
			ultramafic rocks (5%)	

*Granites are distinctly younger than the volcanic-sedimentary-subvolcanic rocks of the greenstone belt.

olite facies (see Grambling and others, this volume), and intruded by several generations of granitic rocks. In addition, both volcanic and sedimentary rocks exhibit marked variation in lithology and thickness over short distances. For these reasons, a detailed stratigraphic column for the greenstone belt has not yet been constructed.

At least two generations of Precambrian granitic rocks intrude the Pecos greenstone succession. The younger of these has been collectively called the Embudo Granite and is extensively exposed west of the Picuris-Pecos fault in an uplifted block forming the Santa Fe Range. Register and Brookins (1979) report a composite Rb-Sr whole-rock isochron age of 1,464 ± 50 Ma for eight samples of Embudo Granite collected along the western margin of the Santa Fe Range north of Santa Fe. Similar granites intrude the greenstone belt in a number of places east of the fault, and one sample, collected along Macho Creek approximately 0.5 km east of the Picuris-Pecos fault, has a U-Pb zircon age of ca. 1,480 Ma (S. A. Bowring, personal communication, 1984). The older granites, represented by a small pluton exposed along the Pecos River just below its confluence with Indian Creek, have apparent U-Pb zircon crystallization ages near 1,650 Ma (S. A. Bowring, personal communication, 1984). Hypabyssal quartz porphyry, exposed in the Santa Fe Range along Dalton Canyon west of the Picuris-Pecos fault (Fig. 2), gives similar U-Pb zircon ages (ca. 1,660 to 1,650 Ma, S. A. Bowring, personal communication, 1985) and appears to "grade northward into a large arcuate batholith of foliated pink biotite granite" (R. H. Moench, written communication, 1987).

U-Pb zircon dates from the Pecos greenstone belt consistently indicate a ca. 1,720 Ma age for the volcanism (and related volcanogenic sulfide deposition) and subvolcanic intrusive activity. A subvolcanic tonalite (PR-1), exposed along the Pecos River about 2 km north of its junction with Macho Creek, has yielded a U-Pb zircon age of 1,718 ± 5 Ma, while a hypabyssal(?) quartz-eye porphyry (MA-25) near the Jones mine has yielded a U-Pb zircon age of 1,720 ± 15 Ma (Bowring and Condie, 1982; S. A. Bowring, personal communication, 1987). Model lead ages on galenas from the Pecos and Jones mines are 1,710 Ma and 1,720 Ma, respectively (Stacey and others, 1976). These model ages are probably appropriate for both the massive-sulfide deposits and enclosing felsic volcanic rocks.

A majority of the volcanic rock protolith names used in this paper were determined using a Zr/TiO_2–Nb/Y diagram of Floyd and Winchester (1978) (Fig. 3). Ultramafic rocks and high-Mg tholeiites are distinguished on the basis of modal mineralogy and whole-rock chemical composition. Felsic members of the subvolcanic suite are named using a normative Ab-An-Or ternary diagram modified by Barker (1979) and the Zr/TiO_2–Nb/Y diagram (Fig. 3). Volcanic and subvolcanic rock names are summarized in Table 2.

PETROGRAPHY

Volcanic and subvolcanic rocks are widely exposed in the southern two-thirds of the area containing the Pecos greenstone belt. They decrease in abundance toward the north, where ca. 1,650 Ma granitic rocks predominate, and toward the northeast, where greenstone-belt-related sedimentary rocks are abundant. Volcanic rocks consist mainly of flows and assorted volcaniclastic rocks that are variably deformed and recrystallized. Mafic volcanics predominate (see Table 1). A subvolcanic complex of concordant to discordant, hypabyssal sheets, dikes, plugs, and small stocks of metamorphosed tonalite-trondhjemite and gabbro or diabase intrudes, and may be locally overlain by, portions of the volcanic pile. Rare ultramafic rocks, some clearly intrusive and some possibly extrusive, are tentatively assigned to the subvolcanic complex. The mineralogy, textures, and tectonite fabrics displayed by rocks of the subvolcanic complex suggest they have experienced the same metamorphism and deformational history as the associated volcanics.

Mafic volcanic and volcaniclastic rocks

The most abundant rock type in the Pecos volcanic suite is grayish green to greenish black, fine- to medium-grained, massive to moderately well foliated metabasalt that locally displays relict amygdules, pillows, and pillow breccia (Fig. 4A). These mafic volcanic rocks are best exposed in a broad north-south–trending belt of outcrops along the east-facing range front between Spring Mountain and Elk Mountain (Fig. 2). Mafic volcaniclastic rocks are not common in the Pecos greenstone belt. They are grayish green to greenish gray, fine- to medium-grained, well-bedded to thinly laminated schistose rocks that commonly exhibit alternating light and dark compositional banding. The volcaniclastic rocks are generally interlayered with mafic volcanic sequences and are most abundant south and southeast of Spring Mountain.

A typical metabasalt, now a fine-grained amphibolite, consists of subparallel, stubby to acicular, subhedral to euhedral laths ($\leqslant$0.5 mm) of amphibole (mostly hornblende, but some actinolite), surrounded by a very fine-grained ($\leqslant$0.1 mm) mosaic of anhedral to subhedral sodic plagioclase (An_{10-20}), along with lesser amounts of quartz, epidote, chlorite, and biotite. Accessory minerals include magnetite, ilmenite, sphene, hematite, and minor

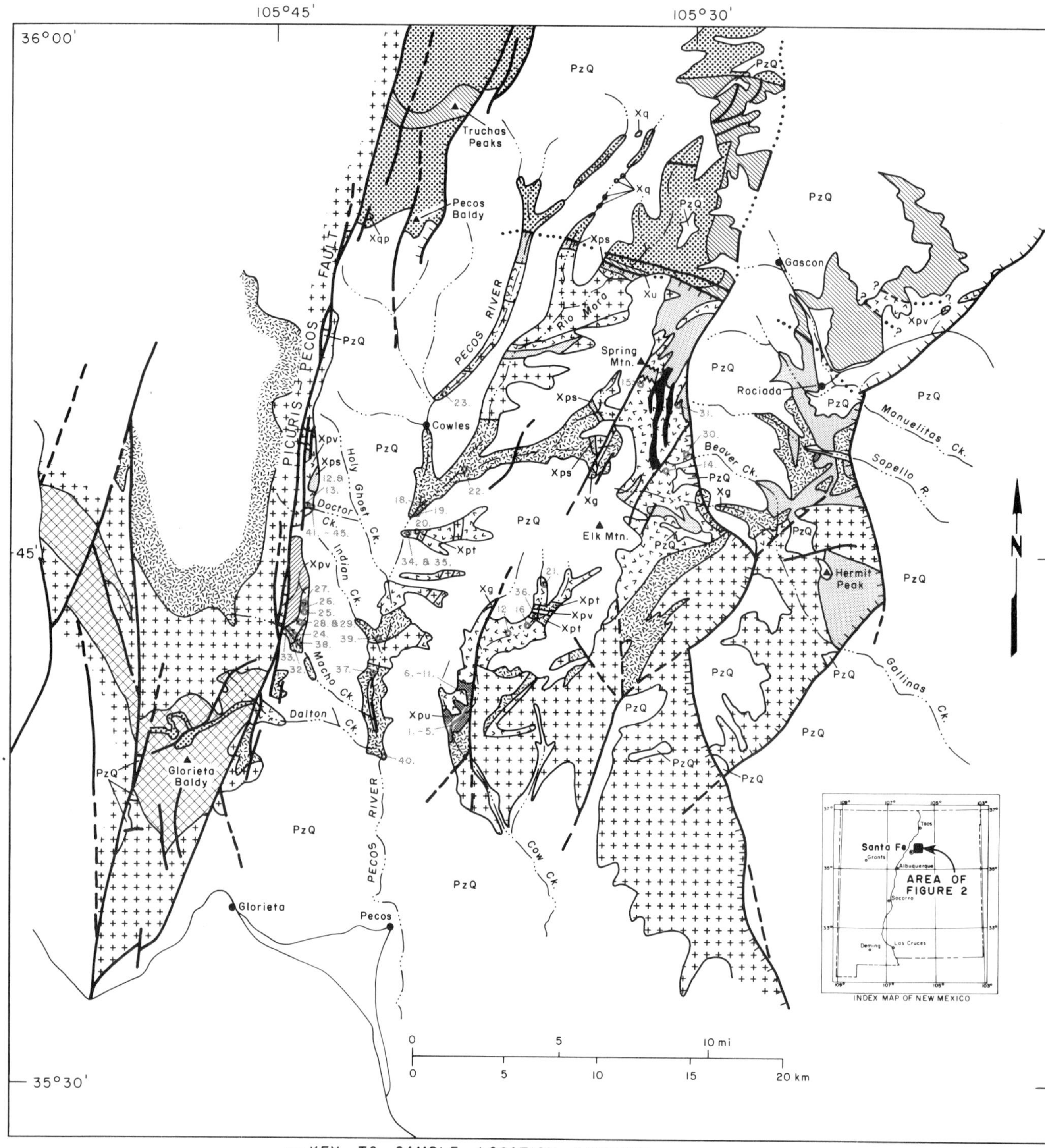

—KEY TO SAMPLE LOCATION NUMBERS—

Ultramafics
1. 79-4-35
2. 79-18-26
3. 79-23-26
4. 80-1-26
5. 80-4-23

High-Mg Tholeiites
6. 79-12-26
7. 79-14-26
8. 79-21-26
9. 79-22-26
10. 80-1-23
11. 80-4-26

Tholeiites
12. DOC-45
13. DOC-46
14. 103-c
15. 113-a
16. CC-2
17. CC-4

Calc-alkaline Basalts
18. PR-4
19. PR-6
20. C-9-17
21. H-7-122
22. C-9-87
23. C-10-24

Pecos Dacites
24. MA-2
25. MA-16
26. MA-18
27. MA-25
28. MA-26
29. MA-27

Pecos Rhyolites
30. 452
31. 557
32. MA-34
33. MA-38
34. PM-8
35. PM-10

Tonalite-Trondhjemite
36. CC-1
37. PR-1
38. MA-1
39. MA-39
40. PEC-R-24

Doctor Creek Dacites
41. DOC-43
42. DOC-49
43. DOC-50
44. DOC-51
45. DOC-54

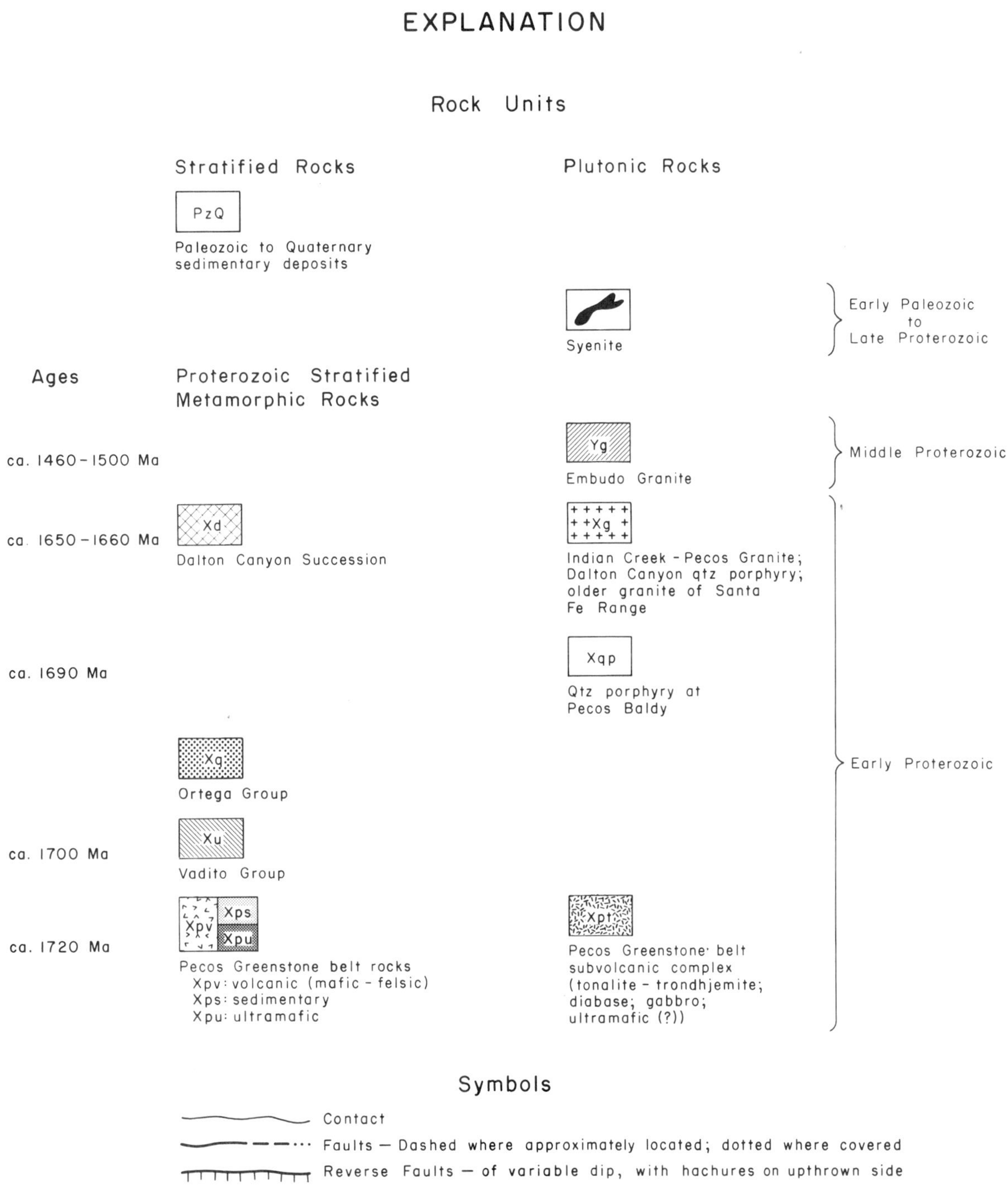

Figure 2. Geologic map of the Pecos greenstone belt and vicinity. Modified from Robertson and Moench (1979), Moench and Erickson (1980), Moench and others (1988), and unpublished mapping of J. A. Grambling, D. B. Codding, I. Klich, D. Mathewson, and J. M. Robertson.

TABLE 2. PECOS VOLCANIC AND SUB-VOLCANIC COMPOSITIONAL GROUPS AND THEIR INFERRED PROTOLITHS

Group	Volcanic protolith	Sub-volcanic protolith
Ultramafic		Ultramafic (pyroxenite?)
Mafic I	High-Mg Tholeiite	?
Mafic II	Tholeiite	Gabbro and Diabase
Mafic III	Calc-alkaline basalt	Gabbro and Diabase
Felsic I	Dacite	Tonalite and Trondhjemite
Felsic II	Rhyolite	Tonalite and Trondhjemite
Felsic III	Doctor Creek dacite	

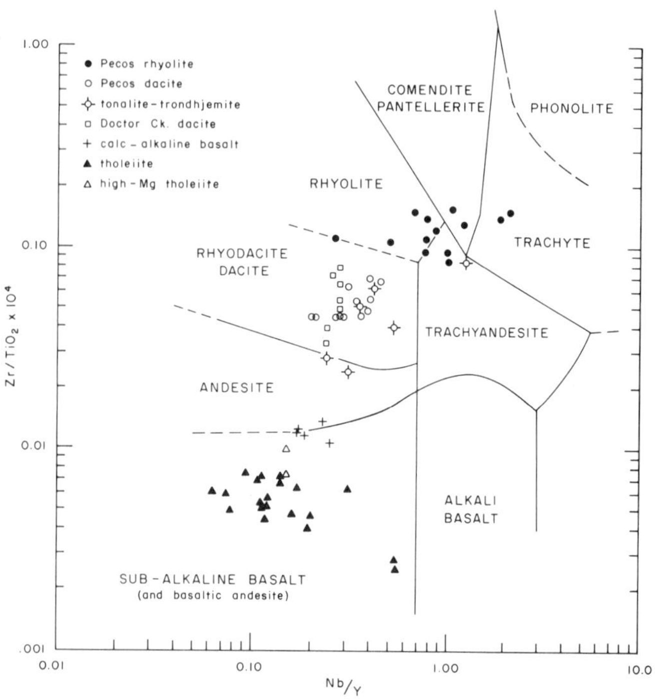

Figure 3. Zr/TiO_2–Nb/Y plot showing distribution of volcanic rocks and associated tonalite-trondhjemite from the Pecos greenstone belt. Classification scheme after Floyd and Winchester (1978).

tourmaline and carbonate. Poorly developed subophitic texture is preserved in a few places, as are relict quartz + epidote–filled amygdules (Fig. 4B). Mafic volcaniclastic rocks are distinguished from mafic volcanic rocks on the basis of quartz and biotite content. A representative volcaniclastic sample contains approximately 30 percent hornblende, 30 percent plagioclase, and 40 percent combined quartz and biotite. Mineralogic banding is defined by alternating layers of hornblende–biotite ± chlorite and quartz–plagioclase ± epidote.

Felsic volcanic and volcaniclastic rocks

Felsic flows and volcaniclastic rocks are locally abundant, especially in the southwestern quarter of the greenstone belt area. These rocks typically form dome-shaped accumulations of interlayered porphyritic flows, volcanic breccias, crystal-rich tuffs, bedded volcaniclastic sediments, and cherty iron-formation. The felsic rocks may represent small (<2 km in diameter) subaqueous eruptive centers developed locally on a regionally extensive basaltic platform. Two felsic centers (in the Macho Creek area and along the Pecos River due west of Elk Mountain) are known to have associated massive sulfide deposits (Fig. 2).

Felsic volcanic and volcaniclastic rocks are typically light gray to orange, massive to well banded, fine-grained to aphanitic, and porphyritic. Two distinct populations are distinguished on the basis of phenocryst mineralogy: (1) metarhyolites that contain both quartz and K-feldspar phenocrysts in subequal amounts; plagioclase is rare (Figs. 5A and 5B); and (2) metadacites that contain mostly plagioclase phenocrysts with only minor quartz; K-feldspar is absent (Figs. 5C and 5D). Phenocrysts typically make up 5 to 25 percent of a given sample. Feldspar phenocrysts are commonly euhedral tabular to lath shaped. Quartz phenocrysts range in shape from euhedral to rounded, and many are embayed. Matrix for both metarhyolites and metadacites consists of fine- to very fine-grained (average ~0.5 mm), anhedral, equigranular feldspar and quartz (90 percent) with minor biotite, chlorite, and muscovite (8 percent), and traces of epidote, iron oxide, and sphene (2 percent). Banding is due to: (a) the presence of abundant, thin (0.1 to 0.2 mm), sub-parallel monomineralic lenses and discontinuous stringers of quartz (Fig. 5A); (b) abrupt changes in phenocryst abundance; (c) alternating mica-rich and mica-poor layers; or some combination of these features. Some samples contain a large number of chaotically distributed phenocrysts of various sizes and shapes, including many feldspar crystals that are clearly broken (Fig. 6A). Others contain abundant phenocrysts displaying a similar range of sizes, but the phenocrysts are sub-rounded to rounded and at least crudely sorted

A B

Figure 4. Mafic volcanic rocks from the Pecos greenstone belt. A. Fine-grained, dark green amphibolite from the Elk Mountain area containing numerous irregularly shaped, angular to amoeboid fragments of light green amphibolite, commonly outlined by greenish black, monomineralic selvages of actinolitic hornblende. These rocks are interpreted as pillow breccia. B. Photomicrograph of metabasalt, now fine-grained amphibolite, composed of actinolite and plagioclase, with relict quartz-filled amygdules. Uncrossed polars. Bar scale is one millimeter long.

(Figs. 6B and C). These rocks are interpreted as crystal-rich tuffs and reworked tuffs, respectively, although in both cases, their fine-grained matrix is so completely recrystallized that no original glass shards have been recognized. In a number of places, felsic volcanic rocks display a well-developed "patchy" or "spotted" texture. The spots make up roughly 50 percent of a typical sample, are lighter colored than the enclosing matrix, exhibit slightly flattened ellipsoidal shapes, and have a limited size range such that the majority average 0.5 to 1.0 cm in their longest dimension. In thin section, the spots consist of aggregates of quartz, feldspar, and sericite in a slightly finer grained matrix of the same minerals plus chlorite. These rocks are interpreted as lapilli tuffs.

Felsic volcanic and volcaniclastic rocks have responded to post-depositional alteration and metamorphism in a wide variety of ways. Some metarhyolites and metadacites show almost no obvious physical (or chemical) change. Thin sections of these relatively unaltered rocks exhibit well-defined phenocrysts of quartz and/or feldspar set in a fine-grained groundmass, which preserves fine textural detail. In some places, felsic volcanic rocks are recrystallized to a slightly coarser grain size, and phenocrysts are broken down into clusters of finer grains. In other places, hydrothermal alteration associated with regional metamorphism has converted metarhyolite to quartz-muscovite schist and metadacite to quartz–muscovite–biotite schist by a combination of recrystallization and metasomatism (cf. Gresens, 1976). Felsic volcanics in hydrothermal alteration zones adjacent to volcanogenic massive sulfide deposits are converted to quartz–muscovite–chlorite or quartz–muscovite–biotite–actinolite schist with their original major-element compositions profoundly modified. In these last two cases, original quartz phenocrysts are commonly preserved through alteration and recrystallization even though original feldspar phenocrysts are completely destroyed. The relict quartz phenocrysts may be broken down into aggregates of finer grains, but the overall ovoid shape of the aggregates, including their distinctive embayments, is typically retained (Gresens and Stensrud, 1974; Vernon, 1986).

Mafic subvolcanic rocks

Mafic rocks in the subvolcanic complex have been metamorphosed to greenish black to black, medium- to coarse-grained, massive to poorly foliated amphibolite and metadiabase. These rocks are commonly enclosed within and mesh-veined by tonalite and/or trondhjemite. Metadiabase displays both concordant and discordant contacts with pillow basalts in the Elk Mountain area. Metadiabase dikes and sub-concordant sheet-like

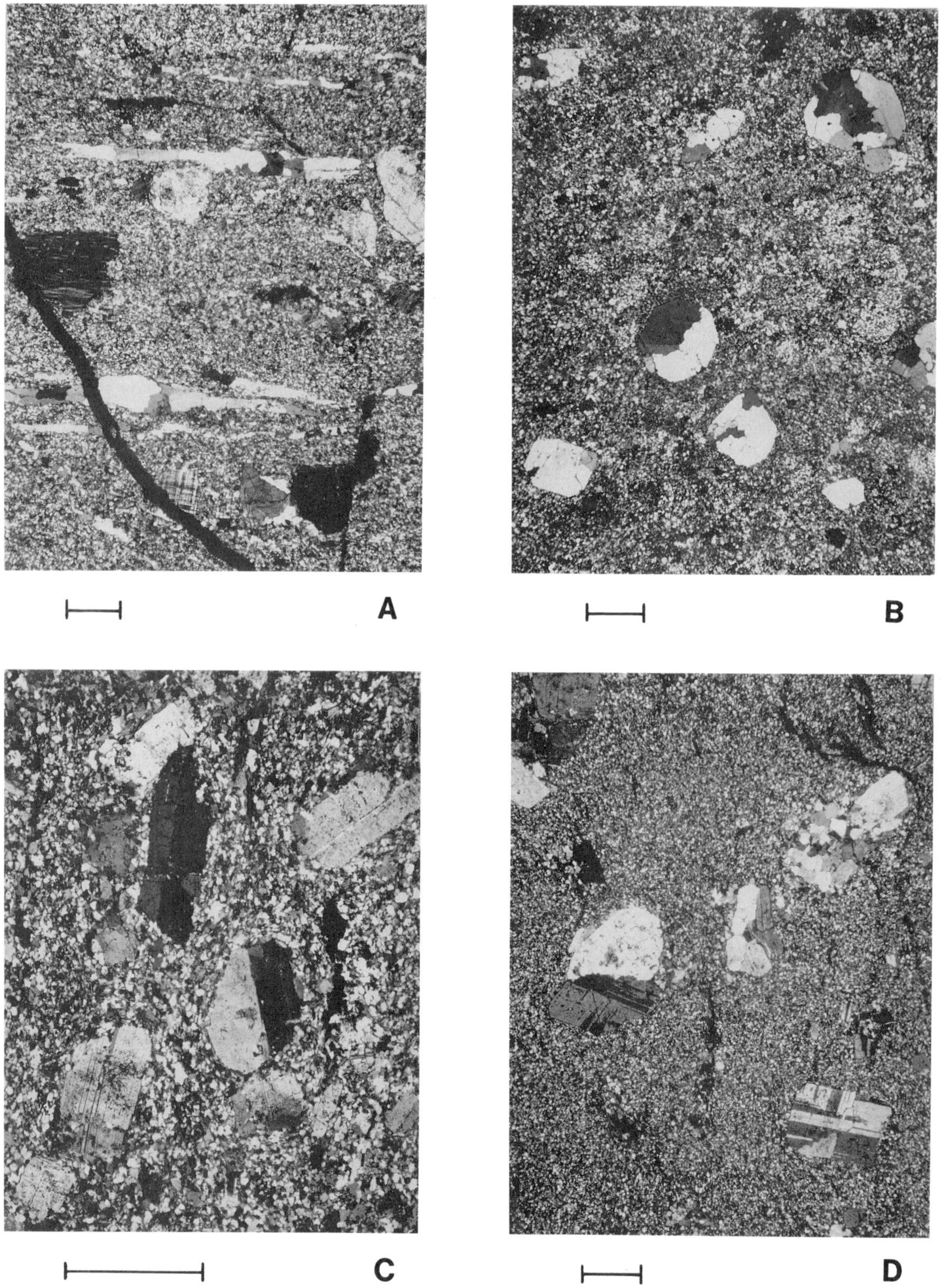

Figure 5. Photomicrographs of selected felsic volcanic rocks from the Pecos greenstone belt. Bar scale is one millimeter long. A. Rhyolite with K-feldspar (microcline) phenocrysts and quartz lenses in a fine-grained matrix of quartz, K-feldspar, and minor biotite. Crossed polars. B. Rhyolite with numerous rounded quartz phenocrysts ("eyes") in a fine-grained matrix of quartz and K-feldspar. Crossed polars. C. Dacite with plagioclase phenocrysts in a fine-grained matrix of quartz, plagioclase, K-feldspar, and biotite. Crossed polars. D. Similar to C.

Figure 6. Selected volcanic and subvolcanic rocks from the Pecos greenstone belt. Bar scale in photomicrographs is one millimeter long. A. Crystal-rich dacitic tuff with zoned plagioclase and quartz phenocrysts and phenocryst fragments in a fine-grained matrix of quartz, biotite, and magnetite. Uncrossed polars. B. Interlayered crystal-rich and crystal-poor dacitic tuff. Crystal-poor horizons contain scattered small quartz fragments in a finer-grained matrix that is biotite-poor. Uncrossed polars. C. Dacitic rock from the Doctor Creek area (re-worked tuff?) containing plagioclase-rich (light colored) and plagioclase-poor (dark colored) horizons that locally display sorting and rudimentary grading. D. Metadiabase showing well-developed sub-ophitic texture. Randomly oriented plagioclase laths with subhedral hornblende filling interstices. Crossed polars.

bodies intrude both the felsic volcanic rocks and felsic members of the subvolcanic complex in the southwestern part of the greenstone area. Metadiabase is distinguished from mafic flows by its coarser grain size, locally discordant contacts, and distinctive relict diabasic to subophitic textures.

In thin section, mafic rocks of the subvolcanic complex contain approximately equal amounts of sodic plagioclase and hornblende and commonly display a moderately to well-preserved subophitic texture (Fig. 6D). Plagioclase (An_{10-20}) forms subhedral to euhedral laths (up to 3 mm in length); hornblende, which appears to have replaced original pyroxene, typically occurs as stubby, subhedral prisms (averaging 1 mm in length) that fill interstices between randomly oriented plagioclase crystals. Intergrown anhedral quartz, epidote, and biotite are typically associated with the hornblende and together make up from 5 to 15 percent of the rock. Accessory minerals are magnetite, sphene, carbonate, and in many samples, tourmaline.

Felsic subvolcanic rocks

Felsic rocks of the subvolcanic complex are represented by light gray to dark gray, fine- to medium-grained, poorly to well-foliated, metamorphosed tonalite and trondhjemite. The larger bodies contain abundant mafic xenoliths, display roughly concordant contacts with the volcanic sequences they intrude, and have undergone varying amounts of ductile deformation. In the southwestern and easternmost parts of the greenstone belt, tonalite-trondhjemite dikes crosscut amphibolites of the subvolcanic complex.

In thin section, the felsic rocks commonly exhibit equigranular, subhedral textures, with a foliation defined primarily by multiple, narrow (0.5 to 1.0 mm), subparallel mylonitic zones. Typical tonalite contains 30 to 40 percent quartz, 40 to 50 percent sodic plagioclase, 5 to 20 percent biotite and/or hornblende, 0 to 5 percent K-feldspar, and minor amounts of magnetite, sphene, epidote, and carbonate. Trondhjemites contain less quartz and mafic minerals and more plagioclase compared to tonalite.

In places along Macho Creek, felsic volcaniclastic rocks contain numerous sub-angular to rounded, pebble-sized tonalite fragments, suggesting that at least some tonalite emplacement and unroofing must have occurred before or during accumulation of the volcanic pile.

Ultramafic subvolcanic rocks

Ultramafic rocks make up a very small percentage ($\leqslant$5 percent) of the greenstone succession. They are exposed along Cow Creek in the south-central part of the greenstone area (Fig. 2) where they are interlayered with and locally intrude fine- to medium-grained amphibolite and minor mafic volcaniclastic rocks and are, in turn, intruded by a tonalite-trondhjemite pluton. The ultramafic rocks also contain numerous discontinuous, concordant to discordant, medium- to coarse-grained, lens-shaped bodies of amphibolite (several meters to several tens of meters thick) that in places display a relict diabasic texture.

Typical ultramafic rocks are dark green to brownish red, medium-grained, massive to moderately foliated, and extensively recrystallized. They consist chiefly of chlorite and tremolite-actinolite, with much smaller amounts of clinopyroxene, calcic plagioclase, serpentine, magnetite, biotite, carbonate, and quartz (Wyman, 1980). Although commonly medium-grained, ultramafic rocks may be much finer grained near their contacts with adjacent fine-grained amphibolites. A typical medium-grained sample contains stubby, subhedral to euhedral crystals ($\leqslant$1 cm in length) of what was probably originally clinopyroxene, now mostly replaced by aggregates of chlorite and tremolite-actinolite, in a finer-grained matrix of the same minerals plus the minor phases listed above. Fine-grained samples consist mainly of intergrown masses of chlorite and subhedral tremolite-actinolite that may contain numerous lensoidal patches of quartz + tremolite-actinolite needles reminiscent of amygdules and/or scattered, randomly oriented, individual tremolite-actinolite crystals $\leqslant$1 cm in length. Rarely, these same fine-grained rocks contain bundles or radiating aggregates of larger (up to 2 cm) tremolite-actinolite blades. Detailed descriptions of the geology and geochemistry of these rocks are given in Wyman (1980).

The field occurrence of the Pecos ultramafic rocks with the interlayered fine-grained amphibolite (metabasalt) and associated medium- to coarse-grained amphibolite (metadiabase and metagabbro dikes and sills) is not unlike that of dismembered Phanerozoic ophiolites (Coleman, 1977; Moores, 1982). The fact that the ultramafic rocks are cut by numerous closely-spaced metagabbro and metadiabase dikes (now amphibolites) and also are cut by minor trondhjemitic (plagiogranite?) dikes, further supports an ophiolite origin for these rocks. However, neither layered gabbros nor sheeted diabase dikes have been found anywhere in the Pecos succession.

GEOCHEMISTRY

Sampling and analytical techniques

Major- and selected minor- and trace-element chemical analyses were made by x-ray fluorescence methods from 101 volcanic and subvolcanic rock samples collected from throughout the greenstone area. An additional suite of trace elements, including rare-earth elements (REE), were determined for 67 of these same samples by instrumental neutron activation analysis (INNA). Analyses of 45 representative samples are presented in Tables 3 through 10, and sample locations are shown in Figure 2.

Major elements (excluding Na) and selected minor and trace elements (Ba, Cr, Cu, Mn, Ni, Pb, Sr, V, Y, Zn, Zr) were determined from pressed pellets by x-ray fluorescence using an analytical procedure modified from that of Leake and others (1969), which utilizes 30 international geochemical standards and published values of Flanagan (1973) for calibration. At typical rock concentrations, analytical error did not exceed 6 percent of the amount present. REE, Na, Co, Sc, Cs, Hf, Ta, Th, and U were determined by INNA using techniques described by Gordon and others (1968) and Lindstrom and Korotev (1982).

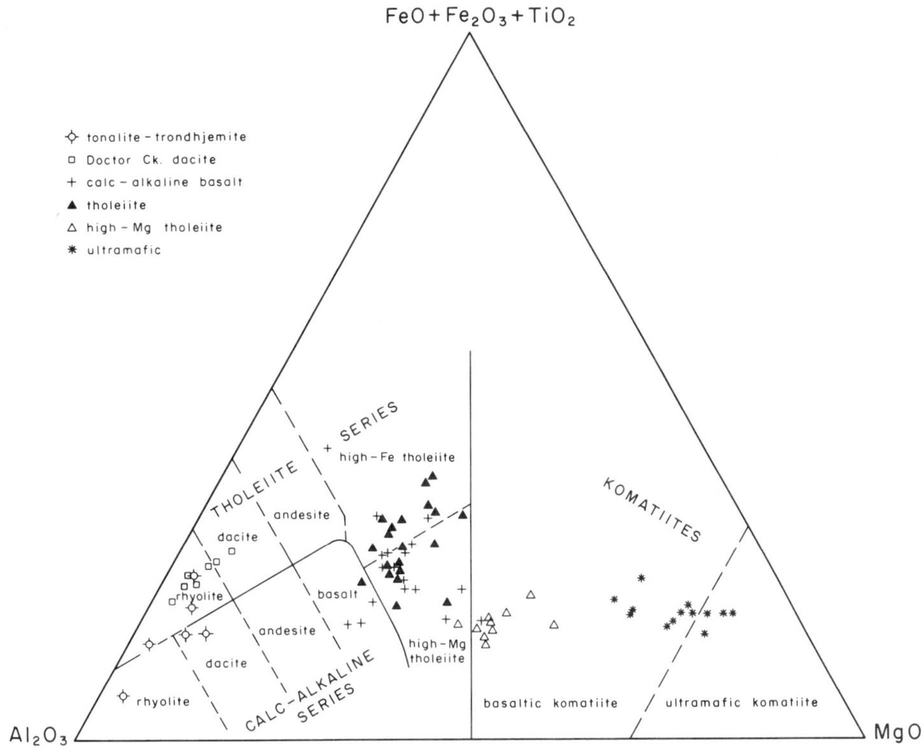

Figure 7. Jensen cation diagram (Jensen, 1976) showing the distribution of volcanic rocks and associated tonalite-trondhjemite from the Pecos greenstone belt.

Diagenesis, alteration, and metamorphism

Greenstone volcanics may be exposed to a number of mineralogically, texturally, and compositionally modifying processes during and after their eruption and accumulation. Among these processes are diagenesis (including hydration and devitrification of volcanic glass); deep-sea alteration (halmyrolysis); hydrothermal alteration associated with volcanogenic massive sulfide formation; contact metamorphism and metasomatism associated with the emplacement of genetically related and/or younger plutonic rocks; and regional metamorphism, involving one or more episodes of deformation, heating, and recrystallization. Volcanic and subvolcanic rocks in the Pecos greenstone belt have probably experienced most, if not all, of these processes, although it is generally not possible to distinguish the separate mineralogical, textural, and/or chemical signatures of each process. Clearly, any assumption that the present composition of these rocks reflects primary igneous compositions must be carefully tested on an element-by-element basis.

Samples analyzed for the present study are classified as "moderately altered" or "severely altered" using a series of petrographic and chemical alteration criteria similar to those used by Gelinas and others (1977) in their work on chemostratigraphic divisions within the Abitibi volcanic belt. These criteria consider: (1) the presence of limpid (clear or transparent) albite in mafic rocks, (2) volatile content as shown by loss on ignition (LOI), (3) the presence of abnormal normative minerals, (4) total alkali content, and (5) K_2O content. Severely altered mafic samples identified by these criteria were eliminated from further genetic consideration in the present study. A number of severely altered felsic samples are included, however, because severely altered samples are generally indistinguishable from their moderately altered counterparts on the trace-element plots. In addition, severely altered felsic samples are commonly associated with volcanogenic massive sulfide deposits and are therefore of economic interest.

Major compositional groups

Using a combination of petrography, whole-rock geochemistry, and selected geochemical discrimination diagrams (Figs. 3 and 7), it is possible to distinguish three general compositional groups of volcanic and subvolcanic rocks from the Pecos greenstone belt: felsic, mafic, and ultramafic. Intermediate rocks of andesitic composition are notably absent. The rocks appear to be mainly sub-alkaline in character, although some rhyolites display alkaline tendencies.

Ultramafic rocks. Ultramafic rocks average $\geqslant$20 percent MgO and have Al_2O_3/TiO_2 ratios of $\leqslant$25 (Table 3). In many respects, these rocks are similar in composition to Archean komatiites. For instance, the samples plot within the komatiite field on the Jensen cation plot (Fig. 7), and largely within the komatiite fields defined by Naldrett and Cabri (1976) on plots of TiO_2–MgO and Al_2O_3–[Fe_2O_3T/Fe_2O_3T + MgO] (Wyman,

1980). They differ from Archean komatiites with similar MgO contents in that the Pecos samples have lower Ni, Cr, and V contents.

REE distributions in the Pecos samples are not well defined because concentrations are near the detectability limits by INNA. Two samples that have measurable quantities of REE exhibit slightly REE enriched patterns (~5x chondrites) and small positive Eu-anomalies (Fig. 8A). The light REE enrichment, which differs from that found in most komatiites and ophiolite-type ultramafic suites, probably reflects secondary enrichment.

Mafic rocks. Pecos mafic volcanic and subvolcanic rocks are further subdivided, on the basis of position on selected geochemical discrimination diagrams (Figs. 7, 9A, and B), into tholeiitic and calc-alkaline populations.

Pecos tholeiites include both normal and high-Mg groups. The high-Mg tholeiites contain <1 percent TiO_2 (Table 4) and have normative olivine contents of 16 to 23 percent. In addition, they are depleted in light REE and do not display conspicuous Eu-anomalies (Fig. 8A). They share in common the low Ti, Zr, Y, Nb, Th, and REE contents characteristic of both Archean and modern high-Mg tholeiites and basaltic komatiites (Fig. 10). Although they fall in the basaltic komatiite field on the Jensen plot (Fig. 7), they differ from basaltic komatiites at the same Mg number by their low SiO_2, Fe_2O_3T, TiO_2, Mn, Cr, V, and Zr and high Al_2O_3, CaO, and Sr (Table 4). Similar differences appear when compared to normal mid-ocean ridge basalt (NMORB). Especially striking are the low contents of Zr, Y, and Ta in the Pecos high-Mg tholeiites (Table 4; Fig. 10). Overall, these rocks are simlar to high-Mg tholeiites from Archean greenstones (Condie, 1981a).

Pecos tholeiites (Table 5) contain <15 percent normative olivine, and four samples are quartz normative. At the same Mg

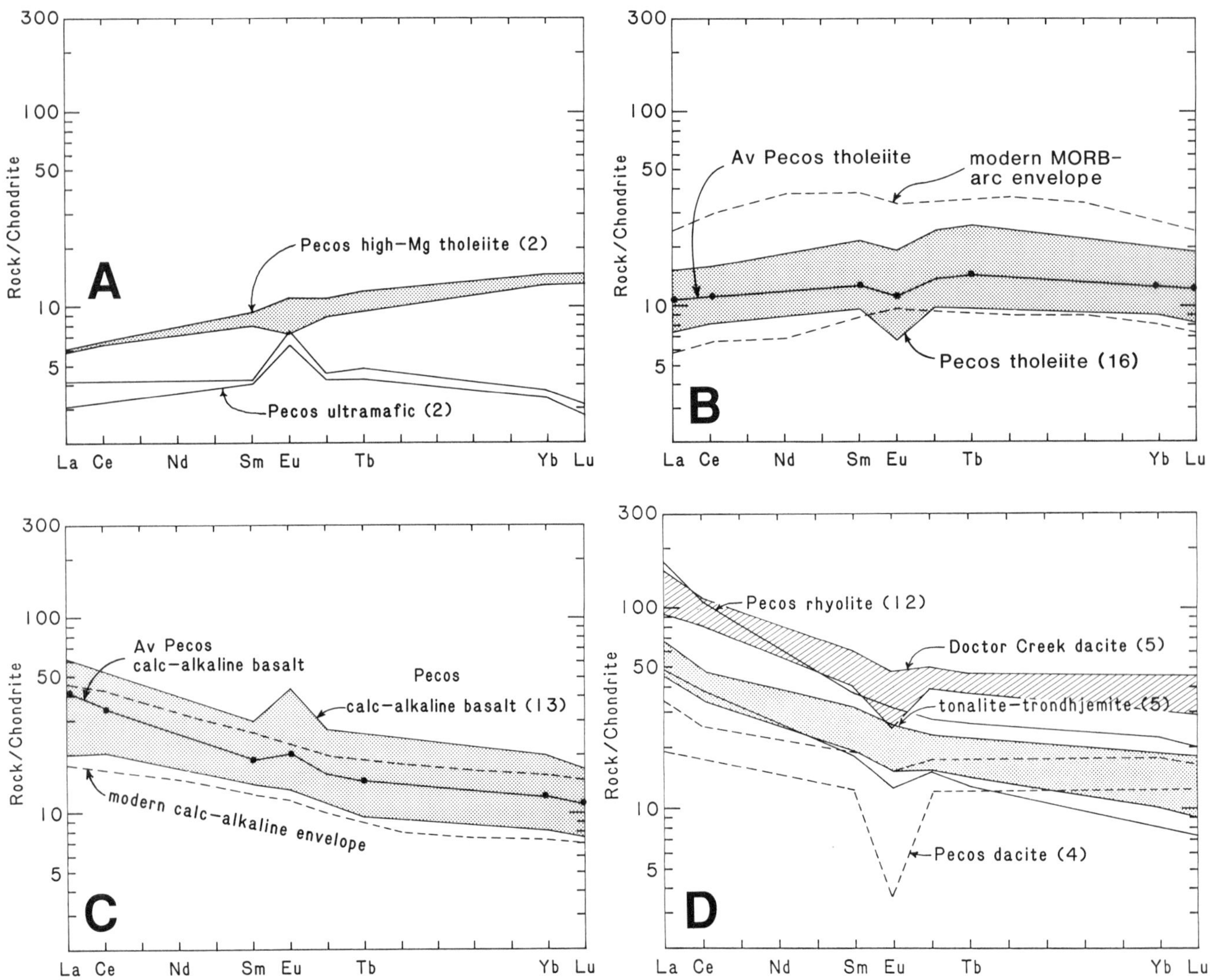

Figure 8. Chondrite-normalized REE distributions in Pecos volcanics and tonalite-trondhjemite. Envelopes of variation and selected mean values are shown for: (A) Ultramafics and high-Mg tholeiites; (B) tholeiites; (C) calc-alkaline basalts, and (D) dacites, rhyolites, tonalite-trondhjemite, and Doctor Creek dacites.

number (55 to 60), Pecos tholeiites are similar to modern basalts from island arcs (IAB; Figs. 11 and 12). They exhibit slight large-ion lithophile element (LILE) enrichment, low Nb and Ta, and relatively unfractionated moderately compatible elements (Fig. 11). REE patterns, relatively flat and 10–15× chondrites, also resemble MORB and IAB (Fig. 8B). The low Sr and negative Eu-anomalies in the Pecos tholeiites probably reflect plagioclase fractional crystallization. As with IAB, Pecos tholeiites exhibit relatively low Zr on a Ti-Zr-Y plot (Fig. 9A) and low Zr and high Ti on a TiO_2-Zr plot (Fig. 9B). In addition, they fall in or near IAB fields on the Zr/Y-Zr, Hf-Ta-Th, MgO-Al_2O_3-CaO, and Th/Yb-Ta/Yb plots (Pearce and others, 1977; Wood, 1980; Pearce, 1983). The average LILE distribution of the Pecos tholeiite is also similar to LILE distributions in IAB (Fig. 11). Although the wide intra-suite variation in LILE in the tholeiites probably reflects secondary remobilization, the scale of such remobilization appears to have been small enough to preserve the original average distribution.

Chemically, Pecos calc-alkaline basalts (CAB) are similar to modern CAB (see Table 6). On an NMORB-normalized plot, they exhibit LILE enrichment, low Nb and Ta, and elevated Ce and Sm (Fig. 11). They also exhibit negative Th, Ta, Nb, Y, and Ti anomalies on a primordial-mantle-normalized plot (Fig. 12). REE distributions are light REE enriched (~50× chondrites) and do not exhibit significant Eu-anomalies (Fig. 8C). Ti contents are low (<1 percent TiO_2) and increase slowly with increasing Zr (Fig. 9B). Samples fall chiefly in the CAB field on the Ti-Zr-Y diagram (Fig. 9A), and Zr/Y ratios are in the range of 3 to 10, similar to modern calc-alkaline basalts. Pecos CAB do not show the LILE or light REE enrichments characteristic of continental-margin arc basalts (Lopez-Escobar, 1984).

Felsic rocks. Pecos felsic volcanic and subvolcanic rocks are subdivided on the basis of petrography and selected geochemical discriminant diagrams (Figs. 3 and 7) into two compositional groups: rhyolite and dacite. For the most part, subvolcanic felsic rocks (mainly tonalite-trondhjemite) fall into the dacite group. Dacites from the Doctor Creek area (Fig. 2) form a compositionally distinct sub-group.

Pecos rhyolites and dacites (exclusive of the Doctor Creek dacites) have major- and trace-element distributions similar to modern counterparts from evolved oceanic arc systems (Tables 7 and 8; Ewart, 1979). Like arc dacites, Pecos dacites are low in Sr, Zr, REE, and Nb. The moderately altered dacites have REE patterns only slightly enriched in light REE (~20× chondrites), while moderately altered rhyolites have light REE concentrations of about 100× chondrites (Fig. 8D). Negative Eu-anomalies and low Sr contents in both rock types probably reflect plagioclase fractionation. In comparison to felsic volcanics from modern continental rift systems, the moderately altered Pecos volcanics (and subvolcanic rocks) have low La/Yb ratios and relatively low Yb contents (Fig. 13). Compared to rift rhyolites, Pecos rhyolites are depleted in Zr, Nb, Ta, and Hf (Table 8 and Ewart, 1979). Intrusive tonalites and trondhjemites are more enriched in CaO, Na_2O, and Sr than the Pecos dacites (Tables 7 and 9), a feature probably reflecting plagioclase accumulation.

Dacitic volcanic and volcaniclastic rocks from the Doctor

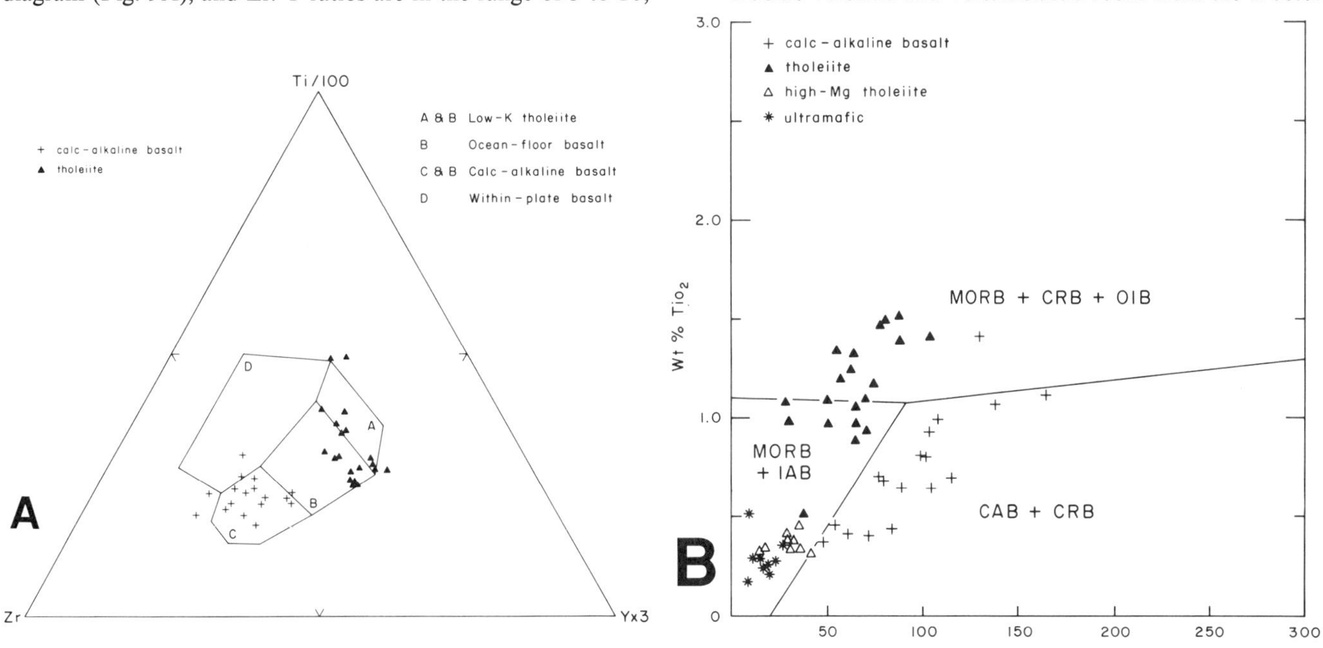

Figure 9. Trace-element plots for Pecos mafic (± ultramafic) rocks. A. Ti-Zr-Y diagram of Pearce and Cann (1973) showing distribution of Pecos mafic volcanic and subvolcanic rocks whose major-element compositions satisfy the restriction: 20 wt percent > CaO + MgO > 12 wt percent. B. TiO_2-Zr plot showing distribution of Pecos mafic and ultramafic rocks. Boundaries after Garcia (1979); MORB, mid-ocean-ridge basalt; CRB, continental rift basalt; OIB, oceanic island basalt; IAB, island-arc basalt; CAB, calc-alkaline basalt.

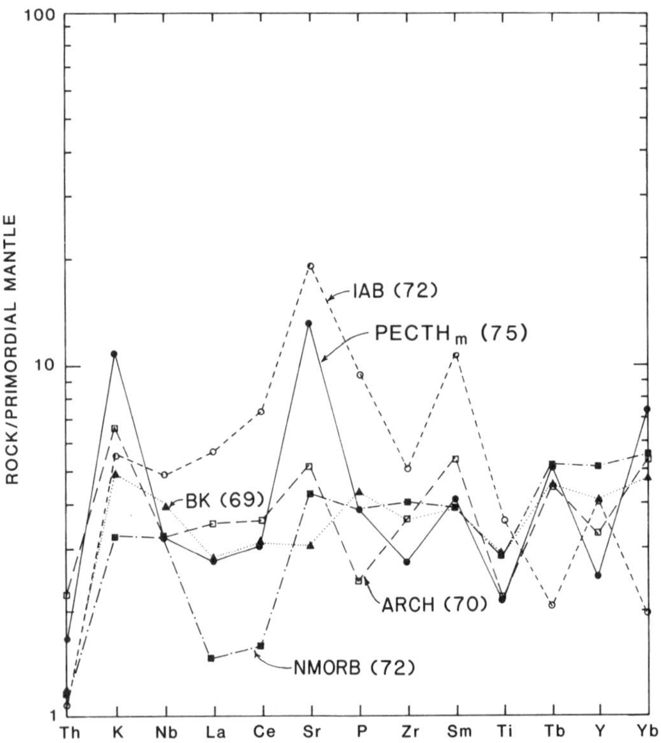

Figure 10. Average high-Mg tholeiites normalized to primordial mantle (after Wood, 1979). Symbols: ARCH, Archean; IAB, island-arc basalt; BK, Archean basaltic komatiite; NMORB, normal mid-ocean-ridge basalt; PECTH$_m$, Pecos high-Mg tholeiite. Principal references: Condie (1981a); Arndt and Nisbet (1982), Basaltic Volcanism Study Project (1981). Mg numbers given in parentheses.

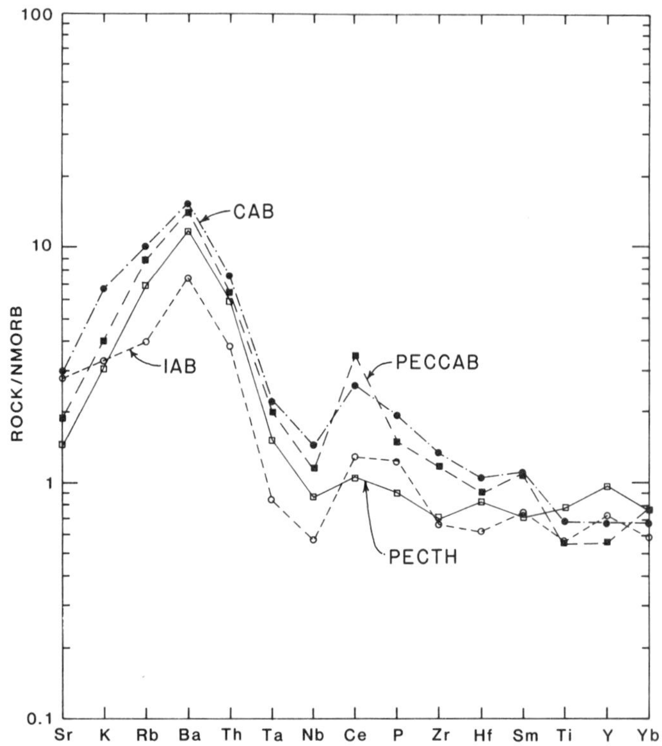

Figure 11. Element distributions in average basalts normalized to NMORB. IAB, island-arc basalt; CAB, calc-alkaline basalt; PECTH, Pecos tholeiite; PECCAB, Pecos calc-alkaline basalt. Normalization values from Pearce (1983). Chief references for IAB and CAB averages: Basaltic Volcanism Study Project (1981), Brown and others (1977), Nicholls and others (1980), Saunders and others (1980), Kay and others (1982).

Creek area are significantly enriched in Fe_2O_3, TiO_2, Sr, and especially Cr and Ni relative to Pecos dacites (Table 10). Doctor Creek dacites form separate populations on a number of trace-element plots (e.g., REE, Fig. 8D, La/Yb-Yb; Fig. 13, Hf-Zr, etc.), and they contain unusually high concentrations of REE and high field strength elements (HFSE). In this respect they are similar in composition to felsic volcanics erupted in continental back-arc basins and continental rifts.

DISCUSSION

Basalts and ultramafic rocks

Numerous models have been tested to evaluate the origin of mafic and ultramafic rocks from the Pecos area. It does not appear likely that the ultramafic rocks and high-Mg tholeiites are related by fractional crystallization, since if this were so, some elements (*viz.,* Sc and heavy REE) require >70 percent fractionation of olivine + clinopyroxene, while other elements permit only moderate amounts of fractional crystallization (Ni and Ti, 20 to 30 percent) or no fractional crystallization (light REE, Zr, Nb, Ta). Problems are also encountered if the high-Mg tholeiites and tholeiites are related by fractional crystallization. Although the tholeiites must have fractionated significant amounts of olivine, the two tholeiite groups do not appear to lie on the same fractionation trend (Fig. 14). Also, if the tholeiites were derived from the high-Mg tholeiites by fractional crystallization, the amounts of olivine removal required are inconsistent between elements (*viz.,* Ni and Sc require <25 percent and Zr, Y, Th, and V require >80 percent). Within either of the tholeiite groups, however, fractional crystallization is capable of accounting for most of the compositional variation.

The two Pecos tholeiite groups also cannot be related to each other by different degrees of melting of the same source. Similar concentrations of heavy REE in both groups cannot be explained by this mechanism. Also, degrees of melting predicted by HFSE and REE are inconsistent and not in agreement with major-element experimental data. Dynamic melting may produce a suite of mafic magmas ranging in composition between low-Mg tholeiite and CAB in incompatible element concentrations (Wood, 1979; Strong and Dostal, 1980). The absence of a complete spectrum of compositions between tholeiite and CAB end members in the Pecos area, however, does not favor such a mechanism. Also, the small amounts of incremental melting required (<5 percent) may present an obstacle to extracting the melts from the source. Although the differences in concentration

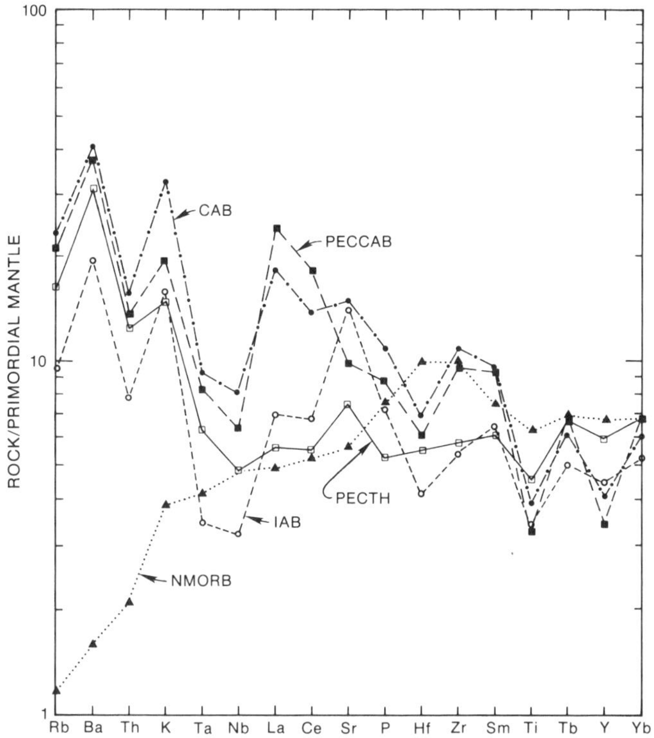

Figure 12. Primordial-mantle normalized graph of average basalts. NMORB, normal ocean ridge basalt; other symbols and references given in Figure 11. Normalization data given in Wood (1979).

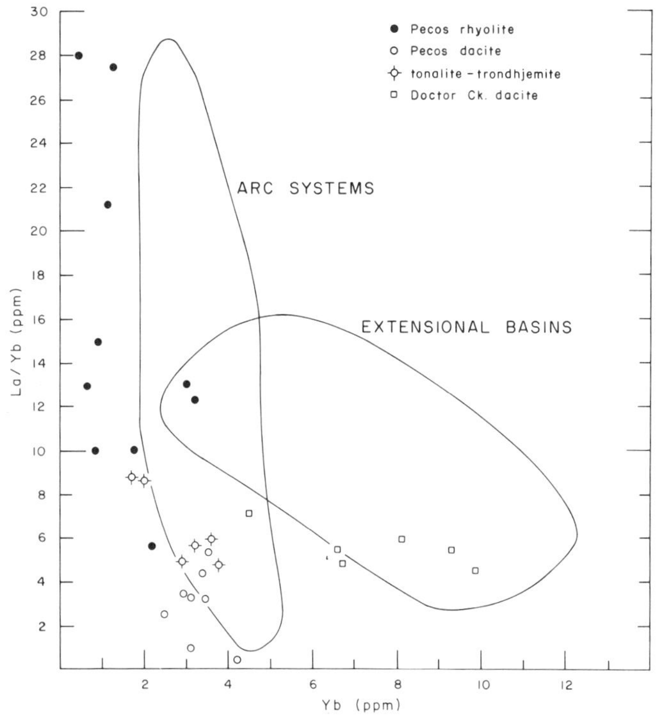

Figure 13. La/Yb-Yb plot of felsic volcanic rocks and tonalite-trondhjemite from the Pecos greenstone belt. Fields defined from modern felsic volcanic populations (Ewart, 1979, and other sources). Arc systems include island arcs and continental margin arcs; extensional basins include back-arc basins and rifts in continental crust.

of most incompatible elements between the tholeiites and CAB may be explained by open-system fractional crystallization (O'Hara, 1977), differences in incompatible element ratios (such as La/Sm and La/Yb, Table 11) are too large to be explained by this process (Pankhurst, 1977) and appear to reflect differences in mantle source composition.

The possible effect of continental contamination is most readily detectable with LILE. Because some LILE (U, Th, Rb, Cs) are typically depleted in the lower crustal terranes (Condie and Allen, 1984), it is unlikely that the lower continental crust can substantially contaminate magmas. Ba and Th enrichment are most characteristic of upper crustal contamination (Weaver and Tarney, 1983; Pearce, 1983). Neither of these elements is enriched significantly above the average IAB content (Fig. 11), and Th exhibits a negative anomaly on a primordial mantle-normalized plot (Fig. 12). The relative enrichment in most LILE in the Pecos high-Mg tholeiites and ultramafics may reflect selective crustal contamination, although our preferred interpretation is that such enrichments occurred during alteration or metamorphism. More definitive evidence against significant contamination by older continental basement is reflected by the positive ϵ_{Nd} values for the Pecos ultramafic rocks and high-Mg tholeiites (Nelson and DePaolo, 1984). These Pecos rocks lie on or very close to the uncontaminated $^{143}Nd/^{144}Nd$ growth curve for IAB.

It would appear that at least three mantle sources are needed for the mafic and ultramafic rocks of the Pecos greenstone belt. Ratios of HFSE and REE, which are least susceptible to secondary remobilization, are particularly sensitive to source compositions. The high La/Yb ratio in the Pecos ultramafic rocks may reflect secondary introduction of light REE, and the slightly high Zr/Y, Zr/Sc, and Al_2O_3/TiO_2 ratios in these rocks may be caused by clinopyroxene fractionation. Many incompatible element ratios in Pecos tholeiites and CAB are similar to corresponding ratios in modern IAB and CAB, respectively (Table 11). As with IAB, the Ba/La ratio in Pecos tholeiites is particularly variable when compared to modern CAB (Hickey and others, 1984). The K/Rb in Pecos low-Mg tholeiite and CAB is lower than IAB and CAB, a feature probably reflecting preferential Rb addition during secondary processes. The implied mantle source for the tholeiites has similar Zr/Nb, Ti/Zr, and Zr/Y ratios to chondrites and primordial mantle. Compared to Ti and Y, Zr is enriched in the CAB source, a feature that may reflect Ti- and Y-bearing minor phases in the residue. Garnet, however, cannot represent the Y-bearing phase because Zr/Y and La/Yb ratios should be higher than observed. Compared to the tholeiite source, the CAB source is also enriched in Ta relative to La. These differences and similarities in element ratios between Pecos tholeiites and calc-alkaline basalts are very similar to those ob-

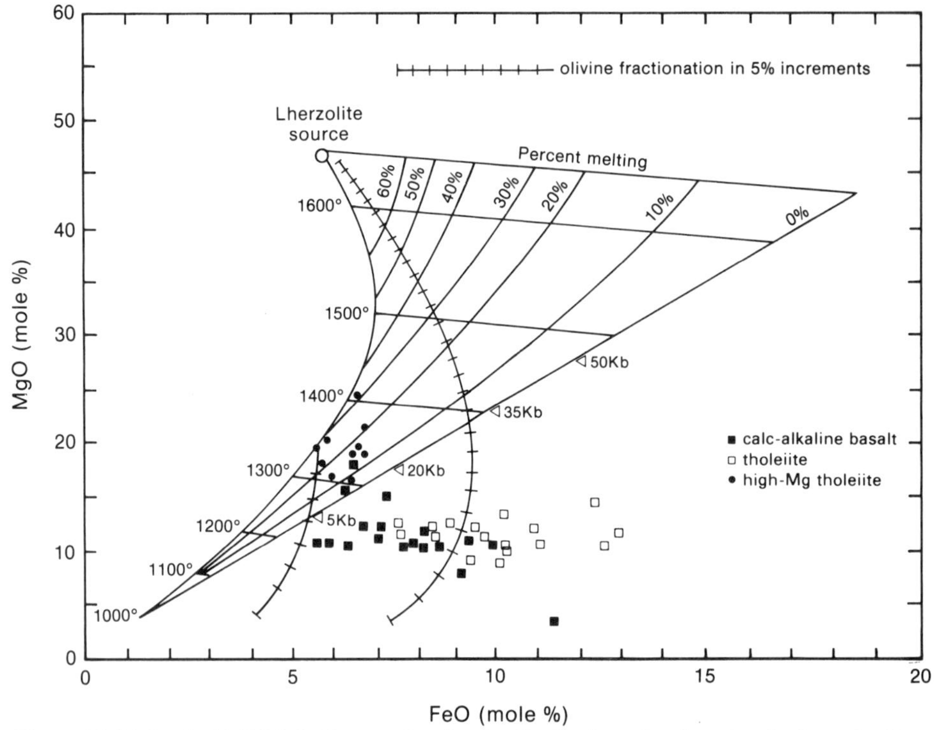

Figure 14. Molecular FeO-MgO diagram showing the distribution of mafic volcanics from the Pecos greenstone belt. Percent of melting of lherzolite source at various depths (in kb) after Hanson and Langmuir (1978) and Rajamani and others (1985). Temperatures given are for 1 atm; temperature at higher pressures can be estimated assuming an increase of 5.5°C/kb.

served between corresponding groups (IAB and CAB) in modern arc systems (Brown and others, 1977; Nicholls and others, 1980; Saunders and others, 1980; Kay and others, 1982; Table 11). It is possible by analogy with modern arcs that the Pecos tholeiites are derived from a more depleted source, perhaps a descending plate, and Pecos CAB from a metasomatized mantle wedge above the plate (Saunders and others, 1980).

The origin of Pecos mafic volcanics can be further constrained when incompatible-element distributions are considered together with compatible- and major-element contents. Particularly valuable in this respect are the molecular FeO–MgO and Zr–Ni relationships proposed by Hanson and Langmuir (1978) and Rajamani and others (1985). The Pecos tholeiites can be produced by 15 to 20 percent batch melting of a garnet lherzolite mantle source at ~100 km depth (30 kb) followed by up to 50 percent fractional crystallization of olivine (± clinopyroxene; Fig. 14). The Pecos CAB require similar amounts of melting of the same source at a depth <100 km. In both cases, neither garnet nor amphibole can remain in the residue. The low MgO and Ni contents of these rocks also necessitate 25 to 30 percent of olivine removal. If minor phases are retained in the tholeiite and CAB sources to explain the negative Th, Ta-Nb, and Ti anomalies (Fig. 12), the degrees of melting were probably less than to 10 percent. The high-Mg tholeiites may be produced by 20 to 30 percent melting of garnet lherzolite at temperatures of 1,400 to 1,500°C and depths of 30 to 40 km (10 to 15 kb). Their relatively high Mg numbers and high Ni contents allow ≤5 percent olivine fractionation. To explain incompatible trace-element distributions, the mantle sources for Pecos tholeiites and CAB must be enriched in LILE and light REE relative to Nb and Ta, a feature referred to a subduction zone geochemical component (SZC; Pearce, 1983). The Pecos high-Mg tholeiites, on the other hand, appear to have been derived from a depleted mantle source, not unlike the source of MORB.

The Pecos ultramafic rocks may represent cumulates, although possible parental magmas have not been recognized in the Pecos mafic rocks. In terms of composition and relict textures (Wyman, 1980), they are similar to pyroxene-rich ophiolite-type ultramafic rocks (Coleman, 1977). The relatively low contents of Ni and Cr in the Pecos ultramafic rocks, however, indicate that olivine and spinel were not important primary minerals.

Felsic volcanics and tonalites

Of the models tested for the origin of Pecos rhyolites, dacites, and tonalites, a model involving fractional crystallization of Pecos CAB results in the best overall agreement. The model is similar to that proposed by Condie and Shadel (1984) for Proterozoic felsic volcanics in southeastern Wyoming. The details of the model, which involves two stages of fractional crystallization, are given in this reference. Liquidus phases include olivine, plagioclase, clinopyroxene, hornblende, quartz, and magnetite. The relatively large variations in heavy REE, Y, Ti, Zr, and Sc in the Pecos dacites can be explained by 15 to 30 percent of hornblende fractional crystallization. The model predicts more light REE in the dacites than observed, a problem that may be explained if trace amounts of allanite (or another phase rich in light REE) crystallize from the melt (Miller and Mittlefehldt, 1982). Variable Eu anomalies and Sr contents in the felsic volcanics and tonalites can be explained by plagioclase crystallization.

Doctor Creek felsic volcanics

Both the petrography and geochemistry of the Doctor Creek felsic volcanic and volcaniclastic rocks set them apart from the other Pecos felsic volcanic rocks. Relict textures reminiscent of welded ash-flow tuffs suggest some Doctor Creek volcanics may have accumulated, or at least erupted, subaerially. In addition, these rocks display numerous chemical similarities with modern continental, rift-related felsic volcanic rocks, especially in terms of REE and HFSE abundances (cf. Ewart, 1979). Interestingly, Doctor Creek felsic volcanics share a number of compositional characteristics with the 1,700-Ma Burned Mountain metarhyolite (J. M. Robertson, unpublished data), a regionally extensive, locally welded ash-flow sheet exposed in the Tusas Mountains 100 km to the northwest. For these reasons, Doctor Creek felsic volcanic rocks are judged not to be part of the Pecos felsic suite but are tentatively assigned to a younger (ca. 1,700 Ma) rhyolite-dominated volcanic succession that is widely exposed in the Tusas Mountains (see Grambling and others, this volume). Similar rhyolitic rocks are present in the Picuris Range (Bauer, 1984), and may overlie portions of the northern margin of the Pecos greenstone belt (Grambling and Ward, 1987).

Tectonic setting

As with other Proterozoic arc associations in the Southwest (Condie and Nuter, 1981; Condie and Shadel, 1984; Condie and McCrink, 1982; Boardman and Condie, 1986; Condie, 1986; Knoper and Condie, 1988), the Pecos succession is characterized by bimodal volcanism. Young bimodal volcanic arc systems have been reported (Whitney and others, 1978; Condie and Shadel, 1984; Condie, 1986), but they appear to be more common in the Proterozoic. The geochemistry and Nd isotopic composition of the Pecos volcanic succession (Nelson and DePaolo, 1984) is similar to that of modern island arcs such as the Aleutian Islands. As in modern oceanic arcs, tholeiites may be produced in the descending plate; as they rise, they undergo only minor LILE contamination in the mantle wedge. Perhaps, as suggested for the Aleutian arc (Kay and others, 1982), tholeiites rise rapidly in regions of extensional stresses. This results in a minimum amount of mantle-wedge contamination and also restricts the opportunities for fractional crystallization to produce felsic magmas. The Pecos CAB, on the other hand, may have been produced at shallower depths in a metasomatized mantle wedge. Their progression to the surface was more sluggish and resulted in greater degrees of fractional crystallization. Just how a depleted mantle source, which appears to have given rise to the Pecos ultramafic rocks and high-Mg tholeiites, is related to the mantle wedge and descending slab is not clear. The ultramafic rocks and high-Mg tholeiites occur in only two relatively small (<5 km^2 in aggregate) areas on the southeast side of the Pecos greenstone. They may represent tectonic fragments of oceanic crust, perhaps from a back-arc basin on the north, in that the Pecos arc seems to continue south into the Manzanita and Pedernal Mountains (Condie, 1980, 1988; McKee, 1988).

The relationship of the Pecos arc to other Proterozoic arc associations in the Southwest is of interest in the tectonic evolution of this area during the early Proterozoic. Arc associations ranging from 1,790 to 1,750 Ma occur in southeast Wyoming, Colorado, and northern New Mexico (Condie, 1982; Condie and Shadel, 1984; Bickford and Boardman, 1984; Condie, 1986, 1988; Bowring and others, this volume). A possible back-arc succession was erupted in a rifted segment of this crust at 1,730 to 1,740 Ma in central Colorado (Condie and Nuter, 1981; Boardman and Condie, 1986; Knoper and Condie, 1988). The Pecos arc in the southern Sangre de Cristo Mountains (~1,720 Ma) appears to represent the youngest addition to these arc successions. It is not yet clear if the Pecos arc represents an evolved island arc system that collided with the edge of the continent in northern New Mexico or a relatively unevolved, back-arc basin associated with a continental-margin arc system. Volcanics from both types of settings are geochemically similar.

ACKNOWLEDGMENTS

The authors acknowledge NSF Grant EAR-8313735, which has partially supported K. C. during his geological and geochemical studies of Proterozoic rocks in the southwestern United States. Samples were irradiated for neutron activation at Sandia National Laboratories (Albuquerque, New Mexico) and we appreciate the able assistance of the reactor staff. We would especially like to acknowledge the mapping contributions of a number of graduate students, including Dave Mathewson, Dunc Riesmeyer, Bill Wyman, Ingrid Klich, and Dave Codding. The original manuscript was greatly improved by critical reviews of J. A. Grambling, R. H. Moench, and M. Garcia.

TABLE 3. CHEMICAL COMPOSITIONS OF PECOS ULTRAMAFIC ROCKS

Sample No*	79--4-35	79-18-26	79-23-26	80-1-26	80-4-23	n	Mean	σ
Map No.	1	2	3	4	5			
SiO_2	44.09	43.32	46.20	44.35	42.17	14	43.93	1.85
TiO_2	0.30	0.18	0.25	0.22	0.29	14	0.27	0.09
Al_2O_3	8.70	3.85	5.53	5.80	6.66	14	6.60	1.62
Fe_2O_{3-T}	11.25	12.93	11.73	12.67	12.52	14	11.97	0.94
MgO	19.70	27.51	23.44	23.61	24.61	14	23.21	2.66
CaO	9.37	3.12	5.47	5.93	7.41	14	6.36	1.87
Na_2O	0.79	0.18	0.47	0.63	0.53	14	0.61	0.17
K_2O	nd	nd	nd	nd	nd	4	0.14	0.04
MnO	0.10	0.02	0.08	0.16	0.18	14	0.19	0.16
P_2O_5	0.06	0.03	0.04	0.04	0.05	14	0.05	0.01
LOI	4.32	8.27	6.04	6.34	5.40	14	5.67	1.21
Total	98.68	99.41	99.25	99.75	99.82			
Rb	nd	2	1	nd	nd	6	4	4
Ba	nd	78	83	nd	nd	2	80	2
Cs	na	0.2	nd	na	na	1	0.2	
Sr	87	4	19	101	89	14	96	66
Pb	nd	9	7	nd	nd	5	12	5
Th	na	0.2	1.0	na	na	2	0.6	0.4
U	na	nd	nd	na	na			
Sc	17	5	15	12	15	12	16	9
V	46	38	55	49	41	14	54	26
Cr	168	238	216	230	84	14	234	103
Co	111	129	113	105	97	14	107	14
Ni	1100	1065	890	860	910	14	827	196
Cu	14	nd	107	70	14	13	60	57
Zn	170	85	56	96	70	14	88	32
Y	2	nd	1	nd	nd	8	5	4
Zr	12	8	15	nd	nd	9	17	6
Nb	na	1	1	na	na	2	<2	
Nf	na	nd	nd	na	na			
Ta	na	0.14	0.11	na	na	2	0.13	0.02
La	na	2.2	1.6	na	na	2	2.0	0.2
Ce	na	5.5	5.3	na	na	2	5.4	0.1
Sm	na	1.1	1.2	na	na	2	1.15	0.05
Eu	na	0.50	0.47	na	na	2	0.49	0.02
Tb	na	0.22	0.20	na	na	2	0.21	0.01
Yb	na	0.67	0.74	na	na	2	0.71	0.03
Lu	na	0.10	0.11	na	na	2	0.11	0.05
Mg Number	80	83	82	81	82	14	81	2

*Representative analyses from a total sample population of 14.

Oxides in weight percent, trace elements in ppm; Fe_2O_{3-T} = total iron as Fe_2O_3; LOI = loss of ignition; Mg number = 100 MgO/MgO + FeO mole ratio where FeO = 0.865 Fe_2O_{3-T}; nd = not detected; na - not analyzed; n = number of samples; σ = one standard deviation.

TABLE 4. CHEMICAL COMPOSITIONS OF PECOS HIGH-MG THOLEIITES

Sample No.* Map No.	79-12-26 6	79-14-26 7	79-21-26 8	79-22-26 9	80-1-23 10	80-4-26 11	n	Mean	σ
SiO_2	46.77	45.71	45.57	46.11	47.09	46.24	11	46.23	0.70
TiO_2	0.35	0.41	2.33	0.39	0.46	0.38	11	0.54	0.57
Al_2O_3	14.24	14.15	12.35	13.81	14.54	14.38	11	13.83	1.03
$Fe_2O_{3\text{-}T}$	8.41	9.27	9.90	9.44	8.30	7.01	11	8.85	0.78
MgO	12.77	12.36	14.17	12.52	11.84	12.30	11	12.73	1.26
CaO	12.60	13.13	11.46	12.45	14.21	15.05	11	13.18	1.18
Na_2O	1.58	1.91	1.72	1.53	2.17	1.63	11	1.72	0.38
K_2O	0.48	0.09	0.05	1.25	0.22	0.15	11	0.34	0.34
MnO	0.11	0.14	0.16	0.15	0.15	0.15	11	0.14	0.01
P_2O_5	0.07	0.07	0.08	0.07	0.09	0.07	10	0.07	0.01
LOI	1.83	1.96	1.54	1.37	0.84	1.53	11	1.64	0.50
Total	99.21	99.20	99.33	99.09	99.91	98.89			
Rb	24	9	28	31	11	10	9	16	9
Ba	nd	nd	183	nd	164	nd	2	173	9
Cs	8.5	na	4.5	3.3	9.4	na	5	6	3
Sr	290	311	233	216	355	327	10	275	74
Pb	10	6	nd	10	12	25	7	13	6
Th	0.1	na	nd	0.2	nd	nd	2	0.1	0.06
U	nd	na	0.30	1.0	0.10	nd	4	0.4	0.35
Sc	30	37	39	39	39	38	10	35	7
V	82	88	123	93	120	91	10	87	24
Cr	429	251	436	302	486	443	11	349	90
Co	96	80	85	67	47	54	11	72	18
Ni	330	250	320	305	223	210	11	338	169
Cu	44	132	95	149	42	81	11	187	336
Zn	59	50	80	52	38	54	11	52	13
Y	13	nd	13	20	17	11	7	13	4
Zr	36	16	32	29	40	32	10	30	8
Nb	2	na	na	3	na	na	2	2.5	0.5
Hf	nd	na	na	nd	na	na	0		
Ta	0.11	na	na	0.20	na	na	2	0.16	0.05
La	1.9	na	na	2.0	na	na	2	1.95	0.02
Ce	5.7	na	na	5.8	na	na	2	5.75	0.07
Sm	1.7	na	na	1.4	na	na	2	1.58	0.13
Eu	0.50	na	na	0.76	na	na	2	0.63	0.13
Tb	0.45	na	na	0.57	na	na	2	0.51	0.06
Yb	2.9	na	na	2.6	na	na	2	2.78	0.17
Lu	0.33	na	na	0.32	na	na	2	0.33	0.01
Mg Number	77	75	76	75	76	80	11	76	2

*Representative analyses from a total sample population of 11.

Oxides in weight percent, trace elements in ppm; $Fe_2O_{3\text{-}T}$ = total iron as Fe_2O_3; LOI = loss on ignition; Mg number = 100 MgO/MgO + FeO mole ratio where FeO = 0.865 $Fe_2O_{3\text{-}T}$; nd = not detected; na = not analyzed; n = number of samples; σ =one standard deviation.

TABLE 5. CHEMICAL COMPOSITIONS OF PECOS THOLEIITES

Sample No.* Map No.	DOC-45 12	DOC-46 13	103-c 14	113-a 15	CC-2 16	CC-4 17	n	Mean	σ
SiO_2	48.80	48.96	46.99	51.03	53.48	50.82	20	49.56	2.15
TiO_2	1.50	0.97	1.41	1.25	1.10	1.35	20	1.16	0.25
Al_2O_3	12.19	14.08	16.31	13.53	14.34	16.06	20	14.75	1.74
$Fe_2O_{3\text{-}T}$	18.12	14.18	13.77	14.25	8.85	10.58	20	13.40	2.58
MgO	6.93	8.57	6.95	6.36	9.60	7.79	20	7.40	1.01
CaO	10.38	9.52	11.02	10.87	9.36	8.62	20	10.47	1.54
Na_2O	2.05	2.18	2.88	2.06	1.85	2.59	20	2.34	0.74
K_2O	0.21	0.60	0.22	0.24	0.60	1.20	20	0.66	0.94
MnO	0.27	0.17	0.26	0.21	0.20	0.22	20	0.25	0.08
P_2O_5	0.16	0.10	0.15	0.14	0.05	<0.05	19	0.11	0.04
LOI	0.38	1.16	0.40	0.60	0.40	1.07	20	0.85	0.87
Total	100.99	100.49	100.36	100.54	99.83	100.30			
Rb	7	18	2	4	21	62	20	17	19
Ba	23	65	nd	nd	342	541	11	229	157
Cs	na	2.2	0.7	2.3	3.7	3.8	13	5.0	4.4
Sr	163	138	109	160	165	192	20	174	65
Pb	6	10	54	53	11	16	20	26	23
Th	2.0	1.0	1.0	0.3	3.0	1.0	14	1.3	1.0
U	1.0	nd	0.5	0.4	nd	1.0	12	1.0	0.8
Sc	57	52	57	34	51	51	20	47	9
V	475	nd	nd	nd	na	na	7	412	54
Cr	166	110	25	43	590	90	20	137	115
Co	nd	55	71	50	47	38	17	48	12
Ni	60	32	nd	nd	na	na	8	60	14
Cu	97	nd	58	47	na	na	14	83	68
Zn	96	125	156	175	na	na	15	148	55
Y	46	29	46	26	17	26	20	29	10
Zr	81	65	104	62	50	54	20	64	19
Nb	5	4	5	2	2	5	20	4	2
Hf	na	2.3	2.9	3.0	<1.4	2.5	9	2.5	0.5
Ta	na	nd	0.5	0.3	0.1	<0.5	10	0.3	0.2
La	na	<3.5	9.5	2.4	4.3	4.4	12	4.6	2.0
Ce	na	<10.7	11.8	<7.3	<6.1	<6.0	3	8.3	3.3
Sm	na	2.25	3.9	2.4	1.8	2.0	17	2.3	0.7
Eu	na	2.20	0.70	0.40	0.89	1.3	17	0.78	0.44
Tb	na	<1.20	1.20	0.60	0.45	0.64	14	0.69	0.24
Yb	na	2.9	4.8	3.1	2.1	2.5	16	2.6	0.9
Lu	na	0.47	0.60	0.30	0.34	0.41	17	0.35	0.11
Mg Number	46	58	53	50	71	62	20	56	6

*Representative analyses from a total sample population of 20.

Oxides in weight percent, trace elements in ppm; $Fe_2O_{3\text{-}T}$ = total iron as Fe_2O_3; LOI = loss on ignition; Mg number = 100 MgO/MgO + FeO mole ratio where FeO = 0.865 $Fe_2O_{3\text{-}T}$; nd = not detected; na = not analyzed; n = number of samples; σ = one standard deviation.

TABLE 6. CHEMICAL COMPOSITIONS OF PECOS CALC-ALKALINE BASALTS

Sample No.*	PR-4	PR-6	C-9-17	H-7-122	C-9-87	C-10-24	n	Mean	σ
Map No.	18	19	20	21	22	23			
SiO_2	53.78	51.87	49.6	50.0	54.0	52.0	18	51.64	2.10
TiO_2	0.81	0.68	0.41	0.44	0.65	1.07	18	0.83	0.44
Al_2O_3	14.29	14.86	15.5	14.0	14.1	14.5	18	14.56	1.52
$Fe_2O_{3\text{-}T}$	10.69	9.86	8.86	8.98	9.49	11.5	18	10.82	2.07
MgO	6.31	7.61	10.4	11.8	7.80	6.37	18	7.16	1.95
CaO	8.89	9.42	11.3	11.0	10.2	9.81	18	10.69	1.63
Na_2O	2.64	3.18	1.55	1.64	2.73	3.49	18	2.62	0.53
K_2O	1.16	0.83	0.58	0.38	0.89	0.51	18	0.59	0.32
MnO	0.19	0.17	0.22	0.21	0.20	0.23	17	0.21	0.03
P_2O_5	0.20	0.14	na	na	na	na	5	0.17	0.02
LOI	1.02	1.27	1.76	0.74	0.78	0.95	13	1.05	0.31
Total	99.98	99.89	100.18	99.19	100.84	100.43			
Rb	35	21	21	11	34	3	17	17	12
Ba	365	559	209	87	409	319	18	288	265
Cs	4.3	3.1	34.1	8.1	2.5	2.7	18	5.9	7.5
Sr	270	225	213	158	248	313	18	225	48
Pb	10	14	11	23	17	20	13	20	13
Th	3.0	1.0	0.8	0.5	2.9	0.4	15	1.3	1.0
U	2.0	nd	na	na	nd	nd	3	1.1	0.7
Sc	35	36	na	na	na	na	5	34	4
V	183	162	na	na	na	na	4	176	26
Cr	100	211	361	914	332	56	18	195	200
Co	40	42	42	49	42	43	18	44	6
Ni	67	161	50	79	33	27	17	52	38
Cu	63	72	112	51	65	67	18	60	26
Zn	86	85	67	66	76	91	18	100	48
Y	21	20	14	9	15	23	18	18	6
Zr	102	79	72	84	104	137	18	99	32
Nb	4	4	na	na	na	na	5	5	1
Hf	2.6	2.2	na	na	na	na	5	2.15	0.24
Ta	nd	0.69	0.01	0.08	0.33	0.77	16	0.36	0.28
La	16.2	10.3	4.0	1.6	15.2	20.0	18	12.2	9.6
Ce	37.7	21.7	8.6	8.8	27.9	45.6	18	28.5	23.5
Sm	2.9	2.6	1.4	1.5	2.7	6.0	18	3.2	1.9
Eu	1.3	1.0	0.23	1.1	1.20	1.36	18	1.4	0.81
Tb	0.58	0.44	0.42	0.44	0.36	1.08	17	0.69	0.33
Yb	1.8	1.8	1.6	1.6	2.1	4.1	18	2.4	1.0
Lu	0.31	0.25	0.26	0.30	0.35	0.59	18	0.40	0.16
Mg Number	57.0	63	72	75	65	55	18	59	11

*Representative analyses from a total sample population of 18.

Oxides in weight percent, trace elements in ppm; $Fe_2O_{3\text{-}T}$ = total iron as Fe_2O_3; LOI = Loss on ignition; Mg number = 100 MgO/MgO + FeO mole ratio where FeO = 0.865 $Fe_2O_{3\text{-}T}$; nd = not detected; na = not analyzed; n = number of samples; σ = one standard deviation.

TABLE 7. CHEMICAL COMPOSITIONS OF PECOS DACITES

Sample No.* Map No.	MA-2 24	MA-16 25	MA-18 26	MA-25 27	MA-26 28	MA-27 29	n	Mean	σ
SiO_2	77.60	77.39	74.28	69.12	73.09	72.33	12	72.95	2.98
TiO_2	0.19	0.23	0.26	0.25	0.22	0.24	12	0.25	0.04
Al_2O_3	13.31	11.55	11.54	14.28	11.68	11.85	12	12.22	1.22
Fe_2O_{3-T}	2.35	5.37	4.92	4.00	6.27	5.59	12	4.93	1.55
MgO	0.41	1.98	3.75	6.10	3.42	3.67	12	4.36	2.19
CaO	2.48	0.06	0.04	0.10	0.07	0.08	12	0.27	0.67
Na_2O	4.05	0.27	0.23	0.24	0.43	0.41	12	0.58	1.05
K_2O	0.21	2.10	1.96	1.99	2.43	1.97	12	1.87	0.65
MnO	0.03	0.07	0.06	0.05	0.07	0.08	12	0.07	0.03
P_2O_5	0.04	0.03	0.03	<0.05	0.03	0.03	11	0.03	0.01
LOI	0.15	1.90	2.53	3.39	2.44	3.21	12	2.67	1.00
Total	100.80	100.95	99.60	99.52	100.15	99.46			
Rb	nd	16	16	53	39	23	11	24	12
Ba	123	927	865	126	1110	1239	12	917	569
Cs	1.1	4.0	na	24.4	9.8	1.7	8	6.3	7.3
Sr	255	6	19	10	7	8	12	32	68
Pb	8	17	18	15	8	11	12	17	9
Th	2	1	2	6	7	5	9	4	3
U	1.0	nd	nd	2	1	1	4	1.1	0.5
Sc	12	12	15	9	10	17	12	14	3
V	9	na	na	na	3	2	4	5	3
Cr	nd	nd	nd	nd	85	71	2	78	7
Co	4	3	nd	4	7	2	8	4	1
Ni	14	6	na	na	<10	<10	8	8	3
Cu	43	65	69	na	<10	11	10	75	70
Zn	12	167	120	na	98	39	11	106	52
Y	30	22	34	29	31	29	12	30	5
Zr	134	104	116	158	150	132	12	130	16
Nb	12	8	7	9	14	12	12	10	2
Hf	3.3	2.4	na	4.8	4.6	4.1	8	3.8	0.7
Ta	0.40	nd	na	0.65	0.52	0.46	6	0.46	0.11
La	14.6	6.3	na	3.0	11.0	19.9	8	9.7	5.5
Ce	16.7	<6.2	na	<6.3	10.9	27.4	5	18.6	5.5
Sm	3.0	2.2	na	1.5	3.6	6.9	8	3.3	1.7
Eu	0.87	0.45	na	0.27	0.12	0.27	7	0.36	0.23
Tb	0.67	0.56	na	0.44	0.80	1.06	8	0.74	0.25
Yb	3.4	2.5	na	3.1	3.5	3.6	8	3.3	0.5
Lu	0.54	0.41	na	0.44	0.59	0.56	8	0.51	0.07

*Representative analyses from a total sample population of 12.
Oxides in weight percent, trace elements in ppm; Fe_2O_{3-T} = total iron as Fe_2O_3; LOI = loss on ignition; Mg number = 100 MgO/MgO + FeO mole ration where FeO = 0.865 Fe_2O_{3-T} ; nd = not detected; na = not analyzed; n = number of samples; σ = one standard deviation.

TABLE 8. CHEMICAL COMPOSITIONS OF PECOS RHYOLITES

Sample No.*	452	557	MA-34	MA-38	PM-8	PM-10	n	Mean	σ
Map No.	30	31	32	33	34	35			
SiO_2	76.59	75.23	75.62	78.62	89.39	72.32	13	78.74	5.99
TiO_2	0.10	0.23	0.09	0.06	0.05	0.08	13	0.09	0.05
Al_2O_3	13.02	13.98	12.38	10.10	5.80	9.11	13	9.82	3.56
$Fe_2O_{3\text{-}T}$	1.81	3.08	1.70	2.28	0.40	5.77	13	2.26	1.34
MgO	0.23	0.43	4.18	1.46	1.12	6.87	13	2.71	2.04
CaO	0.81	0.80	0.18	1.61	0.01	0.12	13	0.71	0.70
Na_2O	4.57	4.72	1.09	3.34	0.37	0.57	13	2.27	1.68
K_2O	2.84	2.54	2.91	0.19	1.54	0.14	13	1.06	1.04
MnO	0.02	0.04	0.04	0.03	nd	0.03	12	0.03	0.01
P_2O_5	0.03	0.06	0.01	0.01	0.01	0.01	13	0.02	0.01
LOI	0.78	0.80	2.04	1.75	1.08	5.50	13	1.80	1.25
Total	100.80	100.91	100.24	99.45	99.77	100.52			
Rb	54	53	45	3	16	3	11	20	19
Ba	850	680	573	2102	310	39	13	599	659
Cs	2.0	2.0	4.7	0.5	0.2	1.6	11	1.3	1.3
Sr	113	101	22	146	3	5	13	77	72
Pb	41	65	9	10	186	43	13	111	157
Th	6.0	8.0	10.2	11.6	4.5	5.3	13	9.3	3.3
U	2.0	3.0	4.1	1.8	1.0	1.0	12	1.8	1.1
Sc	3	3	1	1	1	2	13	2	1
V	na	na	nd	15	19	26	8	22	13
Cr	nd	nd	58	78	50	75	8	72	15
Co	2	3	1	3	1	18	9	6	5
Ni	na	na	<10	<10	<10	42	7	21	14
Cu	<10	<10	<10	<10	70	717	8	240	237
Zn	28	225	45	19	326	3368	13	1790	3751
Y	24	38	10	17	9	24	13	16	9
Zr	148	243	122	93	56	85	13	106	52
Nb	16.0	10.0	19	18	7	12	13	13	5
Hf	3.0	8.0	3.3	2.6	1.5	2.3	11	2.5	2.0
Ta	3.00	1.00	1.10	0.88	0.29	0.62	10	0.86	0.78
La	39.3	40.6	35.9	17.8	12.7	12.7	10	21.0	12.4
Ce	89.2	80.0	67.7	35.7	29.4	25.5	10	43.6	25.0
Sm	3.8	5.4	2.8	2.6	2.2	2.1	10	2.5	1.2
Eu	0.90	1.40	0.34	0.21	0.41	0.61	10	0.48	0.38
Tb	1.00	1.20	0.33	0.44	0.30	0.51	10	0.45	0.35
Yb	3.0	3.3	1.3	1.76	0.86	2.27	10	1.54	0.96
Lu	0.30	0.50	0.21	0.28	0.14	0.33	10	0.23	0.12

*Representative analyses from a total sample population of 13.

Oxides in weight percent, trace elements in ppm; $Fe_2O_{3\text{-}T}$ = total iron as Fe_2O_3; LOI = loss on ignition; Mg number = 100 MgO/MgO + FeO mole ratio where FeO = 0.865 $Fe_2O_{3\text{-}T}$; nd = not detected; na = not analyzed; n = number of samples; σ = one standard deviation.

TABLE 9. CHEMICAL COMPOSITIONS OF PECOS TONALITE-TRONDHJEMITE

Sample No.*	CC-1	PR-1	MA-1	MA-39	PEC-R-24	n	Mean	σ
Map No.	36	37	38	39	40			
SiO_2	74.94	68.61	69.26	72.65	70.24	6	71.32	2.18
TiO_2	0.23	0.39	0.45	0.27	0.39	6	0.31	0.10
Al_2O_3	13.34	14.76	13.39	12.87	15.12	6	13.84	0.81
Fe_2O_{3-T}	3.13	5.33	6.28	3.73	4.13	6	4.01	1.55
MgO	0.32	0.86	0.47	1.28	1.00	6	0.72	0.36
CaO	2.72	3.56	4.38	3.44	2.55	6	2.95	1.05
Na_2O	4.25	3.39	3.83	3.66	4.21	6	4.16	0.71
K_2O	1.07	1.73	0.29	1.25	2.22	6	1.19	0.66
MnO	0.05	0.12	0.09	0.08	0.08	6	0.07	0.03
P_2O_5	<0.05	0.08	0.13	0.06	0.09	6	0.08	0.03
LOI	0.38	0.44	0.24	0.40	0.60	6	0.43	0.11
Total	100.43	99.27	98.81	99.69	100.63			
Rb	22	42	2	40	41	6	27	15
Ba	661	663	173	443	1566	6	795	477
Cs	1.4	2.1	1.1	8.2	1.8	6	2.7	2.5
Sr	213	228	397	235	245	6	244	75
Pb	11	16	6	16	17	6	13	4
Th	1.0	4.0	3.0	4.4	7.0	6	4.6	2.3
U	1.0	2.0	1.0	1.0	2.0	6	1.3	0.5
Sc	10	15	25	10	6	6	13	6
V	na	27	13	38	nd	3	26	10
Cr	nd	6		118	8	3	44	52
Co	3	9	4	6	3	5	5	2
Ni	na	6	6	nd	15	4	10	4
Cu	na	14	21	16	7	5	26	23
Zn	na	53	11	45	53	5	34	20
Y	26	21	32	15	17	6	19	9
Zr	143	109	110	109	199	6	133	32
Nb	11	5	10	8	6	6	7	2
Hf	5.0	3.6	2.5	2.8	5.5	6	4.0	1.1
Ta	0.71	nd	1.17	0.35	nd	3	0.74	0.34
La	14.6	18.4	18.9	15.5	21.9	6	17.8	2.4
Ce	30.1	37.5	42.2	35.3	41.9	6	40.6	8.3
Sm	3.6	3.9	5.7	3.37	3.5	6	4.2	0.8
Eu	1.12	1.49	1.66	0.78	<1.16	5	1.37	0.37
Tb	0.67	1.32	1.80	0.48	<0.40	4	1.07	0.53
Yb	2.9	3.8	3.2	1.75	3.6	6	2.8	0.8
Lu	0.44	0.24	0.28	0.27	0.23	6	0.29	0.07

*Representative analyses from a total sample population of 6.
Oxides in weight percent, trace elements in ppm; Fe_2O_{3-T} = total iron as Fe_2O_3; LOI = loss on ignition; Mg number = 100 MgO/MgO + FeO mole ratio where FeO = 0.865 Fe_2O_{3-T}; nd = not detected; na = not analyzed; n = number of samples; σ = one standard deviation.

TABLE 10. CHEMICAL COMPOSITIONS OF DOCTOR CREEK DACITES

Sample No.*	DOC-43	DOC-49	DOC-50	DOC-51	DOC-54	n	Mean	σ
Map No.	41	42	43	44	45			
SiO_2	71.74	66.16	69.61	65.71	69.57	7	68.84	2.09
TiO_2	0.44	0.97	0.58	0.68	0.51	7	0.67	0.17
Al_2O_3	13.64	14.13	14.07	14.74	13.42	7	14.16	0.49
$Fe_2O_{3\text{-}T}$	5.01	7.95	6.23	7.43	5.89	7	6.65	0.98
MgO	0.39	1.04	0.77	0.80	0.43	7	0.69	0.24
CaO	1.73	3.47	2.25	2.53	2.13	7	2.42	0.53
Na_2O	3.27	3.15	3.69	4.03	3.27	7	3.53	0.29
K_2O	3.50	2.83	2.89	2.71	3.53	7	3.09	0.34
MnO	0.04	0.12	0.10	0.10	0.09	7	0.09	0.02
P_2O_5	0.09	0.25	0.15	0.21	0.14	7	0.17	0.06
LOI	0.71	0.42	0.66	0.94	0.69	7	0.64	0.20
Total	100.56	100.49	101.00	99.88	99.67			
Rb	76	80	67	50	84	7	72	14
Ba	1136	981	909	888	1039	7	1001	79
Cs	0.9	na	3.1	2.9	3.2	6	2.7	0.9
Sr	166	222	157	197	167	7	179	25
Pb	25	22	28	23	32	7	26	3
Th	13.0	7.0	9.0	14.0	9.0	7	10.0	2.8
U	2.0	2.0	1.0	2.0	2.0	7	1.7	0.4
Sc	12	16	13	16	12	7	14	2
V	nd	257	nd	nd	nd	1	257	
Cr	nd	146	nd	nd	nd	1	146	
Co	4	nd	3	2	nd	5	4	2
Ni	16	49	15	20	nd	6	23	12
Cu	nd	13	nd	nd	17	2	15	2
Zn	67	69	114	100	129	7	100	22
Y	72	65	72	64	73	7	69	3
Zr	342	318	375	331	362	7	344	18
Nb	20	17	20	18	19	7	19	1
Hf	9.6	na	9.8	8.7	9.5	6	9.31	0.37
Ta	1.92	na	1.17	1.54	1.61	6	1.39	0.32
La	50.0	na	47.3	44.0	35.1	6	40.22	7.16
Ce	96.4	na	98.1	89.0	71.5	6	83.77	11.18
Sm	10.9	na	9.0	8.6	8.5	6	8.7	1.1
Eu	1.66	na	2.70	5.22	2.33	6	2.54	1.27
Tb	2.22	na	2.02	2.27	2.17	6	1.94	0.34
Yb	9.3	na	8.1	9.7	6.6	6	7.5	1.7

*Representative analyses from a total sample population of 7.
Oxides in weight percent, trace elements in ppm; $Fe_2O_{3\text{-}T}$ = total iron as Fe_2O_3; LOI = Loss on ignition; Mg number = 100 MgO/MgO + FeO mole ratio where FeO = 0.865 $Fe_2O_{3\text{-}T}$; nd = not detected; na = not analyzed; n = number of samples; σ = one standard deviation.

TABLE 11. MEAN VALUES OF SELECTED ELEMENT RATIOS FOR PECOS THOLEIITES AND CALC-ALKALINE BASALTS COMPARED TO YOUNG BASALTS

	Tholeiites			Calc-alkaline basalt	
	Pecos	NMORB	IAB	Pecos	CAB
Ba/La	59	110	15-50	17	10-40
La/Sm	0.93	1.1	1-1.5	2.6	1.7-2.5
La/Yb	1.5	1.3	1-3	6.3	2-15
Zr/Nb	23	26	15-32	26	20-27
Ti/Zr	109	87	70-180	51	50-63
Zr/Y	2.2	3.2	2.1-3.2	6.2	5.5-8.7
La/Ta	20	17	20-50	47	20-50
K/Rb	272	1000	375-750	277	300-500
Mg No.	56	60	50-60	59	50-60

Mg No. = MgO/FeO+MgO molecular ratio; NMORB = normal mid-ocean ridge basalt; IAB = island-arc basalt; CAB = calc-alkaline basalt.
Major references: Brown and others (1977), Nicholls and others (1980), Saunders and others (1980), Kay and others (1982), Pearce (1983).

REFERENCES

Arndt, N. T., and Nisbet, E. G., eds., 1982, Komatiites: Boston, Massachusetts, George Allen and Unwin, 526 p.

Barker, F., 1969, Precambrian geology of the Needle Mountains, southwestern Colorado: U.S. Geological Survey Professional Paper 644–A, 35 p.

—— , 1979,Trondhjemite: Definition, environment and hypotheses of origin, *in* Barker, F., ed., Trondhjemites, dacites, and related rocks: New York, Elsevier, p. 1–12.

Basaltic Volcanism Study Project, 1981, Basaltic volcanism on the terrestrial planets (BVTP): New York, Pergamon Press, 1286 p.

Bauer, P. W., 1984, Stratigraphic summary and structural problems of Precambrian rocks, Picuris Range, New Mexico: New Mexico Geological Society Guidebook to 35th Field Conference, p. 199–204.

Bennett, G. S., 1984, Geochemistry of bimodal volcanic and volcaniclastic metasedimentary rocks, Cochetopa Canyon area, central Colorado [M.S. thesis]: Lawrence, University of Kansas, 82 p.

Bickford, M. E., and Boardman, S. J., 1984, A Proterozoic volcano–plutonic terrane, Gunnison and Salida areas, Colorado: Journal of Geology, v. 92, p. 657–666.

Boardman, S. J., 1986, Early Proterozoic bimodal volcanic rocks in central Colorado, U.S.A.: Part I, Petrography, stratigraphy, and depositional history: Precambrian Research, v. 34, p. 1–36.

Boardman, S. J., and Condie, K. C., 1986, Early Proterozoic bimodal volcanic rocks in central Colorado, U.S.A., Part II, Geochemistry, petrogenesis, and tectonic setting: Precambrian Research, v. 34, p. 37–68.

Bowring, S. A., and Condie, K. C., 1982, U-Pb zircon ages from northern and central New Mexico: Geological Society of America Abstracts with Programs, v. 14, p. 304.

Brown, G. M., Holland, J. G., Sigurdsson, H., Tomblin, J. F., and Arculus, R. J., 1977, Geochemistry of the Lesser Antilles volcanic island arc: Geochimica et Cosmochimica Acta, v. 41, p. 785–801.

Coleman, R. G., 1977, Ophiolites: New York, Springer-Verlag, 229 p.

Condie, K. C., 1980, The Tijeras greenstone; Evidence for depleted upper mantle beneath New Mexico during the Proterozoic: Journal of Geology, v. 88, p. 603–609.

—— , 1981a, Archean greenstone belts: New York, Elsevier, 434 p.

—— , 1981b, Precambrian rocks of the southwestern United States and adjacent areas of Mexico: New Mexico Bureau of Mines and Mineral Resources Resource Map 13, scale 1:1,500,000.

—— , 1982, Plate-tectonics model for continental accretion in the southwestern United States: Geology, v. 10, p. 37–42.

—— , 1986, Geochemistry and tectonic setting of Early Proterozoic supracrustal rocks in the southwestern United States: Journal of Geology, v. 94, p. 845–864.

—— , 1988, Terranes and crustal provinces; Proterozoic collisional growth of southwestern North America: Geological Society of America Bulletin (in press).

Condie, K. C., and Allen, P., 1984, Origin of Archean charnockites from southern India, *in* Kroner, A., Hanson, G. N., and Goodwin, A. M., eds., Archean geochemistry: Berlin, Springer-Verlag, p. 183–203.

Condie, K. C., and McCrink, T. P., 1982, Geochemistry of Proterozoic volcanic and granitic rocks from the Gold Hill–Wheeler Peak area, northern New Mexico: Precambrian Research, v. 19, p. 141–166.

Condie, K. C., and Nuter, J. A., 1981, Geochemistry of the Dubois greenstone succession; An Early Proterozoic bimodal volcanic association in west-central Colorado: Precambrian Research, v. 15, p. 131–156.

Condie, K. C., and Shadel, C., 1984, An Early Proterozoic volcanic arc succession in southeastern Wyoming: Canadian Journal of Earth Sciences, v. 24, p. 415–427.

Ewart, A., 1979, A review of the mineralogy and chemistry of Tertiary-Recent dacitic, latitic, rhyolitic, and related salic rocks, *in* Barker, F., ed., Trondhjemites, dacites, and related rocks: Amsterdam, Elsevier, p. 13–21.

Flanagan, F. J., 1973, 1972 values for international geochemical reference stan-

dards: Geochimica et Cosmochimica Acta, v. 37, p. 1118–1200.
Floyd, P. A., and Winchester, J. A., 1978, Identification and discrimination of altered and metamorphosed volcanic rocks using immobile elements: Chemical Geology, v. 21, p. 291–306.
Garcia, M. O., 1979, Criteria for the identification of ancient volcanic arcs: Earth Science Reviews, v. 14, p. 147–165.
Gelinas, L., Brooks, C., Perrault, G., Carignan, J., Trudel, P., and Grasso, F., 1977, Chemo-stratigraphic divisions within the Abitibi volcanic belt, Rouyn–Noranda district, Quebec, *in* Baragar, W.R.A., Coleman, L. C., and Hall, J. M., eds., Volcanic regimes in Canada: Geological Association of Canada Special Paper 16, p. 265–309.
Gordon, E. E., Randle, K., Goles, G. G., Corlis, J. B., Beeson, M. H., and Oxley, S. S., 1968, Instrumental activation analysis of standard rocks with high-resolution gamma-ray detectors: Geochimica et Cosmochimica Acta, v. 32, p. 369–396.
Grambling, J. A., and Ward, D. B., 1987, Thrusting of the Pecos greenstone belt over younger supracrustal rocks, Rio Mora area, New Mexico: Geological Society of America Abstracts with Programs, v. 19, p. 278–279.
Gresens, R. L., 1976, Geologic, geochemical, and geochronologic investigation of Precambrian metamorphic rocks of the Las Tablas–La Madera Quadrangles and the Picuris Range, northern New Mexico; A summary, *in* Woodward, L. A., and Northrop, S. A., eds., Tectonics and mineral resources of southwestern North America: New Mexico Geological Society Special Publication 6, p. 132–137.
Gresens, R. L., and Stensrud, H. L., 1974, Recognition of more metarhyolite occurrences in northern New Mexico and a possible Precambrian stratigraphy: Mountain Geologist, v. 11, no. 3, p. 109–124.
Hanson, G. N., and Langmuir, C. H., 1978, Modeling of major elements in mantle melt systems using trace element approaches: Geochimica et Cosmochimica Acta, v. 42: p. 725–741.
Hickey, R. L., Gerlach, D. C., and Frey, F. A., 1984, Geochemical variations in volcanic rocks from central-south Chile, *in* Harmon, R. S., and Barreiro, B. A., eds., Andean magmatism; Chemical and isotopic constraints: Cheshire, United Kingdom, Shiva Publications Ltd., p. 72–95.
Jensen, L. S., 1976, A new cation plot for classifying subalkalic volcanic rocks: Ontario Division of Mines Miscellaneous Paper 66, 22 p.
Kay, S. M., Kay, R. W., and Citron, G. P., 1982, Tectonic controls on tholeiitic and calc-alkaline magmatism in the Aleutian arc: Journal of Geophysical Research, v. 87, p. 4051–4072.
Knoper, M. W., and Condie, K. C., 1988, Geochemistry and petrogenesis of Early Proterozoic amphibolites, west-central Colorado: Chemical Geology, v. 67, p. 209–225.
Leake, B. E., Hendry, G. L., Kemp, A., Plant, A. G., Harvey, P. K., Wilson, J. R., Coats, J. S., Aucott, J. W., Lunel, T., and Howarth, R. J., 1969, The chemical analysis of rock powders by automatic X-ray fluorescence: Chemical Geology, v. 5, p. 7–86.
Lindstrom, D. J, and Korotev, R. L., 1982, TEABAGS: Computer programs for instrumental neutron activation analysis: Journal of Radioanalytical Chemistry, v. 70, no. 1-2, p. 439–458.
Lopez-Escobar, L., 1984, Petrology and chemistry of volcanic rocks of the southern Andes, *in* Harmon, R. S., and Barreiro, B. A., eds., Andean magmatism; Chemical and isotopic constraints: Cheshire, United Kingdom, Shiva Publications Ltd., p. 47–71.
McKee, C., 1988, Geochemistry and tectonic setting of some Proterozoic rocks in the Pedernal Hills and Manzano Mountains, New Mexico [M.S. thesis]: Socorro, New Mexico Institute of Mining and Technology, 125 p.
Miller, C. F., and Mittlefehldt, D. W., 1982, Depletion of light rare-earth elements in felsic magmas: Geology, v. 10, p. 129–133.
Moench, R. H., and Erickson, M. S., 1980, Occurrence of tungsten in the Sangre de Cristo Range near Santa Fe, New Mexico: Possible stratabound scheelite peripheral to favorable settings for volcanogenic massive sulfide deposits: U.S. Geological Survey Open-File Report 80–1126, 21 p.
Moench, R. H., and Robertson, J. M., 1980, Geology of the Pecos Wilderness and adjacent areas, Santa Fe, San Miguel, Mora, Rio Arriba, and Taos Counties, New Mexico; Chapter A, Mineral resources of the Pecos Wilderness and adjacent areas, Santa Fe, San Miguel, Mora, Rio Arriba, and Taos Counties, New Mexico: U.S. Geological Survey Open-File Report 80–382, p. 6–41.
Moench, R. H., Grambling, J. A., and Robertson, J. M., 1988, Geologic map of the Pecos Wilderness, Santa Fe, San Miguel, Mora, Rio Arriba, and Taos Counties, New Mexico: U.S. Geological Survey Miscellaneous Field Investigations Map MF–1921B, scale 1:48,000.
Moores, E. M., 1982, Origin and emplacement of ophiolites: Reviews in Geophysics, v. 20, p. 735–760.
Naldrett, A. J., and Cabri, L. J., 1976, Ultramafic and related mafic rocks: Their classification and genesis with special reference to the concentration of nickel sulfides and platinum-group elements: Economic Geology, v. 71, p. 1131–1153.
Nelson, B. K., and DePaolo, D. J., 1984, 1700 Myr greenstone volcanic successions in southwestern North America and isotopic evolution of Proterozoic mantle: Nature, v. 312, p. 143–146.
Nicholls, I. A., Whitford, D. J., Harris, K. L., and Taylor, S. R., 1980, Variation in the geochemistry of mantle sources for tholeiites and calc-alkaline magmas, Western Sundra arc, Indonesia: Chemical Geology, v. 30, p. 177–199.
O'Hara, M. J., 1977, Geochemical evolution during fractional crystallization of a periodically refilled magma chamber: Nature, v. 26, p. 36–38.
Pankhurst, R. J., 1977, Open system crystal fractionation and incompatible element variation in basalts: Nature, v. 268, p. 36–38.
Pearce, J. A., 1983, Role of the subcontinental lithosphere in magma genesis at active continental margins, *in* Hawkesworth, C. J., and Norry, M. J., eds., Continental basalts and mantle xenoliths: Cheshire, United Kingdom, Shiva Publications Ltd., p. 230–249.
Pearce, J. A., and Cann, J. R., 1973, Tectonic setting of basic volcanic rocks determined using trace-element analysis: Earth and Planetary Science Letters, v. 19, p. 290–300.
Pearce, T. H., Gorman, B. E., and Birkett, T. C., 1977, The relationship between major element chemistry and tectonic environment of basic and intermediate volcanic rocks: Earth and Planetary Science Letters, v. 36, p. 121–132.
Rajamani, V., Shivkumar, K., Hanson, G. N., and Shirey, S. B., 1985, Geochemistry and petrogenesis of amphibolites from the Kolar schist belt, south India: Evidence for ultramafic magma generation by low percent melting: Journal of Petrology, v. 26, p. 92–123.
Reed, J. C., Jr., 1984, Proterozoic rocks of the Taos Range, Sangre de Cristo Mountains, New Mexico: New Mexico Geological Society Guidebook to 35th Field Conference, p. 179–185.
Register, M. E., and Brookins, D. G., 1979, Geochronologic and rare-earth study of the Embudo Granite and related rocks: New Mexico Geological Society Guidebook to 29th Field Conference, p. 155–158.
Robertson, J. M., and Moench, R. H., 1979, The Pecos greenstone belt: A Proterozoic volcano–sedimentary sequence in the southern Sangre de Cristo Mountains, New Mexico: New Mexico Geological Society Guidebook to 30th Field Conference, p. 165–173.
Saunders, A. D., Tarney, J., and Weaver, S. D., 1980, Transverse geochemical variations across the Antarctica peninsula; Implications for the genesis of calc-alkaline magmas: Earth and Planetary Science Letters, v. 46, p. 344–360.
Stacey, J. S., Doe, B. R., Silver, L. T., and Zartman, R. E., 1976, Plumbotectonics; IIA, Precambrian massive sulfide deposits: U.S. Geological Survey Open-File Report 76–476, 29 p.
Strong, D. F., and Dostal, J., 1980, Dynamic melting of Proterozoic upper mantle: Evidence from REE in oceanic crust of eastern Newfoundland: Contributions to Mineralogy and Petrology, v. 72, p. 165–173.
Tweto, O., 1979, Geologic map of Colorado: U.S. Geological Survey, scale 1:500,000.
Vernon, R. H., 1986, Evaluation of the "quartz-eye" hypothesis: Economic Geology, v. 81, p. 1520–1527.
Weaver, B. L., and Tarney, J., 1983, Chemistry of the subcontinental mantle: Inferences from Archean and Proterozoic dykes and continental flood basalts, *in* Hawkesworth, C. J., and Norry, M. H., eds., Continental basalts and

mantle xenoliths: Cheshire, United Kingdom, Shiva Publications Ltd., p. 209–229.

Whitney, J. A., Paris, T. A., Carpenter, R. H., and Hartley, M. E., 1978, Volcanic evolution of the southern slate belt of Georgia and South Carolina; A primitive oceanic island arc: Journal of Geology, v. 86, p. 173–192.

Williams, M. L., 1987, Stratigraphic, structural, and metamorphic relationships in Proterozoic rocks from northern New Mexico [Ph.D. thesis]: Albuquerque, University of New Mexico, 138 p.

Wood, D. A., 1979, A variably veined suboceanic upper mantle: Genetic significance for mid-ocean ridge basalts from geochemical evidence: Geology, v. 7, p. 499–503.

—— , 1980, The application of a Th–Hf–Ta diagram to problems of tectonomagmatic classification and to establishing the nature of crustal contamination of basaltic lavas of the British Tertiary volcanic province: Earth and Planetary Science Letters, v. 50, p. 11–30.

Wyman, W. F., 1980, Precambrian geology of the Cow Creek ultramafic complex, San Miguel County, New Mexico [M.S. thesis]: Socorro, New Mexico Institute of Mining and Technology, 125 p.

Manuscript Accepted by the Society September 27, 1988

Printed in U.S.A.

Geological Society of America
Special Paper 235
1989

Geochronologic study of Precambrian rocks of the Sandia Mountains, New Mexico

Douglas G. Brookins
Department of Geology, University of New Mexico, Albuquerque, New Mexico 87131
Arun Majumdar
Department of Geology, Sam Houston University, Huntsville, Texas 77341

ABSTRACT

Precambrian rocks of the Sandia Mountains have been dated previously by the U, Th-Pb methods at about 1.46 Ga. Rb-Sr geochronology yields a similar age, 1.44 Ga, although local open-system conditions in the granite are indicated by data that fall off the whole-rock isochron. It is not known if this scatter reflects multi-domains in the granite or whole-rock-scale incipient alteration. Potassium-argon dates for micas yield 1.3 to 1.4 Ga, and Rb-Sr biotite dates yield 1.35 Ga; this age is commonly noted in north-central New Mexico. We interpret this date as either a thermal event affecting the granite, or possibly extended cooling at temperatures high enough to reset mineral Rb-Sr systematics. The Rb-Sr whole-rock age of the Juan Tabo metamorphic rocks at the north end of the Sandia Mountains is 1.61 Ga. Pegmatites and aplites of the same area yield a date of 1.41 Ga, but with a high initial $^{87}Sr/^{86}Sr$ ratio of 0.712 to 0.714. Granulitic xenoliths from the granite yield a date of 1.48 Ga, and may be older crust sampled by the granite.

Our geochronologic data, and the geochemical data of Majumdar (1985), do not support the hypothesis of Condie and Budding (1979) that there are two major plutons in the Sandia granite. We propose instead one major pluton with some slight and systematic chemical variation within it.

INTRODUCTION

The Sandia granite is one of a series of late mid-Proterozoic granites that are exposed in mountains flanking the present Rio Grande rift in New Mexico (Condie and Budding, 1979). Various aspects of the Sandia granite have been described by Lodewick (1960), Green and Callender (1973), Kelley and Northrop (1975), Enz and others (1979), Berkley and Callender (1979), Condie and Budding (1979), and Majumdar (1985). The granite is a gray to pink, medium- to coarse-grained porphyritic rock. It consists mainly of megacrysts of microcline set in a matrix of quartz, oligoclase, and biotite, with accessory sphene, magnetite, apatite, hornblende, muscovite, pyrite, and less common tourmaline. The average modal analysis of the granite is 35 percent quartz, 15 percent microcline, 35 percent plagioclase, 10 percent biotite, and 5 percent microperthite (Kelley and Northrop, 1975). The Sandia granite actually is more quartz monzonitic and granodioritic in composition, but the name Sandia granite will be used throughout the remainder of this report for consistency with the other literature.

The Sandia granite of north-central New Mexico crops out in a continuous north-south belt, 3 to 6 km wide and 30 km long, just east of the city of Albuquerque (Fig. 1). The granite is exposed along the western escarpment of the Sandia Mountains and as inliers on the eastern side of the range. Overall width of the pluton (including the inliers) is about 14 km (Kelley and Northrop, 1975, p. 23). The northern contact of the Sandia granite with adjacent metamorphic country rocks is in part a fault contact, although a distinct metamorphic aureole is noted near the boundary of the granite (Green and Callender, 1973; Berkley and Callender, 1979). The southern boundary of the granite is mylonitic and gradational with the Precambrian Cibola gneiss. Pennsylvanian sedimentary rocks unconformably overlie the Sandia granite along the crest of the range.

Brookins, D. G., and Majumdar, A., 1989, Geochronologic study of Precambrian rocks of the Sandia Mountains, New Mexico, *in* Grambling, J. A., and Tewksbury, B. J., Proterozoic geology of the southern Rocky Mountains: Boulder, Colorado, Geological Society of America Special Paper 235.

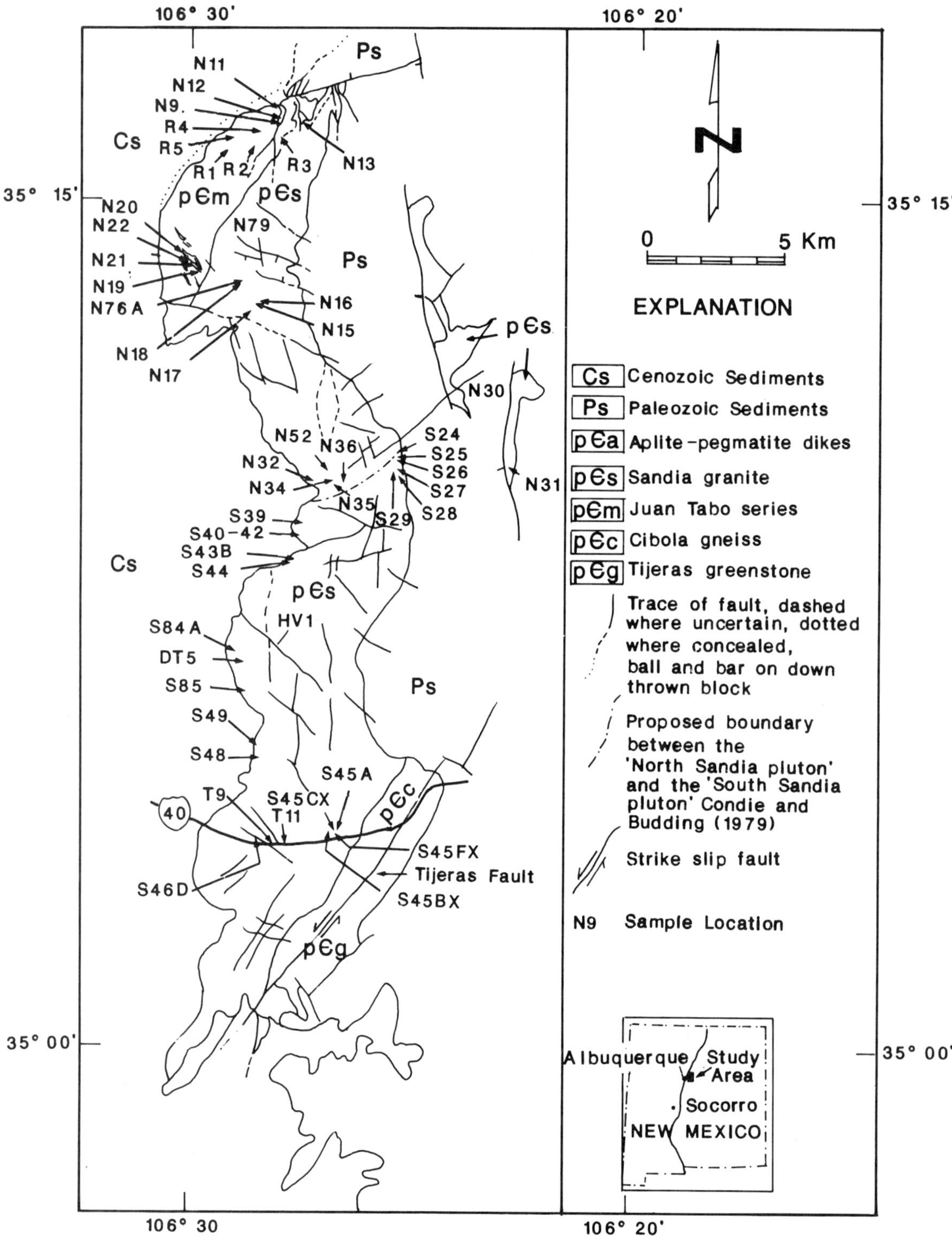

Figure 1. Geologic sketch map of the Sandia granite showing sample locations (from Majumdar, 1985).

Berkley and Callender (1979) have studied the metamorphic history of the Precambrian Juan Tabo series exposed in the Rincon Ridge, north and northwest of the Sandia granite. They identified two episodes of deformation and three episodes of metamorphism. The mineral assemblage of the contact metamorphic aureole, according to Berkley and Callender (1979) implies relatively shallow depth of intrusion (less than 10 km), and the temperature of the contact metamorphism was about 500 to 650°C at a total pressure of 2 to 3 kb.

PREVIOUS GEOCHRONOLOGICAL WORK

The earliest reported geochronological work for the Sandia granite is that by Aldrich and others (1957, 1958) who cited 1.31 Ga for biotite by both the Rb-Sr and K-Ar methods, although no precise sample location is given in their works. Steiger and Wasserburg (1966) and Wasserburg and Steiger (1967) report U, Th-Pb dates for zircons ranging from 0.93 to 1.47 Ga. Tilton and Grunenfelder (1968) report U-Pb dates for sphenes ranging from 1.45 to 1.49 Ga. All of the samples used by the U,Th-Pb workers apparently came from one 5-kg sample from Tijeras Canyon (see discussion in Brookins, 1974). Previous Rb-Sr work by Wasserburg and Steiger (1967) indicated a whole-rock isochron date of 1.41 Ga, and these same authors (Steiger and Wasserburg, 1969) also report a Rb-Sr muscovite date of 1.47 Ga. Brookins (1974) summarizes all geochronological work published between 1957 and 1973.

In the early 1970s, work was started at the University of New Mexico to further investigate the geochronology of the Sandia Mountains (see Brookins, 1974; Taggart and Brookins, 1975; Brookins and Shaffiqullah, 1975; Brookins and others, 1975.) More recently, Brookins and Majumdar (1982a, b) report a Rb-Sr whole-rock isochron for eight of their data plus the three data from Taggart and Brookins (1975) with a date of 1.44 ± 0.04 Ga, although no attempt was made to distinguish between north and south plutons (cf., Condie and Budding, 1979). They also report eight biotite Rb-Sr dates ranging from 1.30 to 1.35 Ga. Brookins and Majumdar (1983) cite a Rb-Sr whole-rock isochron date of 1.61 ± 0.06 Ga for metamorphic rocks from the Rincon area, with initial $^{87}Sr/^{86}Sr = 0.705$.

There has been limited fission track work on the minerals of the Sandia granite. Naeser (1971) cites work by Poupeau resulting in ages of 60 Ma for apatite and 1.4 Ga for sphene from samples from Tijeras Canyon. Brookins (1974) suggests that the granite was heated above 175°C, track retention temperature for apatite, but below 420°C, track retention temperature for sphene. The timing of this heating is unknown.

THE PRESENT STUDY

A scrutiny of the previous work raises a number of research questions that call for further investigation. A detailed study of the Sandia granite, therefore, was undertaken with the following specific objectives:

1. To obtain the Rb-Sr whole-rock age of the Sandia granite by analyzing a large number of samples.
2. To evaluate the Rb-Sr whole-rock age of the metamorphic country rocks surrounding the Sandia granite.
3. To determine the time of metamorphism of the Sandia granite.
4. To determine the whole-rock Rb-Sr age of the deformed pegmatites and aplite dikes cutting across the country rocks.
5. To test the two-pluton hypothesis (Condie and Budding, 1979) on the origin of the Sandia granite by detailed mapping followed by geochemical and petrological investigations.
6. To study the granitoid xenoliths in the Sandia granite.

RESULTS

The data reported here are from the University of New Mexico Geochronology Laboratory, and have been reported in part in Brookins and Majumdar (1982a, b; 1983) and in the Ph.D. thesis of Majumdar (1985). Sample locations are shown in Figure 1.

Analytical procedures, given in detail in Majumdar (1985), are summarized here. Rb and Sr contents were determined by isotope dilution using ^{87}Rb- and ^{84}Sr-enriched spikes, and the $^{87}Sr/^{86}Sr$ ratios were calculated from the experiments. The uncertainty for the reported $^{87}Sr/^{86}Sr$ ratios is ± 0.00005 (one sigma) rounded to ± 0.0001, and for the $^{87}Sr/^{86}Sr$ ratio ± 0.3 percent of the reported ratio (one sigma). Five runs on the Eimer and Amend standard $SrCO_3$ gave 0.70803 ± 0.00007 during the course of the investigation. Standard chemical procedures were followed throughout (Majumdar, 1985), and the isotopic measurements were made using a solid-source, Nier-design NUCLIDE 1290 mass spectrometer. All $^{87}Sr/^{86}Sr$ ratios were normalized to $^{87}Sr/^{86}Sr = 0.1194$. The Rb-Sr data are given in Tables 1 through 4 and previously unpublished isochrons shown in Figures 2 and 3 using the York (1969) model 3 method of isochron construction.

Rb-Sr age of the Juan Tabo series

The strontium isotopic composition and concentrations of Rb and Sr of five fresh samples of the Juan Tabo series (Rincon Ridge area) have been given by Brookins and Majumdar (1982a), who report an age of 1.61 ± 0.06 Ga and $(^{87}Sr/^{86}Sr)_o$ = 0.705 ± 0.001. The rock types analyzed are quartz–mica–microcline schists and hornblende schists. From the study of bulk chemistry and field relations of the Juan Tabo series, Berkley and Callender (1979) conclude that the rocks represent a metamorphosed sequence of clastic sediments with mafic tuffs, flows, and sills. The 1.61 Ga isochron of Brookins and Majumdar (1983) yields the minimum age of sedimentation.

Taggart and Brookins (1975) obtained a date of 1.61 ± 0.07 Ga (initial strontium ratio = 0.703 ± 0.005) for Cibola gneiss, which forms the southern boundary of the Sandia granite, but based on only three data. The significance of the identical age but

TABLE 1. Rb-Sr DATA FOR MAFIC XENOLITHS FROM SANDIA GRANITE

Sample Number	$^{87}Sr/^{86}Sr$	Rb (ppm)	Sr (ppm)	$^{87}Rb/^{86}Sr$
S45CS	0.7339	106.6	226.08	1.37
S46DX	0.7371	168.0	314.21	1.55
S45FX	0.7351	120.7	235.9	1.49
S45GX	0.7531	158.1	200.5	2.30

TABLE 2. Rb-Sr DATA FOR PEGMATITE-APLITE DIKES CUTTING THE JUAN TABO SERIES

Sample Number	$^{87}Sr/^{86}Sr$	Rb (ppm)	Sr (ppm)	$^{87}Rb/^{86}Sr$
N-19	0.8523	268.5	112.0	7.04
N-20	0.8393	214.5	99.80	6.30
N-21	0.8251	265.8	141.6	5.50
N-22	0.7516	196.6	298.5	1.91

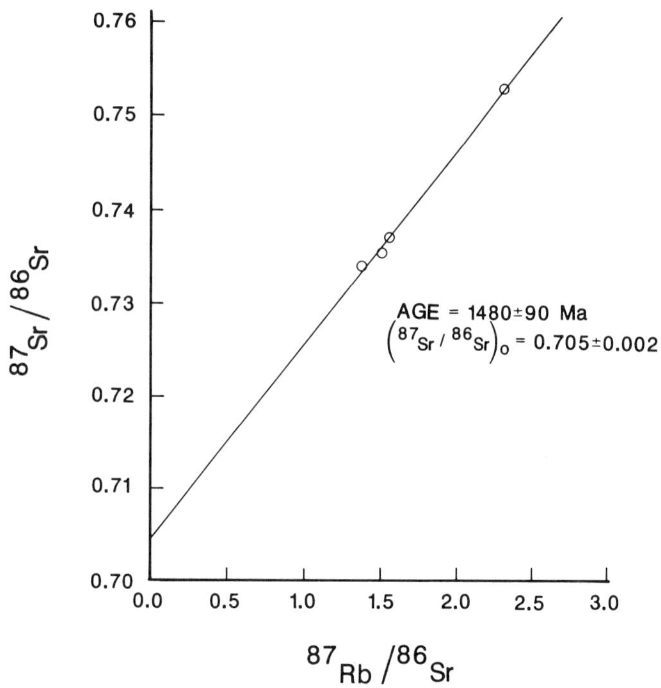

Figure 2. Whole-rock Rb-Sr isochron for the granulitic xenoliths within the Sandia granite.

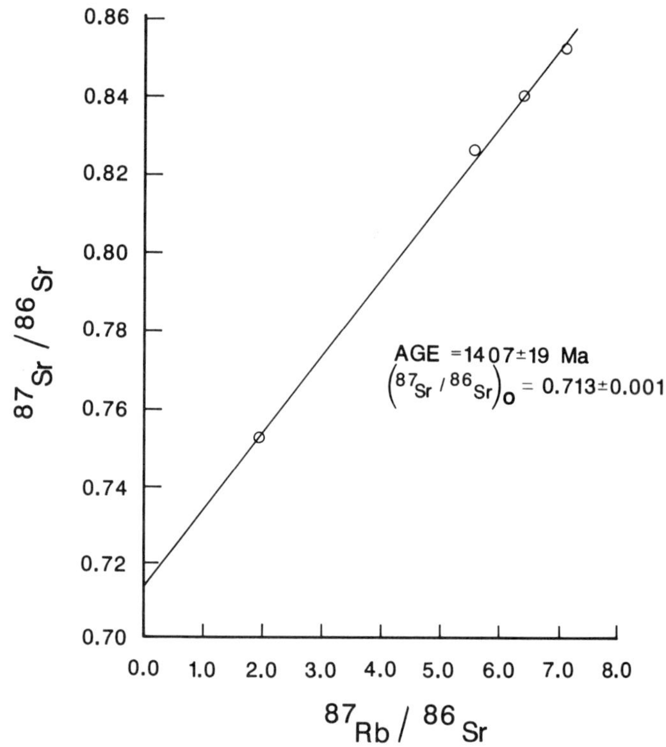

Figure 3. Whole-rock Rb-Sr isochron for the pegmatite and aplite dikes cutting across the Juan Tabo series.

possibly different initial strontium ratios of Cibola gneiss and the Juan Tabo series is unknown.

Rb-Sr age of the pegmatite-aplite dikes cutting across the Juan Tabo series

Three samples of pegmatite and one sample of aplite from the deformed dikes cutting across the Juan Tabo series were analyzed. Data are given in Table 2, and results are plotted on an isochron diagram in Figure 3. The isochron gives an age of 1.41 ± 0.02 Ga with an initial $^{87}Sr/^{86}Sr$ ratio of 0.713 ± 0.001.

Rb-Sr age of the Sandia granite

Rb-Sr data for the Sandia granite are given in Tables 3 and 4, including data from Taggart and Brookins (1975). Previous geochronological work on the Sandia granite has demonstrated that the Rb-Sr systematics of minerals has, in many cases, been perturbed (Wasserburg and Steiger, 1967; Brookins and Majumdar, 1982b).

Wasserburg and Steiger (1967) report a 1.41 isochron with considerable scatter by the Rb-Sr method for the Sandia granite, anchored by four apatites with $^{87}Sr/^{86}Sr$ between 0.7035 and 0.705, at $^{87}Sr/^{86}Sr$ less than 0.01. The apatite compositions argue that the initial $^{87}Sr/^{86}Sr$ of the Sandia granite must fall near 0.704. It is also apparent from the isochron of Wasserburg and Steiger (1967, Fig. 1) that the whole-rock samples have not apparently been closed systems with respect to their Rb-Sr systematics.

Majumdar (1985) has argued for a best isochron of 1.44 ± 0.04 Ga with initial $^{87}Sr/^{86}Sr$ = 0.7054 ± 0.0005. Samples N15, N16, N17, and N18 (Fig. 1) are from the Juan Tabo area and may fall below the isochron. Majumdar (1985) has pointed out that samples N15, N16, N17, and N18 define a 1.35 Ga isochron with initial $^{87}Sr/^{86}Sr = 0.707$. On further consideration of this matter, we cannot unequivocally endorse this as a meaningful date, but we note that this apparent 1.35 Ga date is consistent with mineral dates (below). In addition, Majumdar (1985) used the seven data for the south pluton (Table 4) to calculate model ages. If an initial $^{87}Sr/^{86}Sr$ ratio of 0.704 is assumed (based on Wasserburg and Steiger's, 1967, apatite data), the collective data of Table 4 yield 1.44 ± 0.03 Ga. The model data for the NSP samples (Table 3) with initial $^{87}Sr/^{86}Sr = 0.704$ is also 1.44 ± 0.02 Ga. Hence the isochron for the Sandia granite (Brookins and Majumdar, 1982b) is consistent with all the data north and south. We point out, however, that the scatter in the data (see Majumdar, 1985) is outside analytical uncertainty, a point made by Wasserburg and Steiger (1967) based on their 1.41-Ga isochron for Sandia granite from Tijeras Canyon.

TABLE 3. Rb-Sr DATA FOR 'NORTH SANDIA PLUTON'

Sample Number	$^{87}Sr/^{86}Sr$	Rb (ppm)	Sr (ppm)	$^{87}Rb/^{86}Sr$
N-9	0.75938	191.5	209.1	2.65
N-11	0.75880	174.9	191.7	2.66
N-12	0.75758	183.5	198.2	2.69
N-13	0.75325	155.3	188.3	2.40
N-15	0.7484	175.6	234.0	2.18
N-16	0.7486	181.3	240.9	2.19
N-17	0.7578	200.4	211.0	2.76
N-18	0.7515	192.3	239.3	2.37
3001-S*	0.7758	255.0	216.0	3.49
3002-S*	0.7613	251.0	265.0	2.79
3003-S*	0.7481	184.0	251.0	2.17

*Data from Taggart and Brookins (1975).

TABLE 4. Rb-Sr DATA FOR 'SOUTH SANDIA PLUTON'

Sample Number	$^{87}Sr/^{86}Sr$	Rb (ppm)	Sr (ppm)	$^{87}Rb/^{86}Sr$
S-25	0.7533	196.9	239.7	2.39
S-26	0.7556	210.7	242.3	2.53
S-27	0.7894	256.8	170.2	4.40
T-9	0.7469	145.4	207.5	2.04
T-11	0.7441	148.8	225.0	1.92
HV-1	0.7457	149.6	220.8	1.97
DT-5	0.7449	140.3	212.8	1.92

Rb-Sr age of the granulitic xenoliths

Numerous foliated granitoid xenoliths occur in the Sandia granite around Tijeras Canyon (Majumdar, 1985). The xenoliths are tabular, rounded, oval to lenticular in shape and range from 5 × 10 cm to 50 × 150 cm in size. Four samples (S45 CX, S45 FX, S45 GX, and S46 DX) were collected from larger blocks for Rb-Sr dating. The modal mineralogy and major elements, and REE composition of these rocks are given in Majumbar (1985).

The results of the Rb-Sr analysis on these samples are shown in Table 1. The isochron plot shown as Figure 2 gives an age of 1.48 ± 0.09 Ga and an initial Sr ratio of 0.705 ± 0.002. Within experimental error the age of the xenoliths agrees with that of the Sandia granite.

The xenoliths are found to be granitized to various degrees. Although the samples for dating purposes were collected from the interiors of larger blocks, the effects of granitization could not be avoided. Thus the whole-rock age of 1.48 Ga of the xenoliths and its initial ratio most probably reflect isotopic homogenization with the Sandia granite. Alternatively, the cognate xenoliths may have been formed by partial mixing between mantle-derived mafic melt and granite.

Biotite metamorphic Rb-Sr age

Biotite was studied from eight of the samples from which the whole-rock isochron was obtained. Analyses are given in Brookins and Majumdar (1982b). To obtain mineral ages, $^{87}Sr/^{86}Sr$ and $^{87}Sr/^{86}Sr$ of each biotite sample were used, with the corresponding value of each obtained from its parent whole-rock sample, to draw an isochron whose slope gave the age of Sr homogenization in biotite. For the eight biotite samples this age ranges from 1.32 to 1.35 Ga, with a mean of 1.33 ± 0.02 (one sigma) Ga. This indicates that the biotite samples became closed systems with respect to Rb and Sr at the same time (= 1.33 Ga) throughout the entire Sandia granite. This date may represent the time of cooling of the body through 250°C (estimated strontium retention temperature for biotite; Hart, 1964; Van Breeman and Dallmeyer, 1984). This interpretation is consistent with a ^{39}Ar-^{40}Ar incremental heating experiment on a biotite separate from sample S45A of the SSP by A. K. Baksi of Louisiana State University (see Majumdar, 1985). The results yield medium- and high-temperature plateau ages of 1.35 Ga, showing that the pluton has not been heated above 300° to 350°C (estimated Ar-retention temperature of biotite; Berger and York, 1981) since 1.35 Ga.

DISCUSSION

It is well known that the Rb-Sr systematics of feldspars and, to a lesser degree, biotite are disturbed at low temperatures (e.g., Hart, 1964; Van Breeman and Dallmeyer, 1984). The bulk of the rubidium in these granites is located in the biotite and potash feldspars. These minerals remained partially open to strontium diffusional loss for about 100 m.y. after emplacement of the granite at 1.44 Ga. Lack of good strontium acceptors (minerals) in these granites would lead to disturbance of the Rb-Sr systematics of the granites on a small scale. Such discordant ages obtained from granite bodies, depending on the scale of the outcrops sampled, have been documented in detail by Roddick and Compston (1977). These authors suggested strontium isotopic reequilibration on two scales: (a) from hand-specimen size to outcrop, and (b) for the whole granite body. Case (a) is consistent with Hofmann's (1979) experimental work, which shows that diffusion distance of Sr in a partially melting mantle-type rock is limited to only tens or hundreds of meters. However, case (b) of Roddick and Compston is not at variance with Hofmann's findings. In case (b), the entire granite body would *appear* to be reequilibrated with respect to Sr isotopic composition, as Roddick and Compston (1977) explained, provided that: (1) the granitic body can be subdivided into subvolumes each of which shows the same Rb/Sr ratio as that of the entire body, (2) this subvolume is less than or equal to the minimum volume of Sr reequilibration in the rock, and (3) the Rb/Sr ratio of the subvolume was not altered significantly by any subsequent thermal event.

In the case of the Sandia granite the conditions for (a) may have been obtained after emplacement of the granite. This would explain the local anomalous ages for the NSP and the SSP. However, the conditions for (b) may not have been satisfied so that determination of the *formation age* of the entire Sandia granite has been possible.

Some chemical data for the Sandia granite are summarized in Tables 5 and 6 (from Majumdar, 1985). It is noteworthy that the chemical differences noted by Condie and Budding (1979) between their north and south Sandia plutons, based on only nine samples, are not noted by Majumdar's (1985) much larger study (27 samples). Further, the trace-element data (Table 6; from Majumdar, 1985) also show few differences. The north and south samples show nearly identical values for the trace elements Cr, Co, Rb, Sr, Ba, Cs, and the REE except for Ce. These data support Brookins and Majumdar's (1982a) contention that there is no pronounced chemical difference between the north and south plutons as defined by Condie and Budding (1979). It should be pointed out further that samples N15, N16, N17, and N18 all have somewhat higher REE, and especially Ce, than the other NSP and all the SSP samples.

Dominance of microcline over orthoclase in the NSP, and the reverse in the SSP, suggest (Majumdar, 1985) that the nouthern part of the Sandia granite cooled more slowly than the southern part. The slower rate of cooling in the north may have been responsible for a relatively greater degree of open-system behavior of the Rb-Sr system. This would account for the more disturbed isochrons—with high scattering of points for samples from the NSP as compared to the SSP (see Majumdar, 1985).

Differences in initial strontium isotopic ratio between the pegmatite-aplite dikes and the Sandia granite argue against any simple genetic connection between the dikes and the granite. This conclusion is also supported by the fact that the average total REE of the four dike samples is much lower than that of the Sandia granite (see Majumdar, 1985).

TABLE 5. COMPARISON OF CHEMICAL ANALYSES OF THE 'NSP' AND THE 'SSP'

Column	1*	2*	3*	4*
Oxides	'NSP'	"SSP'	'NSP	'SSP'
SiO_2	69.2	66.3	68.00	66.87
TiO_2	0.76	1.03	0.88	0.95
Al_2O_3	13.7	13.9	14.23	14.24
Fe_2O_3	4.81	6.85	5.35	5.92
FeO				
MgO	1.00	1.33	1.17	1.33
CaO	2.75	3.73	2.82	2.90
Na_2O	3.29	3.11	3.24	3.32
K_2O	3.99	3.25	4.34	3.99
P_2O_5			0.31	0.31
MnO			0.11	0.17
SrO	0.024	0.023	0.022	0.024
Total	99.5	99.5	100.47	99.23

*Column 1: average of 4 samples from 'NSP' (Condie and Budding, 1979).
Column 2: average of 5 samples from 'SSP' (Condie and Budding, 1979).
Column 3: average of 14 samples from 'NSP' (this study).
Column 4: average of 13 samples from 'SSP' (this study).

If the 1.41-Ga age is the true age of crystallization of the dikes, then they may be genetically related to the 1.4-Ga granitic rocks exposed in south-central New Mexico, which show $(^{87}Sr/^{86}Sr)_0$ comparable to the dike rocks (Condie and Budding, 1979). Alternatively, the dikes in the Juan Tabo series could be older than the Sandia granite because they are folded and deformed in many places, whereas the pegmatite and aplites in the Sandia granite are not. In that event, however, the lower age of the dikes could represent a (partially) reset age (metamorphic), as these dikes occur in the contact metamorphic aureole of the Sandia granite.

SUMMARY AND CONCLUSIONS

The major findings of this study are summarized below.

1. The minimum age of deposition of the sedimentary and volcanic protoliths for the metamorphic country rocks (Juan Tabo series) with which the Sandia granite shows an undoubted intrusive contact is 1.61 ± 0.06 Ga (Brookins and Majumdar, 1983).

2. The age of formation of the Sandia granite is 1.44 ± 0.04 Ga, as established by an 11-point whole-rock isochron (Brookins

TABLE 6. COMPARISON OF SOME TRACE ELEMENTS FROM THE 'NSP' AND THE 'SSP'

Element	NSP		SSP	
	1* (ppm)	2* (ppm)	3* (ppm)	4* (ppm)
Cr	9	12	10	15
Co	11	8	18	11
Rb	192	168	150	158
Sr	241	213	228	217
Zr	364	-	365	-
Ba	840	1114	790	1054
Cs	8.3	8.4	7.6	8.4
La	62	64	49	61
Ce	140	166	118	114
Sm	13	13	11	12
Eu	3.1	2.6	2.9	2.5
Tb	2.3	1.7	1.9	1.7
Yb	4.9	7.0	4.1	6.3
Lu	0.79	1.03	0.64	0.92
REE (Total)	226	255	188	198
Th	-	16.17	-	14.69
Hf	-	11.10	-	10.18
Sc	-	13.15	-	15.11
K/Rb	173	206	180	209
Rb/Sr	0.80	0.89	0.66	0.73
La/Yb	13	9	12	9.6
Ce/Yb	-	23.6	-	18.1

*Data in columns 1 (4 samples) and 3 (5 samples) are from Condie and Budding (1979); data in columns 2 (12 samples) and 4 (12 samples) are from this study. Experimental error = ± 10%.

and Majumdar, 1982b). A calculated model age of 1.44 Ga also agrees with the isochron age. The initial $^{87}Sr/^{86}Sr$ ratio of 0.7045 obtained from isochron intercept is approximately the same as the $^{87}Sr/^{86}Sr$ ratio of some apatite samples from the granite (Wasserburg and Steiger, 1967).

3. The dikes of pegmatite and aplite intruding the Juan Tabo series are dated at 1.41 ± 0.02 Ga, with an initial Sr ratio of 0.713 ± 0.001. The relationship of these rocks to the Sandia granite is not clear. It is possible that the dikes are older than the Sandia granite and their age was partially reset during the emplacement of the granite.

4. The following thermal events related to the emplacement of the Sandia granite are envisaged.

(a) The Juan Tabo series of rocks were reheated by intrusion of the granite somewhat before 1.44 Ga when temperature of the country rocks rose to about 500° to 600°C, based on the occurrence of second-generation overgrowth on andalusite and on the occurrence of sillimanite (Berkley and Callender, 1979).

(b) The culmination of the granite emplacement at 1.44 Ga was at a shallow depth (less than 10 km) and temperature of 700° to 800°C (dry granitic magma; Majumdar, 1985). The thermal effect of the intrusion raised the temperature of the country rocks to about 650°C, leading to the development of sillimanite in the Juan Tabo series.

(c) Both Rb-Sr and $^{39}Ar/^{40}Ar$ data for biotite suggest cooling of the Sandia granite to 300° to 350°C at 1.33 Ga. Order-of-magnitude calculations for heat loss following the methods of Jaeger (1964) suggest that cooling of the Sandia granite to below 300°C could have taken about 100 m.y. after its final emplacement (for detailed calculation see Majumdar, 1985).

(d) It is most likely that during this prolonged period of slow cooling, orthoclase was inverted to microcline.

5. In spite of the open behavior of the Rb-Sr system on a local scale, judicious use of the whole-rock isochron, and mineral–whole-rock pairs representative of the entire Sandia granite, has provided meaningful information about the granite emplacement and a subsequent metamorphic event.

6. Condie and Budding (1979) reported that the southern part of the Sandia granite is granodioritic, and the northern part quartz monzonitic. Majumdar's (1985) study reveals that variation from granodiorite to quartz monzonite exists both in the north and south. Both major- and trace-element variation diagrams for both the NSP and SSP show similar trends of chemical differentiation (Majumdar, 1985). The two populations together define a Rb-Sr isochron that gives the absolute age of the Sandia granite as a whole. In the field, the NSP shows petrological and structural continuity into the SSP across the Pino Canyon proposed contact.

All the above findings tend to establish that the NSP and the SSP belong to a single pluton. However, a small isolated outcrop of granite near the northern end of the Sandia granite may represent part of a separate pluton (samples N15, N16, N17, N18).

7. The strontium-initial-ratio oxygen-isotopic data (Majumdar, 1985) and REE distribution of the mafic xenoliths and the Sandia granite suggest that they are genetically linked. From the major-element variation diagrams (Majumdar, 1985) and petrologic and geochemical considerations, it is concluded that the mafic xenoliths represent modified residuum after partial melting of the source rocks of the Sandia granite. Incipient partial melting of these source rocks gave rise to the felsic xenoliths (with two micas and antiperthite), which crystallized at depth and were caught up by the Sandia granite magma during its ascent.

8. Finally, on the question of addition of new mantle-derived materials to the crust of the western North American continent in the middle of late Proterozoic time, DePaolo (1981) concluded that a new segment of continental crust was added about 1.8 Ga; subsequent magmatic events at 1.67, 1.40, and 1.00 Ga were shown to be mainly intracrustal differentiation with little or no addition to the crust from the mantle. DePaolo also suggests that the 1.4-Ga plutons may have formed by remelting *mantle* that was depleted at 1.7 Ga. The results of the present study on the Sandia granite and its cognate xenoliths, although not conclusive, are consistent with DePaolo's (1981) broad thesis.

ACKNOWLEDGMENTS

R. K. Mathaney, M. T. Murphy, M. S. Abashian, and D. Ward helped with the analytical Rb-Sr work. Los Alamos National Laboratory provided the neutron activation analyses. Ms. J. Binder prepared the manuscript.

REFERENCES

Aldrich, L. T., Wetherill, G. W., and Davis, G. L., 1957, Occurrence of 1,350 m.y. old granitic rocks in the western United States: Geological Society of America Bulletin, v. 68, p. 655–656.

Aldrich, L. T., Wetherill, G. W., Davis, G. L., and Tilton, G. R., 1958, Radioactive ages of micas from granitic rocks by Rb-Sr and K-Ar methods: EOS Transactions of the American Geophysical Union, v. 39, p. 1124–1134.

Berger, G. W., and York, D., 1981, Geothermometry from ^{40}Ar-^{39}Ar dating experiments: Geochimica et Cosmochimica Acta, v. 45, p. 795–811.

Berkley, J. L., and Callender, J. F., 1979, Precambrian metamorphism in the Placitas–Juan Taboarea, northwestern Sandia Mountains, New Mexico: New Mexico Geological Society 30th Field Conference Guidebook, p. 181–188.

Brookins, D. G., 1974, Radiometric age determinations from the Sandia granite, New Mexico: Summary and interpretations: Isochron/West, no. 10, p. 11–14.

Brookins, D. G., and Majumdar, A., 1982a, The Sandia granite, New Mexico; Biotite metamorphic and whole rock Rb-Sr ages: Isochron/West, no. 33, p. 19–21.

——, 1982b, The Sandia granite; Single or multiple plutons?: New Mexico Geological Society 33rd Field Conference Guidebook, p. 221–224.

——, 1983, Whole rock Rb-Sr age of the Juan Tabo series, Sandia Mountains, New Mexico: Isochron/West, no. 38, p. 19–21.

Brookins, D. G., and Shafiquallah, M., 1975, K-Ar ages from pegmatitic and metamorphic muscovites, Sandia Mountains, New Mexico: Isochron/West, no. 12, p. 9.

Brookins, D. G., Enz, R. D., Kudo, A. M., and Shafiqullah, M., 1975, K-Ar and Rb-Sr age determinations of orbicular granite, Sandia Mountains, New Mexico: Isochron/West, no. 12, p. 11–13.

Condie, K. C., and Budding, A. J., 1979, Geology and geochemistry of Precambrian rocks, central and south-central New Mexico: New Mexico Bureau of Mining and Mineral Resources Memoir 35, 58 p.

DePaolo, D. J., 1981, A neodymium and strontium isotopic study of the Mesozoic calcalkaline granitic batholiths of the Sierra Nevada and Peninsular Ranges, California: Journal of Geophysical Research, v. 86, p. 10470–10488.

Enz, R. D., Kudo, A. M., and Brookins, D. G., 1979, Igneous origin of the orbicular rocks of the Sandia Mountains, New Mexico: Geological Society of America Bulletin, v. 90, part 1, p. 138–140, part 2, p. 349–380.

Green, J. A., and Callender, J. F., 1973, Hornblende-hornfels facies metamorphism in a contact aureole adjacent to the Sandia Mountain pluton, New Mexico: Geological Society of America Abstracts with programs, v. 5, p. 642–643.

Hart, S. R., 1964, The petrology and isotopic-mineral age relations of a contact zone in the Front Range, Colorado: Journal of Geology, v. 72, p. 493–525.

Hofmann, A. W., 1979, Diffusion experiments in isotope geology, *in* Jager, E., and Hunziker, J. C., eds., Lectures in isotope geology: Heidelberg, Springer-Verlag, p. 189–193.

Jaeger, J. C., 1964, Thermal effects of intrusion: Reviews in Geophysics, v. 2, p. 443–466.

Kelley, V., and Northrop, S. A., 1975, Geology of the Sandia Mountains and vicinity, New Mexico: New Mexico Bureau of Mining and Mineral Resources Memoir 29, 135 p.

Lodewick, R. B., 1960, Geology and petrography of the Tijeras (Cibola) gneiss, Bernalillo County, New Mexico [M.S. thesis]: Albuquerque, University of New Mexico, 63 p.

Majumdar, A., 1985, Geochronology, geochemistry, and petrology of the Precambrian Sandia granite, Albuquerque, New Mexico [Ph.D. thesis]: Baton Rouge, Louisiana State University, 190 p.

Naeser, C. W., 1971, Geochronology of the Navajo–Hopi diatremes, Four Corners area; Journal of Geophysical Research, v. 76, p. 4978–4985.

Roddick, J. C., and Compston, W., 1977, Strontium isotopic equilibration; A solution of a paradox: Earth and Planetary Science Letters, v. 34, p. 238–246.

Steiger, R. H., and Wasserburg, G. J., 1966, Systematics in the Pb^{208}-Th^{232}, Pb^{207}-U^{235}, and Pb^{206}-U^{238} systems: Journal of Geophysical Research, v. 71, p. 6065–6090.

Steiger, R. H., and Wasserburg, G. J., 1969, Comparative U-Th-Pb systematics in 2.7×10^9 yr plutons of different geologic histories: Geochimi Cosmochimica Acta, v. 33, p. 1213–1231.

Taggart, J. E., and Brookins, D. G., 1975, Rb-Sr whole rock age determinations for Sandia granite and Cibola gneiss, New Mexico: Isochron/West, no. 12, p. 5–8.

Tilton, G. R., and Grunenfelder, M. H., 1968, Sphene; Uranium-lead ages: Science, v. 159, p. 1458–1461.

Van Breeman, O., and Dallmeyer, R. D., 1984, The scale of Sr isotope diffusion during post-metamorphic cooling of gneisses in the inner Piedmont of Georgia, southern Appalachians: Earth and Planetary Science Letters, v. 68, p. 141–150.

Wasserburg, G. J., and Steiger, R. H., 1967, Systematics in the Th-U-Pb systems and multiphase assemblages, *in* Radioactive dating and methods of low level counting: International Atomic Energy Agency, p. 331–347.

York, D., 1969, Least-squares fitting of a straight line with correlated errors: Earth and Planetary Science Letters, v. 5, p. 320–324.

Manuscript Accepted by the Society September 27, 1988

Geological Society of America
Special Paper 235
1989

Early recumbent folding during Proterozoic orogeny in central Arizona

Karl E. Karlstrom
Department of Geology, Northern Arizona University, Flagstaff, Arizona 86011

ABSTRACT

Two generations of folds affected the folded unconformity between the 1,750-Ma Brady Butte Granodiorite and overlying Proterozoic Texas Gulch Formation metagraywackes and slates in the Brady Butte area of central Arizona. F_1 folds are isoclinal and were predominantly northwest-verging recumbent folds prior to F_2 folding. They occur at all scales, with macroscopic fold amplitudes exceeding 500 m. F_1 folds were temporally associated with transposition of bedding and mesoscopic thrusting in metasedimentary rocks, and development of mylonitic foliation in both metasedimentary rocks and granodiorite. Both style of structures and asymmetry of fabrics suggest F_1 folding accompanied northwest-directed thrusting. F_2 folds are open to tight and have northeast-striking, steeply dipping, axial plane foliation. F_2 folds are gently northeast plunging in the Brady Butte area, coaxial with F_1 folds, and the enveloping surface of F_2 folds is subhorizontal.

New structural data may have important regional implications. A subhorizontal enveloping surface for F_2 folds implies that stratigraphic units and the basal Texas Gulch Formation unconformity are repeated across strike in F_2 fold hinges. Thus, the metasedimentary rocks of the Crazy Basin area may correlate with the Texas Gulch Formation, and parts of the Spud Mountain Volcanics may correlate with parts of the Iron King Volcanics. However, transposition during F_2 and rotation of fold hingelines toward the subvertical F_2 finite stretching axis in areas of high shortening strain complicate structural geometry such that rootless intrafolial folds are common. Furthermore, F_2 folds affected an already transposed S_1 tectonic layering formed by recumbent folds and thrusts. The complex overprinting of F_1 and F_2 on a regional scale make stratigraphic interpretations tenuous until more detailed structural, geochronologic, and geochemical data are available.

INTRODUCTION

Early Proterozoic (1,800 to 1,600 Ma) rocks in Arizona are exposed in a northwest-trending band in the Transition Zone, between the Colorado Plateau and Basin and Range Provinces (Fig. 1). This band provides semi-continuous exposure for more than 500 km perpendicular to strike of the Proterozoic orogenic belt. The Transition Zone has been affected by Tertiary high-angle faults of moderately large displacement, but no post-Precambrian ductile deformation or large Tertiary extensional strains have been documented, so it is possible to study a cross section of Proterozoic crust that has been largely undisturbed since ca. 1,400 Ma.

This orogenic cross section can be divided into two major provinces (Fig. 1): an older (1,800 to 1,700 Ma) province to the northwest, underlain by oceanic volcanogenic rocks of the Yavapai Supergroup[1], and a younger (1,740 to 1,600 Ma) province to the southeast, underlain predominantly by continentally derived rocks of the Tonto Basin Supergroup and Diamond Rim Intrusive Suite (Conway and others, 1987; Karlstrom and others, 1987). Although the exact geometry and tectonic history of the

[1]This sequence of rocks was named the Yavapai Series by Anderson and others (1971). The chronostratigraphic term "Series" is replaced in this paper by the lithostratigraphic term "Supergroup" in accordance with recommendations of amendments to the U.S. Code of Stratigraphic Nomenclature (Henderson and others, 1980).

Karlstrom, K. E., 1989, Early recumbent folding during Proterozoic orogeny in central Arizona, *in* Grambling, J. A., and Tewksbury, B. J., Proterozoic geology of the southern Rocky Mountains: Boulder, Colorado, Geological Society of America Special Paper 235.

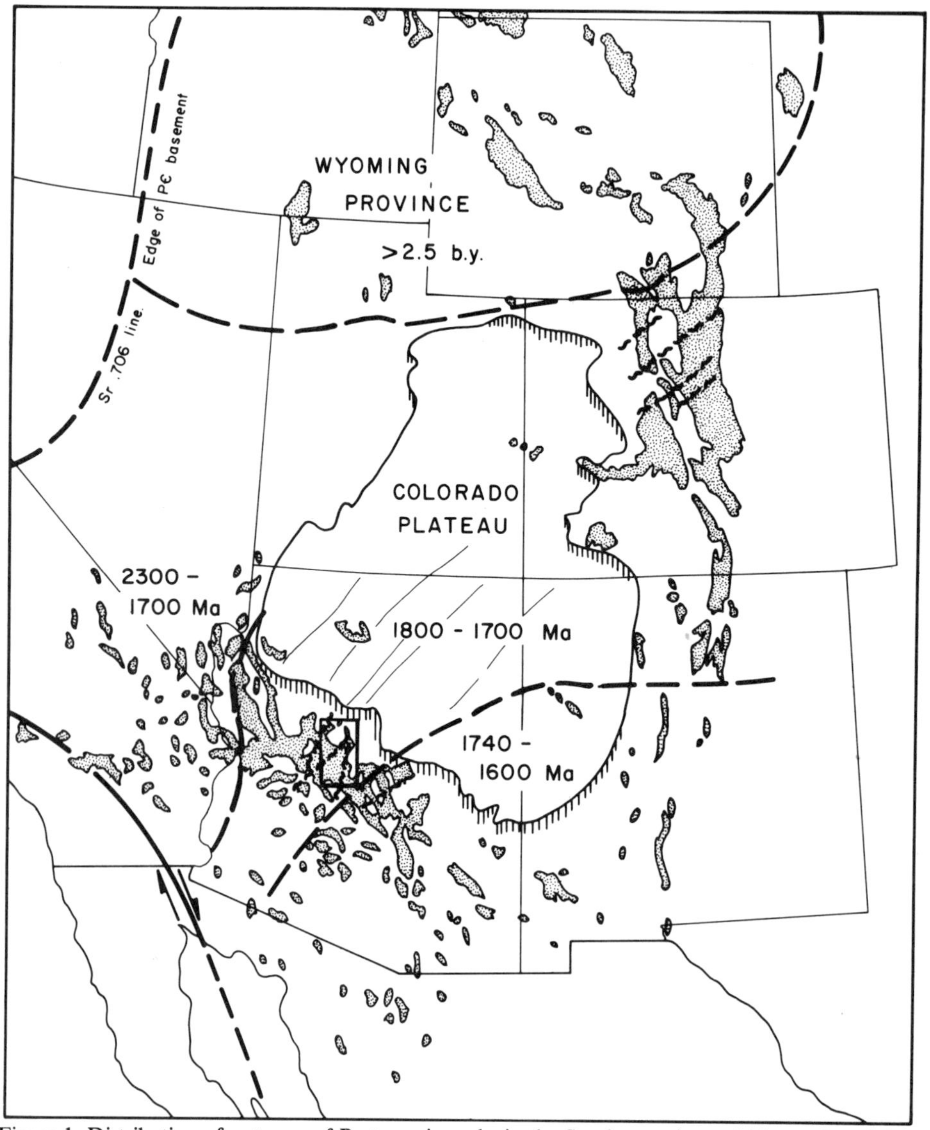

Figure 1. Distribution of outcrops of Proterozoic rocks in the Southwest. Area of Figures 3 and 4 in Arizona is outlined. The Transition Zone of central Arizona exposes a semi-continuous 500-km-long transect northwest-southeast, nearly perpendicular to orogenic strike.

boundary remain controversial, there is a sharp difference in structural style, metamorphic grade, tectonic setting, and metallogenic characteristics between rocks of these two major provinces (Karlstrom and others, 1987). Figure 2 summarizes nomenclature and structural styles of supracrustal rocks of the two major provinces. Rocks of the two provinces in central Arizona are juxtaposed across the Moore Gulch fault, one of a family of north- and northeast-trending major shear zones that divide the orogenic cross section into a series of blocks.

Based on structural studies and U-Pb geochronology of several of the fault-bound blocks, Karlstrom and Bowring (1988) have proposed that the two major provinces are both composites of diverse tectonostratigraphic terranes and that some of the shear zones represent sutures across which terranes were juxtaposed during continental assembly. Assembly of the northwestern province at about 1,700 Ma is referred to as the Yavapai orogeny; assembly of the southeastern province and juxtaposition of the two major composite provinces between 1,700 and 1,630 Ma is referred to as the Mazatzal orogeny. Further testing of this "suspect terrane" working hypothesis requires detailed understanding of tectonic history within each block so that tectonic elements and tectonic events can be correlated and/or contrasted block to block.

This chapter examines the structural geometry and deformational history of highly deformed volcanogenic rocks of the Yavapai Supergroup within one area of the Big Bug block of Karlstrom and Bowring (1988). This is the rectangular area shown in Figure 3 between the Shylock fault zone on the east and the Chaparral fault on the northwest. This area has received a great deal of attention in the literature, largely because of the

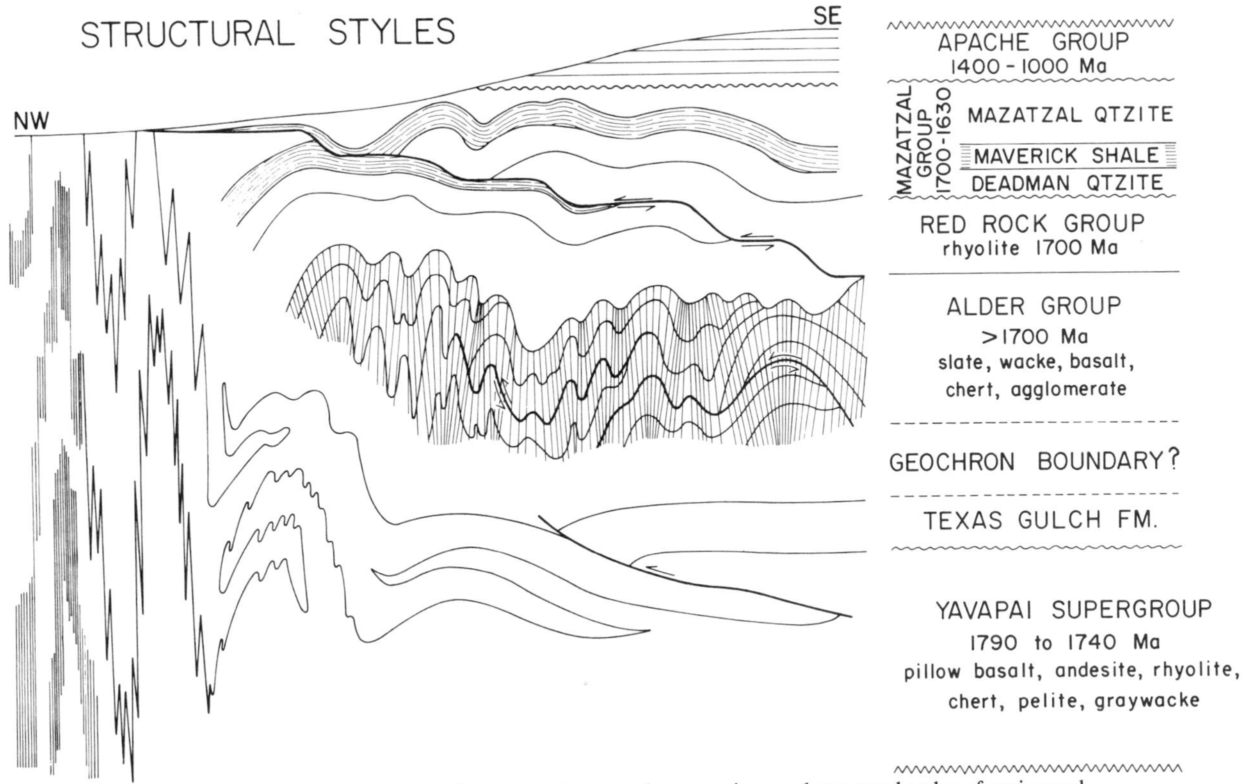

Figure 2. Stratigraphic nomenclature, geochronologic constraints, and structural styles of major rock packages in central Arizona.

presence of an unconformity between metasedimentary rocks of the Texas Gulch Formation and the underlying 1,750-Ma Brady Butte Granodiorite (Blacet, 1966; Anderson and others, 1971), and because of the economic importance of the area for sulfide mineralization.

PREVIOUS WORK AND CONTROVERSIES IN THE YAVAPAI SUPERGROUP

The Yavapai Supergroup of the northwestern province has been described as a Proterozoic greenstone belt (Anderson and Silver, 1976). Like Archean granite-greenstone terranes, this Proterozoic greenstone terrane has roughly equal parts of supracrustal rocks (dominantly mafic volcanics and volcaniclastic sediments) and calc-alkaline batholiths (Fig. 3). More recent workers have suggested a probable island-arc tectonic setting for both supracrustal rocks and batholiths, based on the presence of pillow volcanics, cherts, and massive sulfides, and the calc-alkaline character of the igneous rocks (Anderson, 1978, 1986; Condie, 1986; Vance, 1986). Metamorphic grade is fairly low: greenschist to middle amphibolite facies. Deformation is penetrative and dominantly expressed as north- to northeast-trending, subvertical foliation related to tight to isoclinal folds. Stretching lineations and mesoscopic fold hingelines are also dominantly subvertical.

The stratigraphic nomenclature of Anderson and others (1971) for supracrustal rocks in this area is shown in Figure 4. This stratigraphy was based on nearly three decades of 1:24,000 and smaller scale mapping by the U.S. Geological Survey (Anderson and Creasey, 1958; Krieger, 1965; Blacet, 1968; Anderson and Blacet, 1972a, b, c) and has been the generally accepted nomenclature for the Yavapai Supergroup. The distribution of stratigraphic units was believed by these workers to be strongly controlled by two major faults: the Shylock fault zone and the Chaparral fault. The oldest unit of Anderson and others (1971) was the Ash Creek Group, with a U-Pb zircon age of 1,790 ± 10 Ma from rhyolites near Jerome. The Ash Creek Group was mapped only east of the Shylock fault zone, on Mingus Mountain, and in the Black Hills. The Big Bug Group was believed to be younger, based on U-Pb zircon dates of 1,755 ± 10 Ma[2] on volcanics and 1,750 ± 10 Ma on crosscutting plutonic rocks (Fig. 4). The Big Bug Group was divided into the Green Gulch Volcanics, which crop out northwest of the Chaparral fault; the Spud Mountain Volcanics, believed to stratigraphically overlie the Green Gulch Volcanics; and the Iron King Volcanics, believed to occupy the core of a regional syncline (Fig. 4). The Texas Gulch Formation, now considered to be the youngest unit in the Big Bug block, was originally interpreted by Blacet (1966) and Anderson (1968) to be the oldest. It was considered to be older than adja-

[2]Some ages reported in this paper are 20 m.y. younger than the date originally reported, reflecting the approximate correction needed for older U-Pb dates necessitated by newer decay constants (Steiger and Jäger, 1977). See Karlstrom and others (1987) for a more complete tabulation of available U-Pb zircon dates.

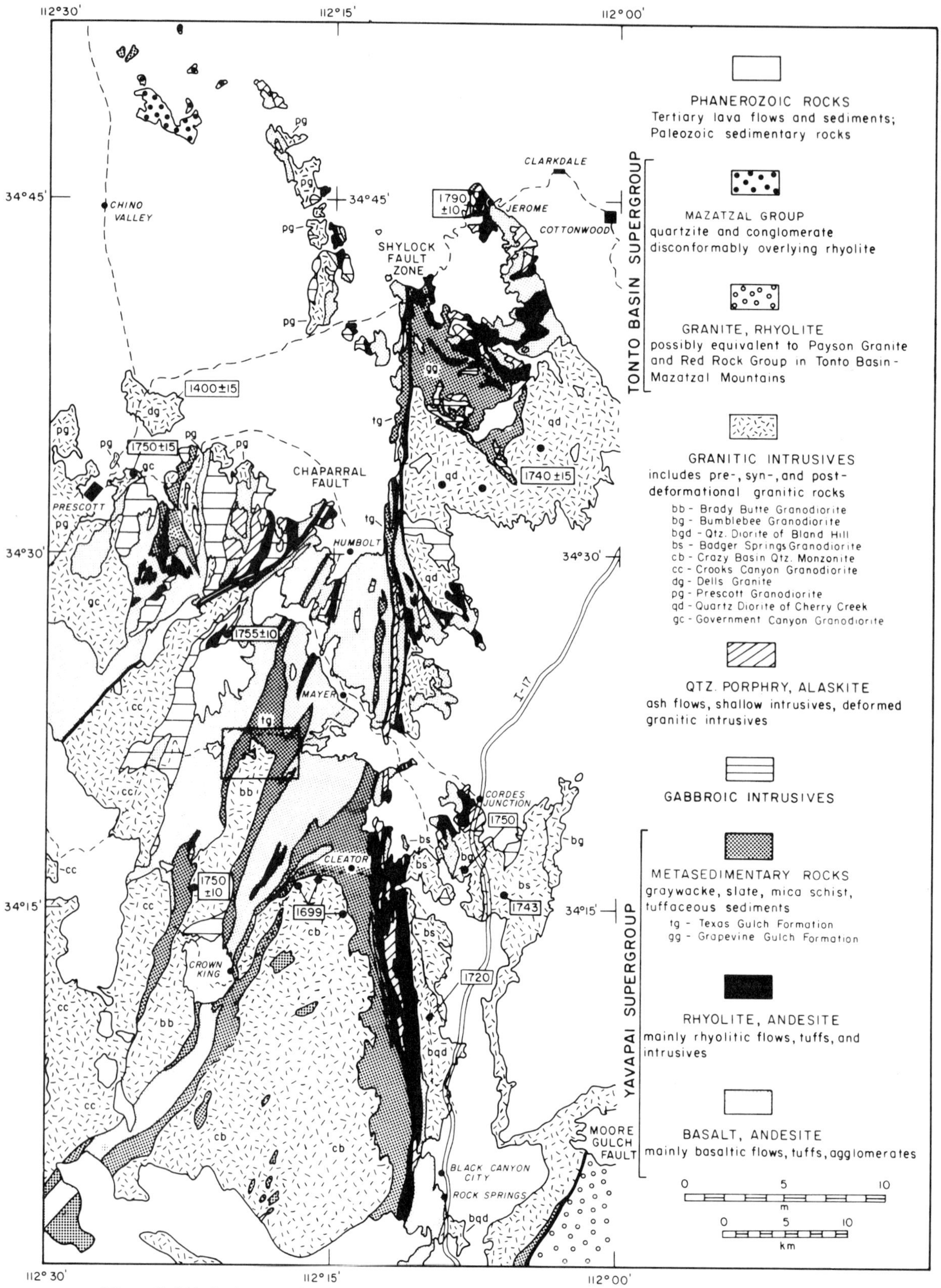

Figure 3. Lithologic map of the Yavapai Supergroup of the Bradshaw Mountains and adjacent areas. Compiled from Jagger and Palache (1905), Jerome (1956), Anderson and Creasey (1958), Krieger (1965), Anderson and Blacet (1972a, b, c), and DeWitt (1976, 1979). Location of map area in Arizona is shown in Figure 1. Box shows area of Figure 5.

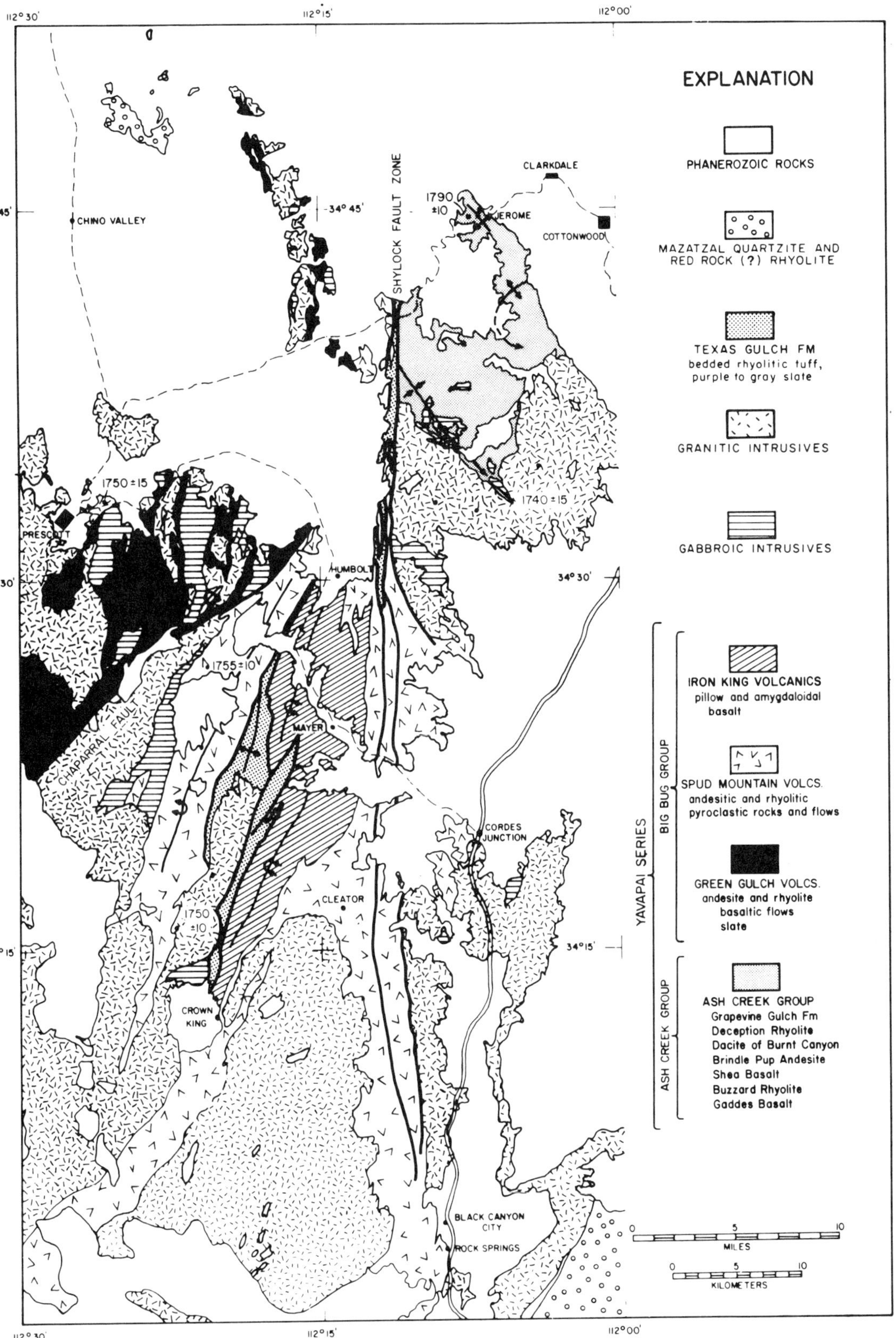

Figure 4. Map showing stratigraphic nomenclature of Anderson and others (1971) for the Yavapai Supergroup of the Bradshaw Mountains and adjacent areas, modified from their Figure 2. Ages and sample locations of U-Pb zircon analyses are shown, but ages here are 20 m.y. younger than those reported by Anderson and others (1971). This is an approximate correction for rocks of ≈1,700 Ma that reflects the change in recommended decay constants (Steiger and Jäger, 1977).

cent volcanic rocks because of its position toward the center of the inferred regional anticline cored by the Brady Butte Granodiorite (Fig. 4), and based on correlation of the Texas Gulch Formation with slates exposed in the core of a major overturned anticline in the Green Gulch Volcanics (Krieger, 1965; Anderson, 1968). Blacet (1966) had documented that basal granodiorite-boulder conglomerate of the Texas Gulch Formation unconformably overlies a northern phase of the Brady Butte Granodiorite (bb in Fig. 3), which he considered at that time to be basement to the Yavapai Supergroup. He also showed that a southern granodiorite intrudes the Spud Mountain Volcanics, and he considered this to be a younger pluton. However, when U-Pb zircon dates showed that the two granodiorites were about the same age (1,750 ± 10 Ma), Anderson and others (1971) concluded that they were probably two phases of a single batholith and that the Texas Gulch Formation must be younger than the Spud Mountain Volcanics. Anderson and others (1971) no longer considered the Texas Gulch Formation to be part of their Yavapai Series (Fig. 4). The unconformity was apparently believed to represent considerable hiatus such that the Texas Gulch Formation did not belong within the 1,800 to 1,750 Ma interval specified for the Yavapai Series.

The structural interpretation of Anderson and his coworkers was that the regional folds are entirely upward-facing[3] and have subhorizontal plunges on a regional scale. They believed that the distribution of stratigraphic units in the Big Bug Group was controlled by a single generation of upright folds. The discovery, from isotopic data, that a unit toward the center of an anticline (Texas Gulch Formation) was younger than a unit farther out (Spud Mountain Volcanics) created problems reconciling structural and stratigraphic interpretations. They proposed that major ductile faults and shear zones entirely bound the Texas Gulch Formation (Fig. 4) and caused this younger unit to be downthrown along "piercement structures" in the core of the regional anticline (Anderson, 1967; Anderson and Blacet, 1972c). Gilmour and Still (1968) questioned the presence of these faults bounding the Texas Gulch Formation, as have other subsequent workers (DeWitt, 1976; O'Hara, 1980; and this chapter).

DeWitt (1976, 1979, 1980), working in the area of the Iron King Volcanics of Anderson and others (1971; Fig. 4), recognized problems with the structural interpretation of Anderson and others (1971). DeWitt reported a preponderance of steeply plunging minor folds and mineral and pebble lineations, and he reinterpreted the stratigraphy in terms of steeply plunging macroscopic folds. This prompted a discussion from Creasey (1980), who pointed out that areas outside DeWitt's study area, such as the Brady Butte area, had predominantly shallow-plunging folds and that it was these folds that controlled regional stratigraphy. This controversy can be reconciled with new structural data that show a much more complicated deformational geometry than envisioned by either worker, as discussed later.

O'Hara and others (1978) and O'Hara (1980) were the first workers to recognize that deformation in the rocks of the Texas Gulch Formation, and by inference older rocks, involved multiple fold generations and that distribution of stratigraphic units was more complex than could be explained by a pattern of macroscopic anticlines and synclines. They noted that Blacet's (1966) map pattern in the Brady Butte area (Fig. 3) suggested fold overprinting relationships (Blacet had interpreted this map pattern to be a topographic effect), and that two generations of mesoscopic folds could be seen in Mule Canyon, at the north end of the Brady Butte Granodiorite. This chapter documents the macroscopic and mesoscopic geometry of multiple fold generations in the Brady Butte area (see Fig. 3 for location), and reevaluates structural and stratigraphic interpretations of the Yavapai Supergroup of the Big Bug block in light of new structural data.

GEOLOGY OF A FOLDED UNCONFORMITY

The Brady Butte area (Fig. 5) is very important for understanding the structure of the Yavapai Supergroup. Most of the Yavapai Supergroup of the Bradshaw Mountains consists of lenticular and gradational volcanic units, many lacking primary layering. In contrast, the Brady Butte area contains a mappable unconformity, and metasediments of the Texas Gulch Formation preserve primary structures and fold overprinting relationships. The preservation of early structures may be due to the location of this area near the hinge of a macroscopic antiform cored by the massive Brady Butte Granodiorite (Fig. 3), where regional shortening strains were apparently lower than in adjacent areas.

The unconformity, seen at several locations (shown with dots in Fig. 5), is marked by basal conglomerate with boulders of granophyre, granodiorite, mafic volcanics, chert, and quartz in an arkosic matrix. This conglomerate is widespread but is not present everywhere at the contact (Fig. 5), suggesting that the conglomerates initially were lenticular debris flows or alluvial channels and/or that the contact has been tectonized. These basal boulder conglomerates reach a maximum thickness of about 10 m, with lenses of pebble conglomerate also occurring higher in the section. In areas where basal boulder conglomerate is absent, the unconformity is marked by pebble conglomerates or metasandstones. Below the contact is highly altered granodiorite in which feldspars have been converted, either partially or completely, to fine-grained muscovite and quartz. In places these altered rocks form a gradational zone between less altered granodiorite and basal conglomerate or arkose. These intensely altered granodiorites may represent regolithic material, as is suggested by their association with the unconformity. Alteration of feldspars may also have accompanied mylonitic deformation of the granodiorite (Knipe and Wintsch, 1985).

The importance of the unconformity from a structural viewpoint is that it is a marker horizon, across which younging is

[3]Shackleton (1958) defined fold "facing" as the younging direction seen in fold hinge areas. Combined with the concept of enveloping surface (Hobbs and others, 1976), facing is useful for reconstructing stratigraphy in isoclinally folded and transposed sequences. Folds can face up or down or in any compass direction.

known, that is nearly continuously exposed in an area of more than 500 m vertical relief. As such, it is an excellent form surface to define macroscopic fold geometries. Although this contact is tectonized, the presence of immature sediments containing clasts of the immediately underlying granodiorite indicates that no major translations occurred along this contact prior to folding. Thus, the contact can be assumed to have been an originally subhorizontal depositional contact. Folds of this unconformity, and by inference the kinematically related mesoscopic folds in the Texas Gulch Formation, cannot be attributed to soft-sediment deformation, which is a possible source of ambiguity in interpreting mesoscopic pre-cleavage folds in deformed graywacke sequences. Likewise, the deformation cannot be attributed to granite emplacement since the granite is strongly foliated and, therefore, was emplaced before the deformation. The Brady Butte area thus records the regional folding deformation.

The Texas Gulch Formation is younger than both the Brady Butte Granodiorite and the nearby metavolcanic rocks of the Yavapai Supergroup. This was documented by Anderson and others (1971) and is further demonstrated by the observation that the unconformity can be recognized above volcanic units as well as granodiorite in the Brady Butte area. Basal sediments contain lithologies derived from the rocks directly below the contact. Thus, the regional deformation that affects the Texas Gulch Formation can be inferred to also have affected older units of the Yavapai Supergroup, including volcanogenic units and massive sulfides that are often harder to study structurally because of the lack of continuous marker horizons.

NEW DATA—MACROSCOPIC FOLD GEOMETRY

Results of new mapping of the Brady Butte area was presented in Figure 5, and structural data are summarized in Figure 6. Mapping was done at scales of 1:6,000 to 1:10,000 by the author and by the 1985 Northern Arizona University geology summer field course. The combination of complex structure and topographic relief create a map pattern where patches of granodiorite are surrounded by Texas Gulch Formation and vice versa. This map pattern differs considerably in detail from that of Blacet (1966, 1968). Interpretation of the structure was aided by a series of cross sections (A through G) into which data were projected both downplunge and up-plunge parallel to the statistically defined F_2 fold axis of 23° to N17E (Fig. 6). These cross sections are shown in Figure 7, as vertical faces of a segmented block diagram; the topography and map contacts are shown on the top of the blocks. Figure 8 shows a three-dimensional diagram of the folded unconformity that was made by "reassembling" the block in Figure 7.

The structural interpretation shows that the Brady Butte Granodiorite occupies the core of a macroscopic anticline, as reported by previous workers. This anticline and associated parasitic folds of the unconformity have subvertical axial surfaces and plunge gently northeast (Figs. 7 and 8). These upright folds, including the macroscopic anticline, are second generation folds (F_2) that locally overprint early recumbent folds (F_1) that also plunge shallowly in a northeasterly direction (Fig. 8).

The documentation of macroscopic F_1 recumbent folds is the most important contribution of the new mapping. At locations 1, 2, 3, and 4 of Figure 5, basal conglomerate was found to be underneath granite, in downward-facing F_2 folds. Since the unconformity was originally subhorizontal, and since upward-facing contacts can locally be mapped continuously into downward-facing contacts (e.g., location 2 of Fig. 5), these downward-facing contacts establish the presence of macroscopic F_1 recumbent folds. The large F_1 fold at location 1 can be traced north in successive cross sections (Fig. 7) to location 2. Here, a tight syncline (half-wavelength of about 20 m) exposed in Wolf Creek can be mapped up the steep north-facing slope of the creek to where both limbs are folded by F_2 folds. Figures 7 and 8 suggest that this large F_1 fold verges and faces west or northwest, and decreases in amplitude toward the north. Northwest-directed material transport taken up by this fold is apparently transferred to other F_1 folds (e.g., location 3) as its amplitude decreases.

F_1 folds were also observed in several locations east of the high-angle Tertiary(?) fault shown in Figure 5. At location 4, east-facing, east-verging recumbent folds with amplitudes of tens of meters were mapped. These are overprinted by steeply dipping S_2 cleavage in both granodiorite and metasediments. The significance of the facing direction of these folds relative to the west-facing folds discussed earlier and macroscopic fold geometry is not understood. West-facing, west-verging structures predominate in the area and are interpreted to record dominantly top-to-the-west movements during F_1.

The geometry of macroscopic F_2 folds of the unconformity is also shown by Figures 7 and 8. The major F_2 anticline cored by the Brady Butte Granodiorite has a broad hinge with parasitic folds that appear to increase in amplitude toward the northwest and have a subhorizontal enveloping surface. Location 5, an apparent synformal closure of Texas Gulch Formation on the west side of the Brady Butte Granodiorite (Fig. 3), is interpreted as an F_2 syncline with the Texas Gulch Formation above an eroded basement that includes granodiorite and Spud Mountain Volcanics. This interpretation differs from the "piercement fault" interpretation of Anderson and others (1971); I found no evidence that the contacts of the Texas Gulch Formation have experienced more movement than other transposed contacts in the area. My interpretation also differs from that of O'Hara (1980), who postulated that this closure (location 5 of Figs. 7 and 8) is an F_1 fold. If so, the fold would have originally faced east (when unfolded about F_2) before being rotated into its present position, and it would have to be a different generation than the F_1 recumbent fold at location 1. This makes a more complicated deformational history than seems necessary to explain the data. The interpretation of Figures 7 and 8 explains the presence of volcanics unconformably beneath Texas Gulch Formation at locations 3 and 5 by proposing that the Texas Gulch Formation in this area occupies tight, upward-facing F_2 folds with a shallowly dipping enveloping surface. This concept of a broadly folded but regionally sub-

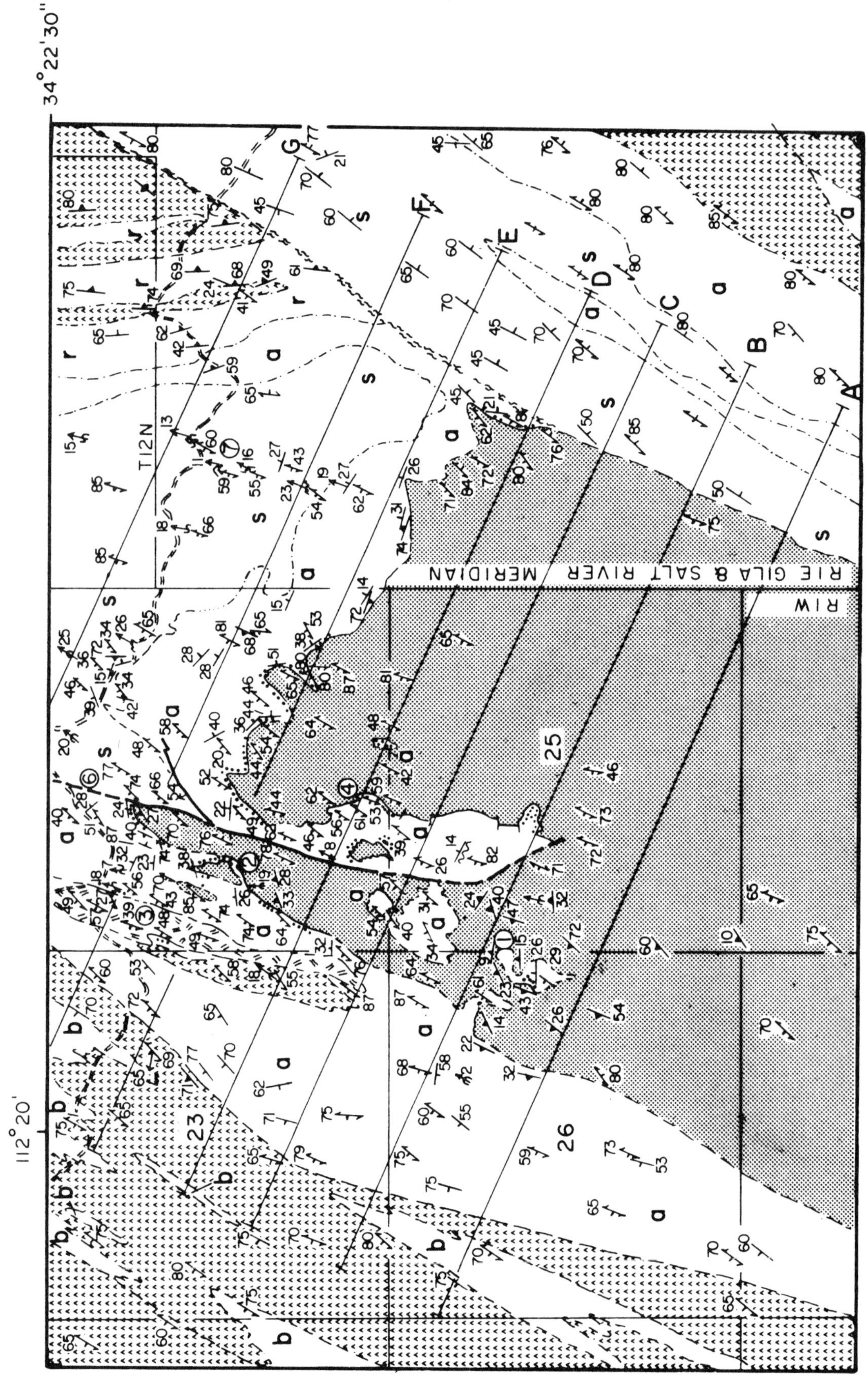

Figure 5. Geologic map of the Brady Butte area showing the trace of the basal Texas Gulch Formation unconformity. See Figure 3 for location.

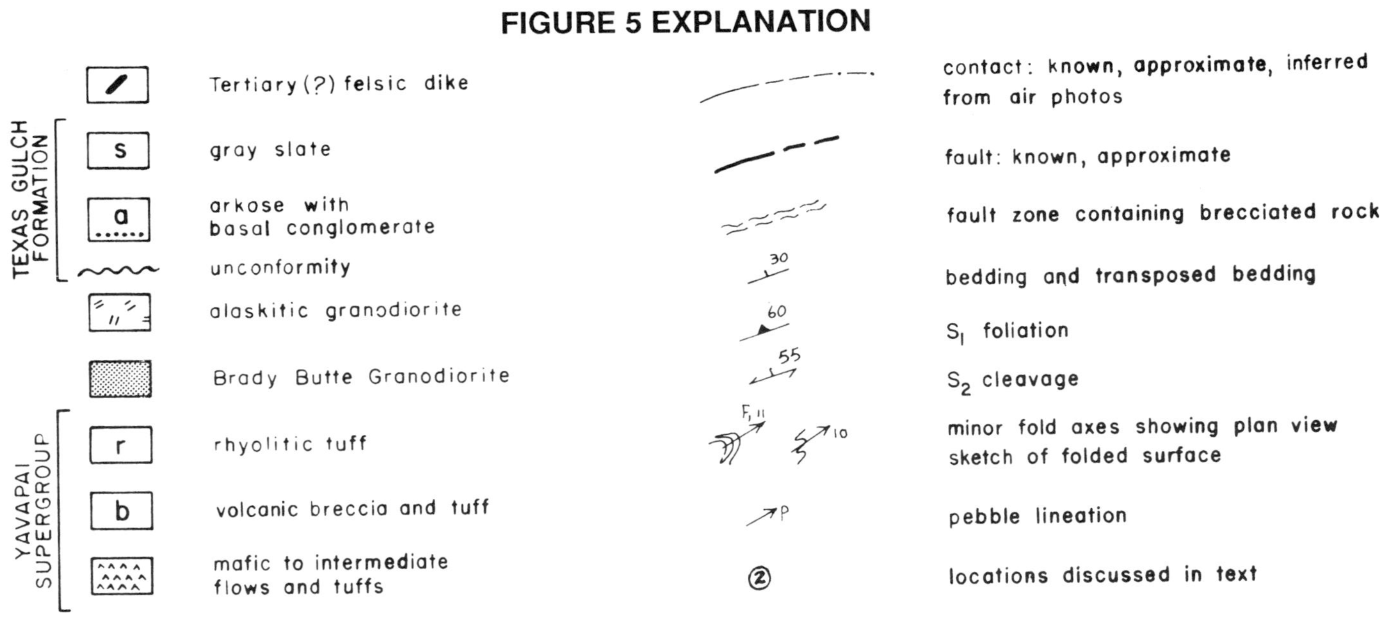

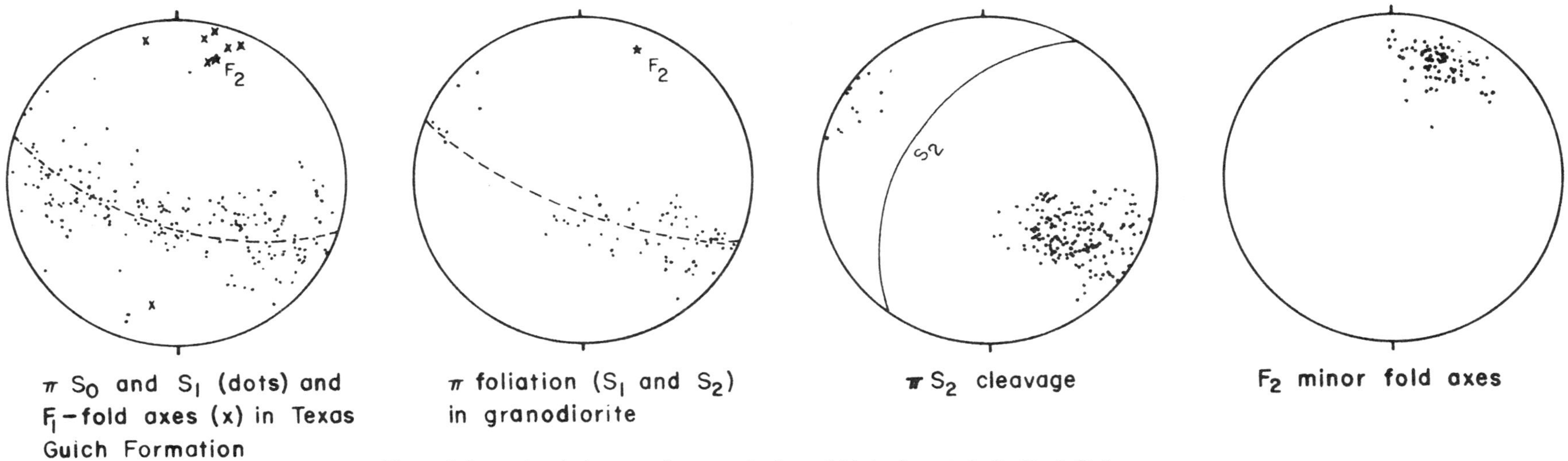

Figure 6. Lower-hemisphere equal-area projections of fabric elements in the Brady Butte area.

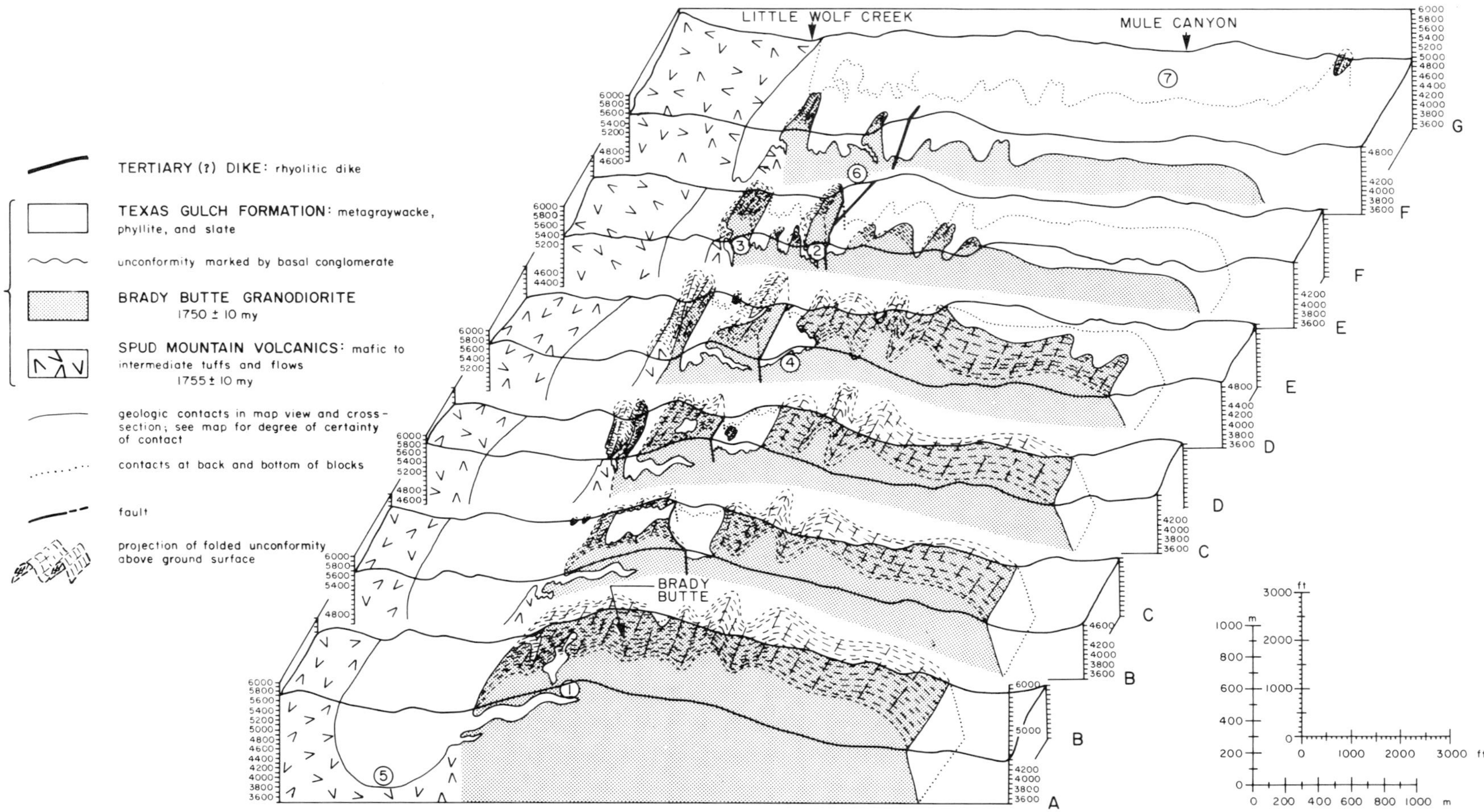

Figure 7. Isometric block diagram of the Brady Butte area, segmented to show structural cross sections A through G (see map for location of cross sections). Surface of block shows mapped contacts and topography, cross sections are based on projection; of data up- and down-plunge of F_2 onto vertical planes. Numbers correspond to locations discussed in text.

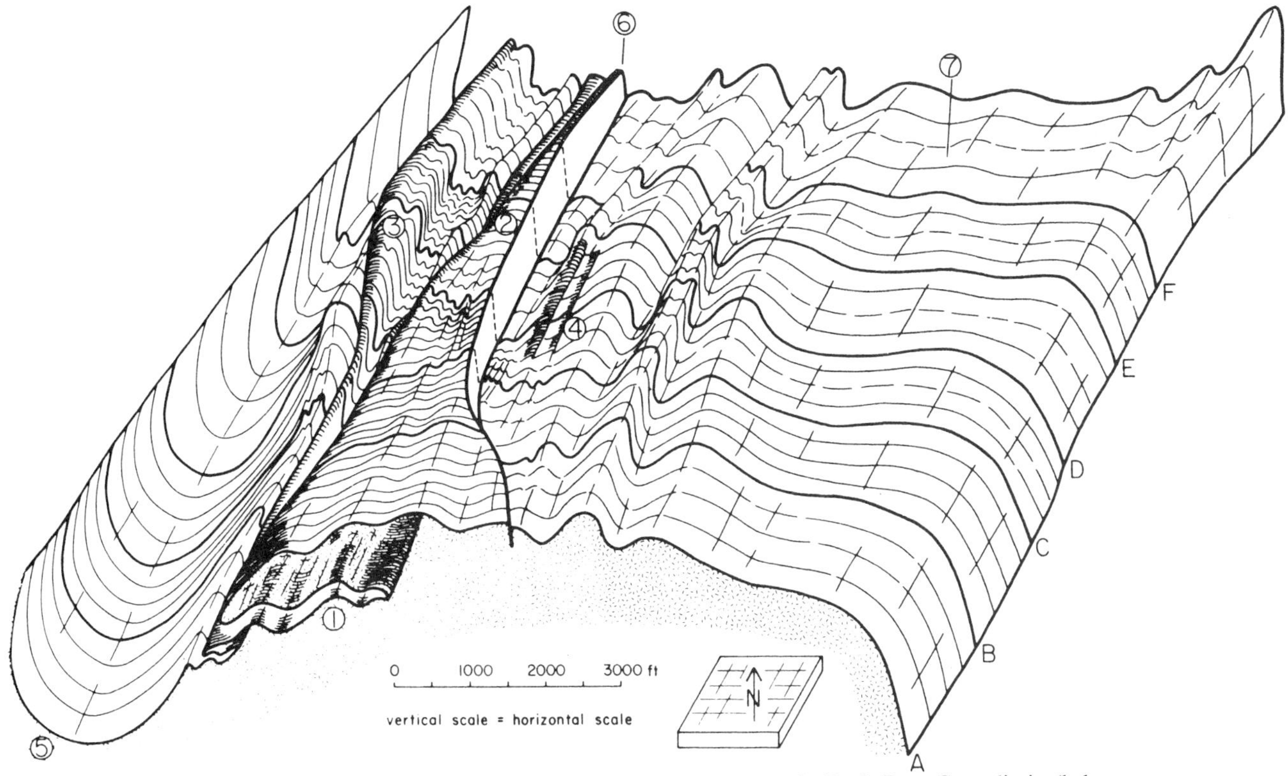

Figure 8. Isometric projection of the folded unconformity between the Brady Butte Granodiorite (below the folded surface) and Texas Gulch Formation (above). This diagram was constructed by "reassembling" the block in Figure 7. Numbers correspond to locations on Figures 5 and 7 that are discussed in the text.

horizontal enveloping surface for F_2 folds also helps explain the distribution of Texas Gulch Formation east of the Brady Butte area (Fig. 3) and has important regional implications, as discussed later.

The youngest structural feature in the area of Figure 5 is a high-angle fault (location 6) that truncates a northwest-dipping rhyolite dike. This dike has flow-banded margins and contains xenoliths of granodiorite, volcanics, and Texas Gulch Formation and is probably of Tertiary age. Post-dike fault movement was down to the east, as shown by fragments of the dike that have dropped down to the level of Wolf Creek. These fragments indicate a minimum throw of 30 m on the fault.

MESOSCOPIC AND MICROSCOPIC STRUCTURAL FEATURES

Mesoscopic examples of F_1 folds refolded by upright or inclined F_2 folds can be seen in the Mule Canyon area (location 7 of Fig. 5). F_1 folds are tight to isoclinal (Fig. 9a, b) and nearly co-axial with F_2 folds (Fig. 6). F_1 folds restore to a recumbent orientation when F_2 folds are "unfolded" by rotation of F_2 limbs to horizontal about the F_2 fold axis. Where asymmetry of F_1 folds can be determined, they verge predominantly west or northwest (Fig. 9a, b, c, d). This relatively consistent vergence, the recumbent style, and the association of F_1 folds with mesoscopic bedding-parallel thrusts (Fig. b, d) suggest that F_1 was related to thrusting.

Mesoscopic F_1 folds affect compositional layering of alternating pelite and psammite, as well as conglomerate layers, all of which presumably represent original bedding. However, bedding was intensely transposed during F_1 (Fig. 9a, b), and most compositional banding in the Texas Gulch Formation, therefore, cannot be used to determine stratigraphic relationships or thicknesses. In many areas, it is difficult or impossible to match up bed segments into contiguous layers because of translation of material parallel to S_1 and drastic attenuation of original layering during transposition (Fig. 9a).

S_1 foliation is present in the granodiorite as a mylonitic foliation defined by shape-preferred orientation of quartz and feldspar grains and grain aggregates, and alignment of phyllosilicates. Various stages of grain-size reduction of quartz can be seen, ranging from undulose extinction and subgrain development to small recrystallized quartz grains as "necklaces" around and along quartz grain and subgrain boundaries (Fig. 10a). S_1 foliation is folded by F_2 folds. At the microscopic scale, strain-free recrystallized quartz grains that developed in quartz porphyroclasts to facilitate dislocation creep during F_1 become plastically deformed in the hinges of F_2 folds (Fig. 10a).

Stretching lineations are locally developed in the mylonitic granodiorite, and in those areas, asymmetric porphyroclasts,

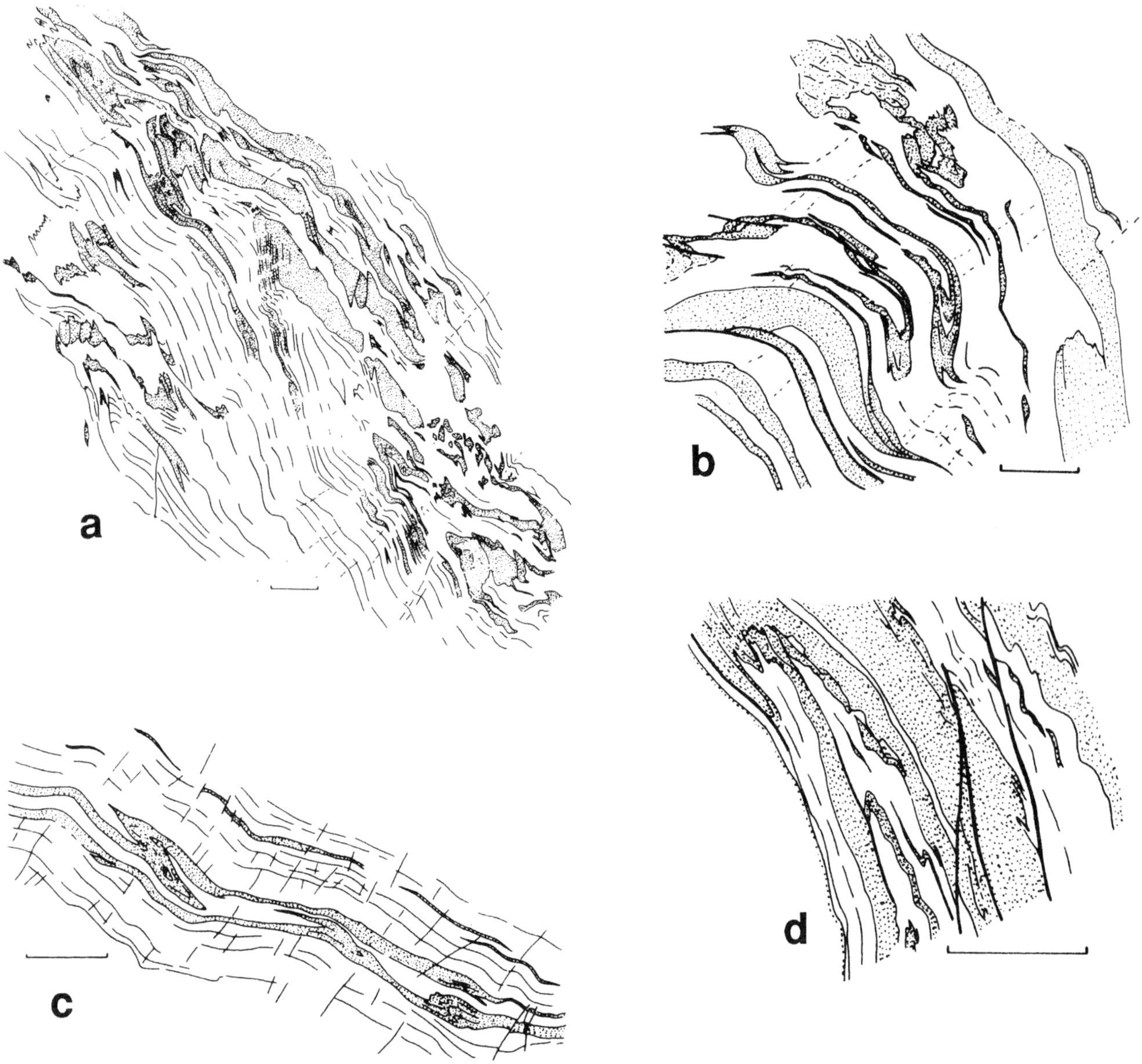

Figure 9. Sketches of photographs of F_1 folds and F_1–F_2 overprinting relationships in the Mule Canyon area. Stippled layers are psammites, white areas are pelites, scale bars are 5 cm long. All sketches are in the profile plane of F_2, looking down the F_2 fold axis, about 20° toward N15E. S_2, a weakly-developing crenulation cleavage, dips 40° to 70° west (left) in all sketches and is shown by short dashed lines. (a) Transposition of bed segments makes it difficult to match once-contiguous layers. Disrupted folds appear to be northwest verging in the upper left and southeast verging in the lower right. Translation parallel to S_1 has disrupted and attenuated layers. (b) F_1–F_2 overprinting. F_1 folds are northwest verging and nearly isoclinal. Thicker psammite layers in the lower left show mesoscopic thrusts. Their geometry suggests northwest vergence. (c) F_1 intrafolial folds within an otherwise uniformly layered sequence of psammite and pelite. Folds verge northwest. (d) Many F_1 fold hinges are truncated by movement surfaces. Thrusts and folds verge northwest after F_2 folds are "unfolded" by simple rotation of limbs to horizontal.

offset feldspar grains, and S-C relationships were observed. These kinematic indicators have not yet been studied systematically, but numerous examples show top-to-the-northwest sense of movement (Figs. 10b, c), compatible with the predominant northwest vergence and facing of F_1 folds. Top-to-the-northwest movement was found on west-dipping limbs of F_2 folds, leaving no doubt that these movements pre-dated F_2 upright folding.

F_2 folds are overturned with steeply west-dipping axial surfaces and fold axes that plunge shallowly northeast (Fig. 6). Folds vary in style from open chevron folds to tight to isoclinal folds. F_2 folds are asymmetric in Mule Canyon and indicate an F_2 antiformal axis toward the west, compatible with the macroscopic geometry shown in Figure 8. S_2 is a differentiated crenulation cleavage in metasedimentary rocks in Mule Canyon.

TIMING OF DEFORMATION

Available U-Pb zircon dates allow some constraints to be placed on timing of deformation. Volcanic rocks of the Spud Mountain Volcanics were emplaced about 1,755 Ma, then intruded by the Brady Butte Granodiorite at about 1,750 Ma. Unroofing of the batholith, presumably from considerable depth, and subsequent deposition of the Texas Gulch Formation may have taken place soon after arc volcanism and plutonism, as part of the progressive evolution of an island-arc complex, as perhaps is suggested by the character of graywackes and slates of the Texas Gulch Formation. Alternatively, the unconformity at the base of the Texas Gulch Formation may represent a considerable hiatus.

Deformation, both F_1 thrusting and recumbent folding and F_2 shortening by upright folding, post-dated deposition of the Texas Gulch Formation. Both F_1 and F_2 took place at moderate depths in the crust, as shown by evidence of crystal plastic deformation of quartz during both F_1 and F_2, brittle deformation of large feldspar porphyroclasts, and by upper-greenschist-facies metamorphic mineral assemblages in metasedimentary rocks (garnet, biotite, chlorite, clinozoisite, epidote) that grew during deformation. Biotite, chlorite (primary and retrograde), and muscovite define the S_1 layering and are folded and kinked by F_2. Garnet overgrew S_1 layering, as shown by inclusion trails of aligned elongate oxide minerals. Some oxide inclusions are broadly warped, suggesting that garnet growth may have been synchronous with the early stages of F_1 folding (Fig. 10d). However, garnet apparently was also synchronous with F_2 folding, as shown by oxide inclusions that are kinked at the margins of garnet grains, with the kink slightly overgrown by garnet rim material (Fig. 10e). Thus, it appears that garnet growth (and peak metamorphism?) was pre- or syn-F_1 and syn-F_2 (Bell, 1985), suggesting that F_1 and F_2 were part of a single progressive deformation. This deformation took place at depths in the crust corresponding to upper-greenschist-facies conditions. F_1 thrusting is a probable mechanism for burying the Texas Gulch Formation to moderate crustal levels after its deposition.

Younger age brackets on timing of deformation are not firmly established in the Brady Butte area. However, in the area near Cleator, at the northern end of the Crazy Basin Quartz Monzonite (Fig. 3), an upright isoclinal fold, which probably correlates with F_2 folds or rotated F_1 folds of the Brady Butte area, is crosscut by aplites related to the 1,700-Ma Crazy Basin Quartz Monzonite. Hence, F_2 is pre- or syn-1,700 Ma. Argenbright and Karlstrom (1986) and Conway and others (1987) suggested that the 1,700-Ma Crazy Basin Quartz Monzonite was in fact emplaced synchronously with regional shortening by upright folding, and suggested that the vertical fabric of the Big Bug block was established at ca. 1,700 Ma.

IMPLICATIONS OF NEW STRUCTURAL DATA FOR REGIONAL INTERPRETATIONS

Several aspects of this study may have important regional implications. First, the unconformity at the base of the Texas Gulch Formation bevels volcanics as well as granodiorite, and the character of the sediment changes depending on the nature of the basement below the unconformity. This unconformity is not an angular unconformity between a previously deformed basement and overlying cover, but it is an important marker horizon for defining structure and possibly stratigraphy. If it is of more than local extent, it may be identifiable elsewhere. The folded unconformity mapped by DeWitt (1976, 1979) and Argenbright (1986) at the base of pelitic schists and graywackes in the Crazy Basin area to the southeast may represent the same unconformity. This suggests a correlation between the two major bands of metasedimentary rocks in Figure 3: Texas Gulch Formation near Brady Butte and metasedimentary rocks near Cleator.

Second, if the enveloping surface of F_2 folds is subhorizontal on a regional as well as local scale, stratigraphic units may reappear across strike in successive macroscopic fold hinges. Transposition associated with F_1 and F_2 makes recognition of these folds difficult. The correlation of metasedimentary packages mentioned above is compatible with this regional subhorizontal F_2 fold envelope. Furthermore, the geometry of structures in the Crazy Basin area supports the idea of a regionally subhorizontal fold envelope (Karlstrom and Conway, 1986; Conway and others, 1987; Darrach, 1988). By this interpretation, the Spud Mountain Volcanics of Anderson and others (1971) west of the Texas Gulch Formation are believed to correlate, at least in part, with the Iron King Volcanics, east of the Texas Gulch Formation. Digitations of Texas Gulch Formation east of Brady Butte are interpreted to be transposed fold hinges rather than fault slivers (Fig. 3). If this is correct, it implies that present regional stratigraphic nomenclature (Anderson and others, 1971) should be applied with caution. The good quality of regional maps by Anderson and co-workers is such that published stratigraphic names can probably be modified and revived, but only with due consideration of the complex nature of regional structural geometries. In the meantime, a combination of lithologic mapping and detailed structural studies is needed in many areas of the Yavapai Supergroup.

New structural data is more compatible with the interpreta-

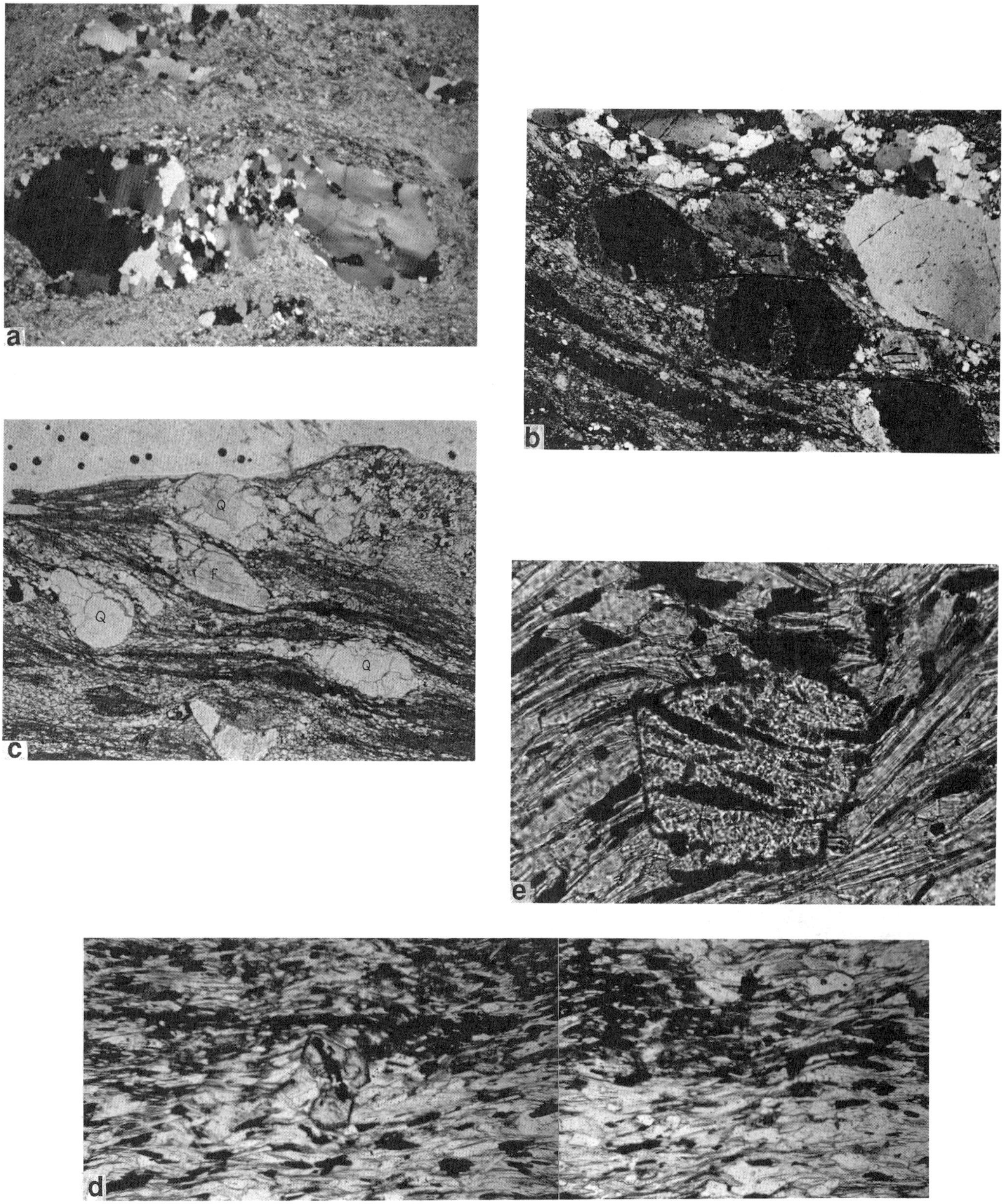
a
b
Q
F
Q
Q
c
e
d

tion of Anderson and others (1971) than DeWitt (1976), in suggesting that stratigraphy is basically controlled by folds that have shallow plunge on a regional scale. However, DeWitt's (1976) observations that fold hingelines are subvertical throughout much of the Yavapai Supergroup is correct. This apparent contradiction can be reconciled by new mapping of strongly curvilinear hingelines in the Crazy Basin area (Conway and others, 1987), which suggests that hingelines rotated toward the finite stretching direction in areas of high strain (cf. Cobbold and Quinquis, 1980; Ramsay and Huber, 1983). Thus, F_2 shortening was characterized by tight to isoclinal folding of subhorizontal layering, with subvertical extension. Areas that experienced lower shortening strains (e.g., Brady Butte) preserve shallow F_1 and F_2 fold orientations, whereas in higher strain areas (for example, between granitoid batholiths), both F_1 and F_2 fold hingelines rotated to steep orientations, toward the subvertical finite stretching lineation. The implication for massive sulfide exploration and for correlating strata is that, whereas lateral continuations of individual outcrops and ore bodies are likely to be subvertical (DeWitt, 1979), in boudins or rootless fold hinges, regional continuations may also be found across strike.

A second aspect of the Creasey (1980) and DeWitt (1980) discussion/reply is resolved by the new structural data presented here. Since deformation of the Yavapai Supergroup involved multiple fold generations, it is inappropriate to try to describe lineations as kinematic "a" versus kinematic "b" lineations. That nomenclature for kinematic axes (from Sander, 1930) is meaningful only for relatively simple displacement fields involving simple shear (Hobbs and others, 1976; Ramsay and Huber, 1983).

The regional significance of F_1 recumbent folds and thrusts to distribution of lithologies in the Yavapai Supergroup is not yet known. Folds that pre-date the upright folds (F_2) are apparently of regional extent; they are present at Brady Butte and near Crazy Basin (Conway and others, 1987). Since both the Brady Butte and Crazy Basin areas are dominated by upward-facing F_2 folds, there is little evidence at present for regional-scale recumbent folds. However, regionally important thrusts (now folded by F_2) may be present. The problem for much of the Yavapai Supergroup is that the F_2 shortening deformation was intense and F_1-related transposed layering was transposed again during F_2 shortening. In areas where isoclinal folding and transposition related to F_2 took place, all linear elements were rotated to subvertical attitudes, toward the vertical F_2 extension direction. In these areas, earlier F_1 fold hingelines and axial surfaces have been rotated into parallelism with F_2 folds and are difficult or impossible to identify. In such areas, which make up a majority of the area of the Yavapai Supergroup, the regional foliation is a composite foliation resulting from F_1 and F_2 transposition.

SUMMARY

This paper provides structural evidence for early recumbent folds in the Yavapai Supergroup with amplitudes exceeding half a kilometer. These F_1 folds verge and face northwest and are associated with development of mylonitic layering in the Brady Butte Granodiorite. The geometry and style of F_1 folds and kinematic indicators in mylonitic granodiorite suggest F_1 was related to northwest-directed thrusting. These early subhorizontal fabrics were overprinted by F_2 upright folds that became more intensely developed away from the Brady Butte area. The subvertical foliation and lineation characteristic of most of the Yavapai Supergroup is interpreted as a composite fabric developed by F_2 folding and transposition of earlier (F_1) tectonic layering. The enveloping surface for F_2 is broadly folded but subhorizontal in the Brady Butte area and adjacent areas of the Yavapai Supergroup. As a

Figure 10. Photomicrographs of rocks of the Brady Butte area. (a) Undulose extinction shows subgrain development in large quartz grains within mylonitized Brady Butte Granodiorite. Feldspars have been entirely altered to white mica and quartz in this sample from the regolithic boundary of the granodiorite. Small recrystallized grains are present along grain and subgrain boundaries and cracks. These recrystallized grains become strained in the hinge of the F_2 fold in the center of the photo. The long dimension of the photo is 8 mm across. Crossed-nicols. (b) Synthetic offsets of an optically continuous feldspar grain shows top-to-the-northwest (top to left) sense of movement in mylonitic Brady Butte Granodiorite. The long dimension of the photo is 4 mm across. Crossed-nicols. (c) Sigmoidal porphyroclasts in conglomerate of the Texas Gulch Formation indicate top-to-the-northwest (top to left) movement during development of S_1. Quartz pebbles (Q) have tails of dynamically recrystallized quartz; feldspars (F), are also sigmoidal with dynamically recrystallized, quartz-filled pressure shadows. The long dimension of the photo is 4 mm across. (d) Transposed F_1 fold defined by folded compositional layer rich in oxides, biotite, and muscovite. The cluster of garnets overgrew a slightly warped oxide grain that was aligned parallel to earlier layering; this layering was intensely folded and transposed by F_1 folding so that S_1 now is nearly at right angles to its earlier orientation as preserved in the garnet. Other garnet grains in the slide preserve oxides of the same orientation suggesting that the S_1 layering, not the garnets, rotated during folding. This suggests that F_1 was syn- or post-garnet growth (Bell, 1985). F_2 folds are broad warps of S_1 layering, with axial surfaces inclined steeply from upper right to lower left in this photo. The long dimension of the photo is 1.5 mm across. Plane polarized light. (e) Elongate oxide grains overgrown by garnet are kinked at the garnet rim by F_2 folds of S_1 layering that are traceable away from the garnet. The rim of the garnet appears to have slightly overgrown the F_2 kinked oxides. This texture suggests that F_2 was syn- or post-garnet growth. The long dimension of the photo is .4 mm across. Plane polarized light.

consequence, the distribution of lithologic units is strongly controlled by shallowly plunging regional structures. In areas of high shortening strains, F_1 and F_2 fold hingelines have rotated toward the subvertical F_2 stretching lineation so that the distributions of individual units and ore deposits are controlled by steeply plunging folds on more local scales. It is postulated that F_1 thrusting and F_2-upright folding were both part of a progressive deformation related to convergence of volcanic arc terranes during accretion of these arc terranes to North America at about 1,700 Ma.

ACKNOWLEDGMENTS

Acknowledgment for support of this research is made to the National Science Foundation (Grant EAR-8411998), donors of the Petroleum Research Fund, which is administered by the American Chemical Society (Grant number PRF 16199-GB), and the Northern Arizona University Organized Research Fund. Significant help in mapping the Brady Butte area came from John Davis, Gary Axen, and the students and teaching assistants of the Northern Arizona University summer field camp of 1985. Patrick O'Hara introduced me to the superimposed folding of the Mule Canyon area. Thanks are due to the following people for helpful reviews of this paper: Jeff Grambling, Patrick O'Hara, John Spencer, Clay Conway, and Gary Axen.

REFERENCES CITED

Anderson, C. A., 1967, Precambrian wrench fault in central Arizona: U.S. Geological Survey Professional Paper 575–C, p. C60–C65.

——, 1968, Metamorphosed Precambrian silicic volcanic rocks in central Arizona: Geological Society of America Memoir 116, p. 9–45.

Anderson, C. A., and Blacet, P. M., 1972a, Geologic map of the Mount Union Quadrangle, Yavapai County, Arizona: U.S. Geological Survey Geologic Quadrangle Map GQ–997, scale 1:62,500.

—— , 1972b, Geologic map of the Mayer Quadrangle Yavapai County, Arizona: U.S. Geological Survey Quadrangle Map GQ–996, scale 1:62,500.

—— , 1972c, Precambrian geology of the northern Bradshaw Mountains, Yavapai County, Arizona: U.S. Geological Survey Bulletin 1336, 82 p.

Anderson, C. A., and Creasey, S. C., 1958, Geology and ore deposits of the Jerome area, Yavapai County, Arizona: U.S. Geological Survey Professional Paper 308, 185 p.

Anderson, C. A., and Silver, L. T., 1976, Yavapai Series, a greenstone belt: Arizona Geological Society Digest, v. 10, p. 13–26.

Anderson, C. A., Blacet, P. M., Silver, L. T., and Stern, T. W., 1971, Revision of the Precambrian stratigraphy in the Prescott–Jerome area, Yavapai County, Arizona: U.S. Geological Survey Bulletin 1324–C, p. 1–16.

Anderson, P., 1978, The island arc nature of Precambrian volcanic belts in Arizona: Geological Society of America Abstracts with Programs, v. 10, p. 156.

——, 1986, Summary of the Proterozoic plate tectonic evolution of Arizona from 1,900 to 1,600 Ma, *in* Beatty, B., and Wilkingson, P.A.K., eds., Frontiers in geology and ore deposits of Arizona and the Southwest: Arizona Geological Society Digest, v. 16, p. 5–11.

Argenbright, D. N., 1986, Geology and structure of Proterozoic rocks of the Yavapai Supergroup in Crazy Basin, central Arizona [M.S. thesis]: Raleigh, North Carolina State University, 81 p.

Argenbright, D. N., and Karlstrom, K. E., 1986, Revised structural and stratigraphic interpretation of the Proterozoic Yavapai Supergroup, Crazy Basin, Arizona: Geological Society of America Abstracts with Programs, v. 18, p. 338.

Bell, T. H., 1985, Deformation partitioning and porphyroblast rotation in metamorphic rocks; A radical reinterpretation: Journal of Metamorphic Geology, v. 3, p. 109–118.

Blacet, P. M., 1966, Unconformity between gneissic granodiorite and overlying Yavapai Series (older Precambrian), central Arizona: U.S. Geological Survey Professional Paper 550–B, p. B1–B5.

—— , 1968, Precambrian geology of the southeast quarter of the Mount Union Quadrangle, Bradshaw Mountains, central Arizona [Ph.D. thesis]: Palo Alto, California, Stanford University, 244 p.

Cobbold, P. R., and Quinquis, H., 1980, Development of sheath folds in shear regimes: Journal of Structural Geology, v. 2, p. 119–126.

Condie, K. C., 1986, Geochemistry and tectonic setting of Early Proterozoic supracrustal rocks in the southwestern United States: Journal of Geology, v. 94, p. 845–864.

Conway, C. M., Karlstrom, K. E., Silver, L. T., and Wrucke, C. T., 1987, Tectonic and magmatic contrasts across a two-province Proterozoic boundary in central Arizona, *in* Davis, G. H. and VandenDolder, E. M., eds., Geologic diversity of Arizona and its margins; Excursions to choice areas: Arizona Bureau of Geology and Mineral Technology Special Paper 5, p. 158–175.

Creasey, S. C., 1980, Discussion "New data concerning Proterozoic volcanic stratigraphy and structure in central Arizona and its importance in massive sulfide exploration": Economic Geology, v. 75, p. 943–944.

Darrach, M. E., 1988, A kinematic and geometric structural analysis of an Early Proterozoic crustal-scale shear zone; The evolution of the Shylock fault zone, central Arizona [M.S. thesis]: Flagstaff, Northern Arizona University, 000 p.

DeWitt, E., 1976, Precambrian geology and ore deposits of the Maye–Crown Ding area, Yavapai County, Arizona [Ph.D. thesis]: Tucson, University of Arizona, 150 p.

—— , 1979, New data concerning Proterozoic volcanic stratigraphy and structure in central Arizona and its importance in massive sulfide exploration: Economic Geology, v. 74, p. 1371–1382.

—— , 1980, Reply to discussion *on* "New data concerning Proterozoic volcanic stratigraphy and structure in central Arizona and its importance in massive sulphide exploration": Economic Geology, v. 75, p. 944–945.

Gilmour, P., and Still, A. R., 1968, The geology of the Iron King mine, *in* Ridge, J. P., ed., Ore deposits of the United States, 1933–1967: American Institute of Mining, Metallurgy, and Petroleum Engineers, p. 1238–1257.

Henderson, J. B., Caldwell, W.G.E., and Harrison, J. E., 1980, North American Commission on Stratigraphic Nomenclature; Report 8, Amendment of code concerning terminology for igneous and high-grade metamorphic rocks: Geological Society of America Bulletin, part 1, v. 91, p. 374–376.

Hobbs, B. E., Means, W. D., and Williams, P. F., 1976, An outline of structural geology: New York, John Wiley, 571 p.

Jagger, T. A., and Palache, C., 1905, Description of Bradshaw Mountains Quadrangle, Arizona: U.S. Geological Survey Folio 126, 11 p.

Jerome, S. E., 1956, Reconnaissance geologic study of the Black Canyon Schist belt, Bradshaw Mountains, Yavapai and Maricopa Counties, Arizona [Ph.D. thesis]: Salt Lake City, University of Utah, 139 p.

Karlstrom, K. E., and Bowring, S. A., 1988, Early Proterozoic assembly of tectonostratigraphic terranes in southwestern North America: Journal of Geology, v. 96, p. 561–576.

Karlstrom, K. E., and Conway, C. M., 1986, Deformational styles and contrasting lithostratigraphic sequences within an Early Proterozoic orogenic belt, central Arizona, *in* Nations, J. D. and others, eds., Geology of central and northern Arizona: Geological Society of America Rocky Mountain Section Guidebook, p. 1–25.

Karlstrom, K. E., Bowring, S. A., and Conway, C. M., 1987, Tectonic significance of an Early Proterozoic two-province boundary in central Arizona: Geological Society of America Bulletin, v. 99, p. 529–538.

Knipe, R. J., and Wintsch, R. P., 1985, Heterogeneous deformation, foliation development, and metamorphic processes in a polyphase mylonite, *in* Thompson, A. B. and Rubie, D. C., eds., Metamorphic reactions: Advances in Physical Geochemistry, v. 4, p. 180–210.

Krieger, M. H., 1965, Geology of the Prescott and Paulden Quadrangles, Arizona: U.S. Geological Survey Professional Paper 467, 127 p.

O'Hara, P. F., 1980, Metamorphic and structural geology of the northern Bradshaw Mountains, Yavapai County, Arizona [Ph.D. thesis]: Tempe, Arizona State University, 117 p.

O'Hara, P. F., Yoder, M. P., Stamm, C. A., and Niver, R., 1978, The Precambrian Texas Gulch Formation boundary fault system, Yavapai County, Arizona; A folded unconformity?: Geological Society of America Abstracts with Programs, v. 10, p. 140.

Ramsay, J. G., and Huber, M. I., 1983, The techniques of modern structural geology: New York, Academic Press, 307 p.

Sander, B., 1930, Gefügekunde der Gesteine: Vienna, Springer, 352 p.

Shackleton, R. M., 1958, Downward facing structures of the Highland border: Geological Society of London Quarterly Journal, v. 17, p. 168–272.

Steiger, R. H., and Jäger, E., 1977, Subcommission of geochronology: Convention on the use of decay constants in Geo- and Cosmochemistry: Earth and Planetary Science Letters, v. 36, p. 359–362.

Vance, R. K., 1986, Geochemistry and tectonic setting of the Yavapai Supergroup, west central Arizona [Ph.D. thesis]: Socorro, New Mexico Institute of Mining and Technology.

Manuscript Accepted by the Society September 27, 1988

Printed in U.S.A.

Index

[Italic page numbers indicate major references]

Typeset by WESType Publishing Services, Inc., Boulder, Colorado
Printed in U.S.A. by Malloy Lithographing, Inc., Ann Arbor, Michigan